William R. Culbertson | Dennis C. Tanner

D0220387

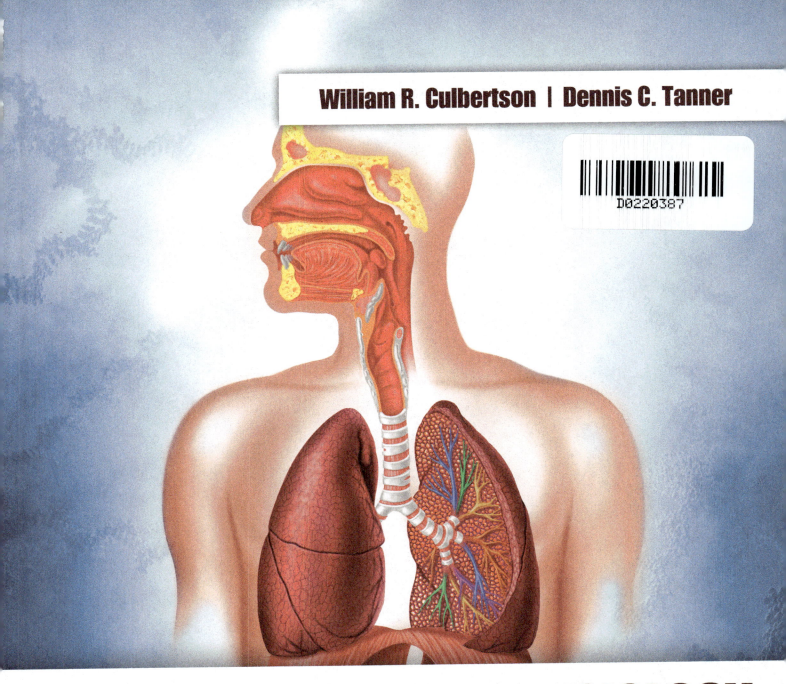

THE ANATOMY AND PHYSIOLOGY OF SPEECH AND SWALLOWING

Revised printing

Kendall Hunt
publishing company

Cover image © Shutterstock

Kendall Hunt
publishing company

www.kendallhunt.com
Send all inquiries to:
4050 Westmark Drive
Dubuque, IA 52004-1840

Copyright © 2011 by William R. Culbertson and Dennis C. Tanner

ISBN 978-1-4652-7742-8

Kendall Hunt Publishing Company has the exclusive rights to reproduce this work, to prepare derivative works from this work, to publicly distribute this work, to publicly perform this work and to publicly display this work.

All rights reserved. No part of this publication may be reproduced, stored in a retrieval system, or transmitted, in any form or by any means, electronic, mechanical, photocopying, recording, or otherwise, without the prior written permission of the copyright owner.

Printed in the United States of America
10 9 8 7 6 5 4 3 2 1

DEDICATION

For Kris Culbertson and Jody Tanner

Contents

Preface

The study of speech anatomy and physiology is a journey into the core of the human ability to communicate. The muscular adjustments necessary to create the acoustic energy for speech sound production provide the foundation for the complex, interwoven process that is human speech production. The ability to communicate is one of the highest evolved functions for the human animal, and it has enabled truly remarkable advances for civilization. Human speech anatomy and physiology lie at the nucleus of this remarkable ability to communicate. This textbook explores speech anatomy and physiology in all its intricacy and complexity.

We have written *The Anatomy and Physiology of Speech and Swallowing* for serious students from all walks of academia. Whether you are a first-year undergraduate testing the waters to see if speech-language pathology and audiology are potential careers worthy of commitment or a doctoral student reviewing the literature for a dissertation, this book provides current, relevant, and accurate information about human speech anatomy and physiology. This book is also appropriate for students, clinicians, scientists, and scholars from neurology, neuropsychology, laryngology, phonetics, virtually any discipline involved in the study of human communication and myriad of diseases, disorders, defects, and disabilities that can lay waste to it.

We have devoted several years to writing *The Anatomy and Physiology of Speech and Swallowing* and our professional lifetimes preparing for it. In addition, the publisher, Kendall Hunt, has generously provided the necessary support, research and illustration funding, counsel, and know-how to help us in our attempts to reach our goal; the creation of an authoritative, accurate, and clinically relevant speech and swallowing anatomy and physiology textbook. While this book certainly has limitations and drawbacks, we have given it out best shot. We hope you will not be disappointed with our efforts.

William R. Culbertson
Dennis C. Tanner
Flagstaff, Arizona
June 15, 2015

To the Reader

Anatomy is the study of the structures of living organisms and *physiology* is the branch of knowledge dealing with anatomical functions. In this book, we address anatomy as it relates to speech production. As for physiology, we examine not only the normal essential process of the speech production mechanism, but in some instances, how disease affects it. Although this book is not about speech pathologies, some communication disorders are addressed where they can edify salient anatomical and physiology concepts. The goal is to provide a thorough and complete treatise on human speech and swallowing anatomy and physiology. The approach of this textbook is based on the *Basel Nomina Anatomica* conceived in 1895 in Basel, Switzerland. The system represents western European tradition with terms originally given in Latin. In this book, English words and terms from other languages are used where they enhance clarity.

Several features of this textbook distinguish it from other speech anatomy and physiology books. First, each chapter is a one-stop resource for its topic. Subsections of chapters include gross structure, physiology as it relates to communication, cytology and histology, embryology, and ontogeny. There is also a chapter devoted to deglutition. Second, scientific and clinical sidebars add depth to the information contained in the main text. Third, chapters have introductions that address the subject matter in condensed form and conclusions that bring closure to the subject matter. In addition, the chapters have glossaries for easy review of salient terms.

This textbook is written to serve students from the beginning of their undergraduate educations to well into their professional practices. It is designed to be useful as a primary text in basic speech anatomy and physiology courses, and it can also serve as an auxiliary text in neuroanatomy, speech and hearing sciences, dysphagia, embryology, and communication disorders.

We believe this textbook provides a comprehensive and complete examination of the anatomy and physiology of the speech mechanism. It should provide the reader with a basic and fundamental understanding of the human ability to communicate through speech.

Acknowledgment

The authors gratefully acknowledge the exceptional assistance and support provided by Angela Lampe and Linda Chapman, and Elizabeth Cray for the second printing. The authors wish to express their appreciation to Ms. Ashley Hepperle for her editorial assistance in the preparation of the second printing.

About the Authors

William R. Culbertson is a Professor of Health Sciences at Northern Arizona University in Flagstaff, Arizona where he teaches anatomy and physiology in the Speech-Language Sciences and Technology program. Dr. Culbertson received his Ph.D. from Michigan State University in Audiology and Speech Sciences and was a clinical practitioner for twenty years.

Dennis C. Tanner received the Doctor of Philosophy degree in Audiology and Speech Sciences from Michigan State University. He has been named "Outstanding Educator" by the Association of Schools of Allied Health Professions and the College of Health Professions' "Teacher of the Year." He has published widely in academic and professional journals and is the author of books in communication sciences and disorders. Dr. Tanner is currently Professor of Health Sciences in the Speech-Language Sciences and Technology program at Northern Arizona University.

CHAPTER 1
Introduction to Human Anatomy and Physiology

CHAPTER PREVIEW

This chapter explores the language of human anatomy and physiology. It comes in the form of a brief history of conceptual revolutions in humankind's attempt to understand itself and the creation of words to represent them, posited in the context of a condensed history of anatomical science. Derivations and combining forms of principal speech and hearing anatomical and physiological terms are discussed, as are major nomenclature systems, including the *Basel Nomina Anatomica* and the *Terminologica Anatomica*. Major subdivisions of speech and hearing anatomy and physiology are examined, and there is a review of anatomical systems, positions, planes, directions, and orientations.

CHAPTER OUTLINE

The Language of Human Anatomy and Physiology

Human anatomy specifically refers to the collective forms, configurations, and structures of human beings. Coupled to the study of human anatomy is the study of physiology. Physiology addresses how living organisms work and includes the effects of disease and other agents influencing them. For example, whereas anatomy examines form, physiology addresses function, and a fitting clinical report describing the unremarkable examination of a patient's peripheral speech mechanism might simply report, "normal in form and function."

For the beginning anatomy student, the language of anatomical and physiological science is, perhaps, its most daunting aspect. Once the vocabulary is assimilated, the student will find the sciences of anatomy and physiology much easier to study.

Yet, the language of anatomy may, in some sense, be a fundamental aspect of mankind's journey toward self-awareness. A compendium of names, that is, a nomenclature, is fundamental to understanding and exploration in any arena. The nomenclature of mankind's physical presence is something particularly compelling to us. Certainly, a nomenclature of the human body is fundamental to mankind's self awareness.

Fundamental to mankind's journey toward self-awareness has been the discovery of human anatomy and the application of nomenclature to its concepts. In the *Art of Awareness*, the noted semanticist, J. Samuel Bois, discussed the periods of philosophical stability and the conceptual revolutions occurring in humankind's attempt to understand itself. The first period of relative conceptual stability extended from the appearance of homo sapiens to the arrival of the Greek philosophers. During this time, humans had no meta-concept of logic, and interpretation of the world was animistic and mythological. Gods were everywhere. Mankind continued to learn more about the science of anatomy and physiology, discovering underlying structures and functions along the way, and ironically settling old superstitions while uncovering vast new questions with every step.

Speech and Hearing Epistemology

Epistemology is the branch of philosophy that addresses the origin, nature, and limits of human knowledge. The study of speech and hearing anatomy and physiology is at the core of humankind's attempt to understand itself in general, and, in particular, how humans communicate.

Whether it involves the study of the primary language centers in the brain, research into the neurochemistry of verbal memory, or exploration of the aerodynamic forces involved in voice production, the study of speech and hearing anatomy and physiology has been a dynamic epistemological process. The language of speech and hearing anatomy and physiology changed, evolved, and adapted to cultural shifts, religious revelation, technological advances, scientific discoveries, and other innovations. New words were coined, and the meanings of existing words underwent metamorphosis, as new explanations changed perceptions about long-observed phenomena.

The language of human speech and anatomy and physiology is no exception to the ever-changing nature of language, and reflects human evolution in thought processes. As is the case with all languages, a group of individuals must, at least tacitly, agree upon the structure of the language. Readers of this chapter are taking the first steps toward inclusion in that group that uses the language of anatomy.

A good starting point for the evolution of the language of anatomy and physiology is found with the Greek philosophers. Aristotle, Plato, Socrates, and Pythagoras challenged the mythological view that human beings were simply "playthings of the gods," and embraced the idea that there is a "nature" of things and that all existing human beings were identical to their "natural selves." Aristotle, in particular, set the groundwork for science with his introduction of words to describe humankind including key anatomical and physiological notions such as category, form, and principle. Aristotle and other early philosophers established the rules of deductive logic, common sense conclusions about the nature of things, based on observations of the real world largely unfettered by mythological beliefs.

Copernicus, Galileo, Newton, Descartes, Spinoza, and Bacon harkened a second conceptual revolution by moving beyond simple "deductive" observations to "inductive" logic. This was a fundamental change, a reversal in the mental process involved in understanding humankind. Rather than to assume that the nature of things was already known, these theorists and philosophers proposed that knowledge had to be discovered through careful observation and systematic experimentation.

Support for this change in epistemology came in the form of linguistic terms to organize the new array of concepts in the human psyche. The revolution in mental processing attributed to the Greek philosophers directly led to the scientific method and systematic investigation into the laws of nature. Many new terms appeared and were added to those coined by Aristotle. "We find them not only in scientific treatises but also in everyday conversations, such as factor, variable, attraction, repulsion, analysis, field of forces, dynamics, progress, evolution, interaction, vector, environment, and many others" (Bois, 1966, pp. 7–8).

Ancient humankind struggled to discover itself through science and to discover its relationship to a higher spiritual power through religion. At times, science and theology clashed, and, as is still the case, long periods passed during which discovery gave way to reconciliation. Initially, taboos against the handling of human corpses, apparently inherent in human nature, required science to reconcile detailed examination of cadavers with closely held spiritual stricture.

Anatomical study in the Middle East seems to have fallen into a dark age after the appearance of Islam, which forbade the touching of human corpses. While much was learned from the study of the living, detailed examination of internal structures was limited for obvious reasons. It remained for the people of the northern Mediterranean to continue the science albeit at a painfully slow pace.

A similar situation existed in Western Europe. To western science, the most ancient study of anatomy was founded, for religious reasons as well as for convenience, upon the study of animal corpses. It was thus that confusion between animal organs and human counterparts inhibited the growth of anatomical science. Since animals did not and do not communicate to the same complex extent as do human beings, the functions of the organs used for communication remained undiscovered or, at least, not detailed.

The Science of Anatomy Begins

Mankind's quest for a relationship with the spiritual did not always inhibit scientific discovery. Ancient Egyptians, with their practices of embalming the human body in preparation for the afterlife, discovered and reported the earliest details of human anatomy. The *Ebers Papyrus*, for example, was written some 1550 years before the current era (BCE). In it are described vessels that conducted fluids and gases throughout the body. The *Smith Papyrus*, written about the same time, contains the first known description of cerebrospinal fluid (Wilkins, 1964).

Ancient Greek scientists struggled with anatomy and produced literature that today is considered somewhat fatuous. It was, however, the next disciplined step in the discovery of the body. Thus, in spite of his position for the ages as the so called "Father of Medicine," Hippocrates, and those who may have written under his name, are seen today as no great founders of the science as we know it, and misconceptions persisted.

All this erroneous thinking was to end, or begin to end, when Aristotle of Stagira was hired by Philip of Macedon to be a pedagogue for his son, Alexander. The hire took place some 345 years before the current era (BCE).

Aristotle's two greatest realizations may have been that there was a structural hierarchy of anatomical components and that anatomical similarities exist between human beings and other animals. Although many of Aristotle's conclusions, such as those that placed the heart at the center of the nervous system, have been generally refuted, his contributions to the science are inarguable. He remains the earliest known writer to engage the study of anatomy in a thorough and systematic manner. Aristotle founded the science of *comparative anatomy*, the study of anatomical similarities between lower animals and mankind.

The ancient study of anatomy concerned basic life-supporting processes, and these processes included respiration and digestion. It was not until the seventeenth century that the organs of speech and hearing were studied particularly for their communicative functions.

ARISTOTE

(Polygraphe).

To the extent that the organs of speech are also organs of other vital systems, the discovery of speech, language, and hearing anatomy and physiology may have been concurrent with that of the respiratory and, although to a somewhat lesser extent, digestive systems. Aristotle distinguished for the first time, the trachea from the esophagus noting that the introduction of food into the trachea resulted in much discomfort. He also recognized the larynx, but particularly for its role as protector of the airway. It is not surprising that Aristotle missed the full recognition of the speech function of the larynx, given his practice of studying animal larynges. He did, however, recognize the pharyngotympanic or Eustachian tubes, connecting the middle ear cavities to the nasopharynx. These tubes were named some 1900 years later for Bartolomio Eustachi, also known as Eustachius (Eustachi, 1562).

Aristotle's conclusions regarding communication components fell short in several ways. For example, he concluded that the purpose of breathing was to cool the body and that the heart was the seat of the intellect.

Lack of governmental support and social unrest apparently led the peoples of the northern Mediterranean regions to become disinterested in anatomical science. About fifty years after Aristotle's work, the capital of human knowledge had migrated to Egypt, where, nurtured by the enlightened Egyptian rulers known to history as the Ptolemys, a group of eminent Greek scientists created what was to become known as the Alexandrian School of Medical Science.

The most distinguished anatomists of the Alexandrians were Erasistratus (c. 280 B.C.E.) and Herophilus (c. 290 B.C.E.). They are generally attributed with being the first to conduct scientific dissections of the human body, along with many others who are recorded as having done so during that period (Longrigg, 1988). Their most cogent conclusions concerned the nervous system, for it was they who discovered the true nature of nerves.

CLAUDE GALIEN.

Four and a half a centuries later, Claudius Galenus, known to posterity simply as "Galen," elevated the study of anatomy into a distinct science. Galen studied and wrote during the second century of the Christian era, and what he left us is valuable in two ways. First, as a combination of his original research and second, as a compendium of anatomical knowledge gathered to his time. Most of the names Galen gave to anatomical structures are still in use to this day.

Of the names coined by Galen, many are still in use today. The foremost of these names are those of the bones of the skull and thorax. Galen's best work concerned osteology, or the study of bones, but his names of certain muscles of the head and neck are also used to this day. These muscles include muscles of facial expression, mandibular elevation and depression, the tongue and pharynx, as well as the six extrinsic ocular muscles.

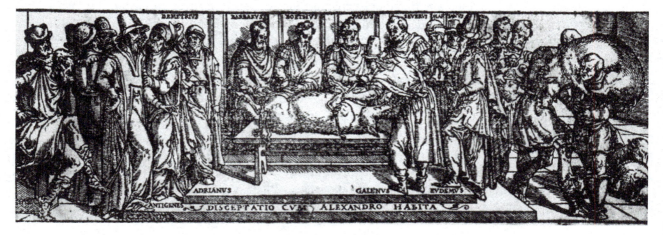

Galen used living animals to demonstrate his discoveries among the scientific community. One such demonstration included severing various nerves of a live pig. Galen demonstrated that cutting a certain nerve in the pig's cervical region caused it to stop squealing, even though it was still living. This nerve is today identified as a communicating branch connecting the superior and inferior laryngeal nerves and is also known as *Galen's nerve*.

After Galen, anatomical science, and, one might argue, science in general, passed into an historical "coma." Little progress occurred in any endeavor except warfare until the end of the Middle Ages.

The works of the Greek Alexandrians spread to the Arab world. In about 850 (CE), Baghdad physician, Hunayn ibn Ishaq, translated and annotated the works of Galen for his colleagues (Tshanz, 2003). Ishaq was particularly interested in ocular anatomy. Thus, in this era, anatomical knowledge was based largely on that discovered or inferred by the Greeks, particularly Galen, until about 1500.

In the Far East, the early approach to anatomy was much different from that in the West. The study of medicine was more closely related to philosophy than to natural science. Medicine was approached by gathering knowledge from clinical experience rather than from experimentation (Loo, 2000).

It was not until the sixteenth century that any serious further development of anatomical science occurred, and even this progress was inhibited by prejudice and dogma. During that era, two figures with confusingly similar names emerged, one in France and one in Holland. Both adopted the Latin name, "Sylvius," loosely translated to English as "Woods." Although they wrote about a hundred years apart, the two "Sylvii" are confused by many to this day.

The Dutch Sylvius was a medical doctor who, among other monumental medical exploits, did much study on the brain. His Dutch name was Franz de le Boe (1614–1672), or Woods, in English. Parts of the brain bear his name today, including the Fissure of Sylvius, also called the lateral fissure, (Collice, Collice and Riva, 2008) and the Aqueduct of Sylvius, also called the cerebral aqueduct.

The other Sylvius was a French anatomist, whose French name was Jacques du Bois (1478–1555), or Woods, in English. The French Sylvius was an admirer of Galen, and spent most of his professional career translating and promoting the ancient Greek anatomist's work with all its important discoveries and all its unfortunate misconceptions. His real claim to fame was as mentor of Andreas Vesalius.

All anatomists owe a debt of homage to Andreas Vesalius. It is he who is generally credited with authorship of the first anatomy text, *De Humanis Corporis Fabrica* or *On the Workings of the Human Body*, published in 1543. Where his mentor, the French Sylvius, was devoted to perpetuating the works of Galen, Vesalius seemed devoted to correcting it. For this, Vesalius earned the mean and constant persecution of the devotees of the earlier anatomist, ultimately highlighted by collegial accusations that he had dissected a living body. The penalty for this was a cruel death. Vesalius evaded the Inquisition only by virtue of royal patronage. His contributions to the anatomy of human communication, although he did not specifically note communicative functions, included an elucidation of brain anatomy, an early description of the pleural membranes, and a detailed description of the sphenoid bone.

By the time of the seventeenth and eighteenth centuries, the study of anatomy was back on track, and the science again proliferated. Like a bird released from a cage, anatomic science was freed by the age of reason. Although still influenced by prejudices founded on past "truth," new discoveries emerged, nevertheless, from many quarters.

Among the high points in the discovery of the anatomy and physiology of human communication were the contributions of Italian anatomists. Perhaps most important to the present work was that of Giulio Cesare Casseri (also called Cassserius), who, in 1601, published the first specialized illustrated text on speech and hearing anatomy and physiology: *De Vocis Auditusque Organis Historia Anatomica* (Kent, 1997). Casserius, professor of anatomy at Padua, also produced much work on the tracheostomy. In the same country, Columbus, a student of Vesalius, described laryngeal ventricles. His compatriot, anatomist Giovanni Battista Morgagni, described the space between

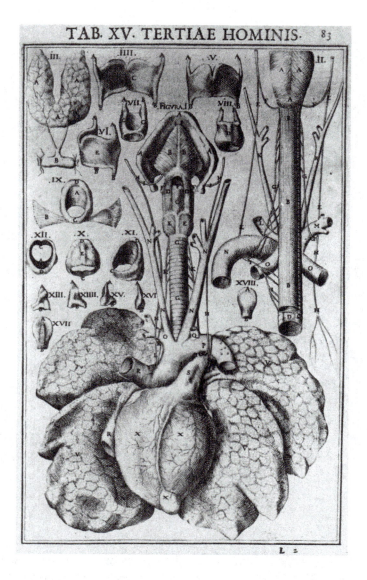

the Vestibular and vocal folds (Morgagni's Ventricle) and the space between the superior border of the levator veli palatini muscle and the base of the skull (Morgan's Sinus). Eustachi, noted previously, reported the pharyngotympanic tubes bearing his name. Santorini described the muscles of facial expression in detail. The first detailed accounts of the hearing organs were attributed to French anatomist Joseph Guichard Duverney, but Italian Valsalva elaborated in this description and lent his name to the *Valsalva* maneuver, still recommended today for clearing the eustachian tubes.

Huge strides were also made in brain research during this period. Thomas Willis first reported the details of the cranial arterial circle that bears his name: the Circle of Willis. This was only one of his important discoveries, many of which advanced the art of microanatomy.

·The almost simultaneous development of injection and microscopy led to detailed descriptions of the more minute details of anatomy. Following the leads of Leeuwenhoek and Willis, anatomical science was extended in many areas. Tissues and smaller vessels were distinguished, including those of the lungs.

Anatomy in the nineteenth century showed no slowing of discovery, and the centers of anatomical science grew strongest in Western Europe. One of the greatest anatomical works to emerge was Henry Gray's *Gray's Anatomy: Descriptive and Surgical,* first published in England in 1858. This work has undergone many editions, including a version specifically designed for students in the United States in 1901 (Pick and Howden, 1901/1977).

During this period, the first attempts were made to standardize anatomical nomenclature. The great expansion of anatomic science resulted in the situation where some structures had as many as twenty names. In Basel, Switzerland, 1895, the German Anatomic Society granted final approval of the Basle Nomina Anatomica, by which the names of anatomical structures were to be in Latin (Encyclopedia Britannica, 2005).

The twentieth century saw a third and current conceptual revolution in understanding the nature of humankind involved general semantics and the notion of semantic relativity. Alfred Korzybski detailed semantic relativity in his landmark book, *Science and Sanity: An Introduction to Non-Aristotelian Systems and General Semantics* in 1933. This view of humankind's journey to knowing itself goes beyond deductive and inductive logic, and includes symbolism on many levels. "We have such words as *multiordinality, indeterminacy, process, multidimensionality, transaction, self-reflectiveness, semantic reaction, thinking models, noise and redundancy in communication* and the introduction of the hyphen in such compound words as space-time, body-mind, and the like" (Bois, 1966, p. 9).

In anatomical nomenclature, both the British and the Germans, in separate nationalistic movements that perhaps reflected the political turmoil of the time, developed their own nomenclature standards. In 1935, the British established a British revision of the BNA, the *British Nomina Anatomica,* while the Germans established the JNA, or *Jena Nomina Anatomica* (Zemlin, 1998).

In an effort to establish worldwide acceptance of an anatomical nomenclature system, the 5th International Congress of Anatomists in Oxford, England, in 1950, appointed the IANC, or International Anatomical Nomenclature Committee. The committee's work was accepted at the 6th International Congress of Anatomists in Paris in 1955. The IANC approved the *Paris Nomina Anatomica,* by which all anatomical terms would be in Latin for international conventions and publications, with "vernacular" terminology acceptable for local meetings and journals, as well as for instructional purposes (Nomina anatomica, 1956).

In the United States, the science of human communication and its disorders gained great impetus in the twentieth century. Lee Edward Travis (1896–1987) was among the first to use electrophysical measurements to study brain functions, particularly during communication. From these studies, he concluded that stuttering was a result of a lack of cerebral hemisphere dominance. His *Handbook of Speech Pathology and Audiology* (1957) was an essential part of the book collections of practitioners in the field for many years.

No discussion of the anatomy and physiology of human communication would be complete without mentioning the work of the American Willard Zemlin (1929–1998). His text, *Speech and Hearing Science,* was first published in 1968, and reigned for the last part of the twentieth century as the preeminent work on the subject. Its last edition was in 1998, the year of its author's death.

Anatomical Science is by no means a static science, and so it goes that, in 1998, the *Terminologica Anatomica,* or *TA,* superseded the International Anatomical Nomenclature Committee. The 56-member Federative Committee on Anatomical Terminology created the Terminologica Anatomica, motivated by the same impetus that led to the British Nomina Anatomica, that is, to promote the study of anatomy without international bias. Terminologica Anatomica contains terms in Latin and the equivalent English, suggesting that, at least in an epistemological sense, two languages, Latin and English, may be considered to have worldwide acceptance in the scientific arena.

For the present work, the authors use Terminologica Anatomica terminology, and supply other nomenclature as required. It is hoped that the reader will recognize and assimilate the variations on some anatomical terms, saving them for practical application when needed.

How to Study Anatomy

Speech and hearing anatomy and physiology specifically addresses the collective forms, configurations, structures, functions, and processes involved in the act of communication. The serious student of anatomy and physiology will recognize and seek out various means of extending this study.

In the past and today, two methods have been used to ascertain how human cells, organs, and systems work, and how they might malfunction. In the first method, simple observation is used to examine the alterations in body functions that accompany changes in nutrition, disease, poisoning, or organ removal. The second method employs controlled experimentation and manipulation of variables to lead to inferences about general phenomena. Both methods provide objective perspectives from which to study structures and functionality of the human body. Human speech and hearing anatomy and physiology are interdependent disciplines, and the study of one cannot be comprehensively addressed without heeding the other.

As the term "anatomy" implies, dissection of cadavers is fundamental to the art and science of human anatomy and physiology. "Anatomy is raised from the dead by the study of the living (Lockhart, Hamilton, and Fyfe, 1959, p. 1). Certainly, knowledge of the human body can be obtained from examining the living. Nevertheless, cadaver dissection, historically and today, provides the most detailed, meticulous, and exacting strategy to knowing the morphology, or the configuration, of the structure of the human body.

Although cadaver dissection provides the bases for the study of human anatomy and physiology, viewing speech and hearing structures from deceased individuals has limitations. Cadavers are in a process of deterioration, and the preservatives used to slow the process causes changes in the tissue appearance. Of course, the causes of death may also leave its marks on the appearances of the specimens. Regardless of the limitations, Lockhart, Hamilton, and Fyfe (1959) term dissection as the "royal road" to the art of anatomy.

> The word "anatomy" is derived from the early Greek philosophers and literally means "cutting up."

As the art of dissection evolves, technology provides continued advancement into viewing and examining the living body. Contemporary speech and hearing scientists have gone beyond gross X-rays to imaging technology, such as computed tomography (CT), magnetic resonance imaging (MRI), single photon tomography (SPECT), positron emission tomography (PET), and functional magnetic resonance imaging (fMRI). From computed tomography of diaphragmatic tissue to functional magnetic resonance imaging of the major language centers of the brain during speech production, technology is opening new horizons for speech and hearing anatomy and physiology. It provides a window into the living human body that was beyond comprehension a few decades ago.

In place of cadaver study, use of accurate artificial models is also very useful. Modern technological advances in molding and fabrication have allowed production of very useful models. Some of these extend the study of anatomy beyond what can be attained in dissection, so that models serve as a study form in their own rights, and are not merely poor substitutes for cadavers.

Although still very two-dimensional, a readily available means of appreciating variations in anatomical and physiological study is by using alternative texts. No matter how thorough an author tries to be, he or she is still bound by editorial constraints to a particular point of view. Perusal of other works adds another point of view from which students can derive their own. A recent and huge advance toward dissemination of diverse academic concepts is provided by the Internet, or World Wide Web. This recently developed resource offers a plethora of materials and viewpoints, and use of one of the popular search engines brings them to the student's desk with a few manual keystrokes.

The language of speech and hearing anatomy and physiology embraces both semantic and iconic references. Descriptive terminology specifies points of muscular origin and attachment, describes structure and function of muscles and other tissues, and defines anatomical and physiological systems. Pivotal to the descriptive terminology are accompanying figures and illustrations visually depicting salient aspects of the human speech and hearing mechanism. The language of speech and hearing anatomy and physiology is a blend of semantic and iconic representation, and a prime example of relativistic semantics. It is an amalgamation of art, science, and epistemology, the branch of philosophy addressing the origin, nature, and limits of human knowledge.

Table 1.1 shows common combining forms of compound speech and hearing anatomical and physiological terms (Daly, 1968). A dash indicates whether the form usually appears before, between, or after the Greek or Latin word from which it is derived.

TABLE 1.1: Common Combining Forms of Compound Speech and Hearing Anatomical and Physiological Terms

a- negative prefix (without)	e- out from	neur- nerve
ab- away from	eso- inside	odont- tooth
acou- hear (also acu-)	exo- outside	onc- bulk, mass
ad- toward	extra- outside of	oss- bone
aden- gland	faci- face	ot- ear
aer- air	fasci- band	per- through
angi- vessel	fiss- split	pha- say, speak
ankyl- crooked, looped	gangli- swelling	pharyng- throat
ante- before	gen- be produced, originate	phon- sound
antr- cavern	-glia glue	phrax- fence, wall off
-aph- touch	gloss- tongue	phren- mind, midriff
arachn- spider	glott- tongue, language	platy- broad, flat
aur- ear	gyr- ring, circle	pne- breathing
bar- weight	hemi- half	pre- before in time or place
bi- two, life	hom- common, same	prosop- face
brachi- arm	hydra- water	pseud- false
brachy- short	hyper- above, extreme	pulmo- lung
bucc- cheek	hypo- under, below	quadr- four
caud- tail	idi- peculiar, separate	re- back, again
cav- hollow	infra- beneath	retro- backward
cephal- head	inter- among, between	rhin- nose
cerebr- cerebrum	intra- inside	sclera- hard
chord- string	is- equal	semi- half
chro- color	kine- move	sens- perceive, feel
chrondr- cartilage	labi- lip	sin- hollow, fold
circum- around	later- side	somat- body
contra- against, counter	lig- tie, bind	stom(at)- mouth, orifice
cortic- bark, rind	lingu- tongue	sub- under, below
cost- rib	lip- fat	super- above, beyond,
crani- skull	loc- place	syg- yoke, union
cune- wedge	macr- long, large	tel- end
de- down from	mal- bad, abnormal	tele- at a distance
dendri- tree	medi- middle	throac- chest
di- two	mega- great, large	thyr- shield
dipl- double	megal- great, large	tors- twist
dis- apart, away from	mening- membrane	trache- windpipe
dors- back	morph- form, shape	tri- three
dur- hard	nas- nose	un- one
dys- bad, improper	necr- corpse	-yl- substance

Anatomy Defined: Subdivisions of Anatomy

Human anatomy is the description of the collective forms and morphologic structure of the human body. Over the years, several specializations within the general field of anatomy have emerged. Below are those pertinent to the study of human speech and hearing.

Cytology and Histology

Cytology, the study of cells, and histology, the study of tissues, frequently use enlarging devices to make observable those objects and substance not visible to the naked eye. Concerning speech and hearing, cytology and histology address the minute anatomy of cells, tissues, and organs involved in human communication. The importance of such study lies in the understanding of the foundations for organ and system function.

Descriptive Anatomy

Descriptive speech and hearing anatomy is the examination of the functional relationships of the individual systems involved in human communication. A system is a collection of organs unified in a particular function, and implicit in the discussion of any system is the fact that one system may share organs with another.

The human speech and hearing system consists of several collections of organs functionally interacting during the act of communication. Most of these organs are parts of other systems with other functions. The other functions are often described as "biological" or "primary," since they are functions essential to life, itself.

For example, descriptive speech and hearing anatomy involves the examination of the respiratory system and its relationship to the phonatory mechanism. While phonation is essential to production of most phonemes, it is not essential to life. Descriptive speech and hearing anatomy examines system relationships and is sometimes called "systemic anatomy."

Developmental Anatomy

This anatomical specialization deals with the development of the human organism from conception to birth. The study of developmental anatomy reveals facts about the arrangement and functions of anatomical components not attainable in any other way.

As for speech and hearing, embryonic and fetal development anatomy examines the morphology of the communication structures in the prenatal human. Since so much ontogeny of communication occurs in the period between birth and twelve years, a mastery of its prenatal foundations is of no small consequence.

Gross Anatomy

Equated with cadaver dissection, gross anatomy is the study of the morphology of the human body without the aid of a microscope or other magnifying devices that enlarge images. "Gross" refers to being viewable by the naked eye as opposed to microanatomy.

Practical Anatomy

Practical speech and hearing anatomy is the study of body structures and relationships involved in communication with particular application to a specialized discipline. This specialized branch of anatomy evolved as refinements in the arts of surgery made precision a necessity. Detailed knowledge of the relationships of superficial structures with deep ones gives the modern surgeon a means to minimize iatrogenic damage during the course of the cure.

Practical anatomy is applicable to several clinical disciplines besides surgery, including speech-language pathology and audiology, neuropsychology, otology, laryngology, neurology, pediatrics, and gerontology, to name a few. Practical anatomy is sometimes referred to as "applied anatomy" because the knowledge base derived from the study of the speech and hearing mechanism directly pertains to a clinical discipline as distinguished from its relevance as an academic endeavor.

Topographical Anatomy

Disciplines and texts addressing the head, neck, and trunk, or the anatomy of the hearing mechanism, are examples of topographical anatomy. This is a specialized area of anatomy focusing on one definite area of the body or the relationship of two or more systems.

Speech and hearing anatomy is a prime example of this topographical anatomy as it focuses on specific structures involved in communication. These structures include the lungs, larynx, ear, brain, and so forth. While studying topographical anatomy, students should bear in mind that normal communication depends upon the concerted actions of various component organs.

Topographic anatomical discussions often create some confusion among students because of the difficulty in determining where one structure ends and another begins. For the present purposes, a clear distinction between the phonatory mechanism and the respiratory mechanism is not possible, and neither is one between the articulatory system and resonance systems. The larynx functions both as the generator of the phonatory sound source and as a glottal consonant articulator.

Physiology Defined: Subdivisions of Physiology

Distinguishing between anatomical science and physiological science is useful, though it is not at all unusual to study the two sciences simultaneously. Whereas anatomy is the study of structure or form, and relationships of body parts, physiology is the study of the functions of those parts. A common phrase written in clinical evaluations of the speech and hearing structures reports that the mechanism is, "Normal in form and function." This phrase succinctly suggests that the examiner observed the structure and its suitability for communicative acts.

General Physiology

This is the study of functional activity and the normal vital processes of cells, muscles, organs, and other structures. In the broadest sense, it includes all of the processes of animal and vegetable organisms. As in anatomical science, several specialized branches within the general field of physiology have emerged. Below are those pertinent to the study of human speech and hearing.

Applied and Experimental Physiology

According to Zemlin (1998), applied physiology is the application of physiological knowledge to problems in medicine and industry. Empirical research investigates phenomena involving the complexities of human communication to enable us to understand its vicissitudes better.

Speech and hearing experimental physiology involves experiments conduced in a laboratory to learn the functional activity of some aspect of communication. For example, speech and hearing physiologists may study auditory physiology as it pertains to noise pollution. Experiments have been carried out to determine the levels of sound pressure required to elicit hearing threshold shifts to protect individuals who may risk exposure to such levels. Once the levels are determined, measures can be undertaken to reduce these damaging levels.

Cellular Physiology

All tissues are composed of microscopic cells. Cellular physiology addresses their vital processes and functional activities individually or collectively. Speech and hearing cellular physiology is the study of individual or small groups of cells involved in the act of communication. Cells of special interest to speech and hearing physiology include neurons, epithelial cells, especially those forming sensory end-organs, skeletal connective tissue cells, and muscle cells, to name a few.

Special Physiology

Zemlin (1998) noted that special physiology addresses the functional activity of particular organs, such as the heart in cardiology, or the glands in endocrinology. The structures and organs involved in communication are the targets of special physiology in speech and hearing science. Of these, the brain is the organ that has received the most interest from special physiologists studying communication.

Anatomical Terms: Position, Orientation, Planes, and Regions

The focus of this introductory chapter has been the language and epistemology of anatomy, and its sister science, physiology. As with any language, the purpose of anatomical language is to provide a conventional system or code for information transmission. For this reason, the first and most important step in the study of anatomy and physiology is to master the basics of the language. Many anatomical terms are derived from Latin or Greek, and will seem difficult at first. However, diligence and study will make the Latin or Greek roots, prefixes, and suffixes familiar and the study of anatomy and physiology more enjoyable.

One of the great ongoing undertakings and conventions in speech and hearing sciences has been the attempt to standardize the anatomical and physiological terminology of location. Although far from complete, scientists worldwide typically use the same terminology to describe postures of the human body and relative positions of the speech and hearing structures.

Anatomical Position is a convention of posture established to avoid confusion in anatomical descriptions. It will be as clear to the beginning student, as it has been to the great anatomists mentioned earlier in this chapter, that when one is standing, the nose is in front of the face, and when one is lying down (supine), the nose is on top of the face. Thus, a conventional posture is needed to enable us to describe the relative positions of body structures no matter what posture the body assumes.

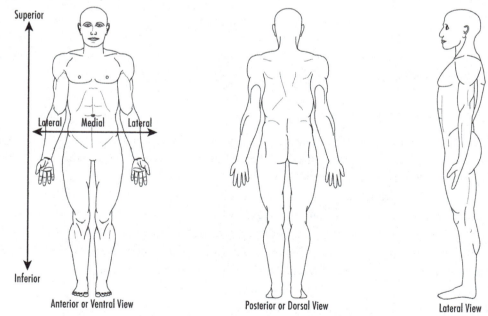

In *Anatomical Position:* 1) the body is erect, 2) the face is forward, 3) the upper extremities (arms) are extended downward and by the sides, 4) the palms are turned forward and fingers extended, 5) the lower extremities ("legs")

are extended, and the feet are forward. This convention assures that anatomical terms describing positioning always describe the same relationship. "The student must use this convention from the outset and think of the subject erect irrespective of its actual position during dissection" (Lockhart, Hamilton, and Fyfe, 1959, p. 2). In addition, the head is in such a position that the lower border of the eye socket is on the same horizontal plane as the ear canal. This is sometimes described as the Frankfort plane (Zemlin, 1998).

> Without such consistency in terminology, a description of the nose on a body lying down might be described as "on top of" the face, while if standing up, that same anatomical structure would be described as "in front of" the face.

Specialized Anatomical Orientation Terms

Anatomists describe the location of anatomical parts and their orientation to the general body by special contrasting terms. Below are the major position terms and examples of their usage, presented in pairs of antonyms.

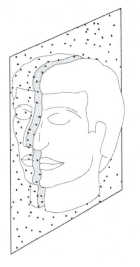 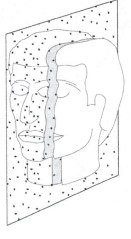 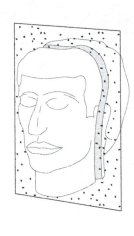 

Anterior	**Posterior**
Situated at or directed toward the front:	Situated at or directed toward the back:
"The nose is an anterior to the eyes."	"The brain is posterior to the eyes."
Inferior	**Superior**
Lower:	Upper:
"The feet are inferior to the legs."	"The legs are superior to the feet."
Central	**Peripheral**
Located at the midpoint or pertaining to the center:	Located away from the center:
"The brain is central to the eyes."	"The external ear is peripheral to the cochlea."

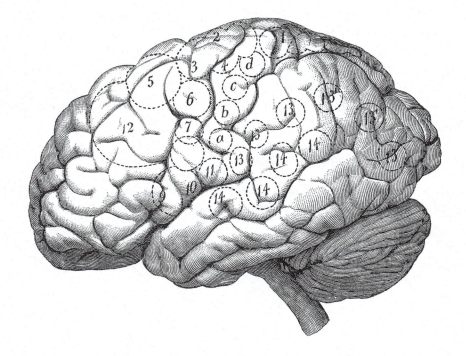

Rostral

Toward the nose or beak:

"The frontal lobes of the brain are rostral to the spinal cord."

Caudal

Toward the tail, or away from the nose or beak:

"The pelvis is caudal to the skull."

Deep

Located near an inner part or internal surface:

"The thalamus is deep to the cerebral cortex."

Superficial

Toward the outer part or external aspect:

"The cortex is a superficial structure of the brain."

Internal

Toward the inner surface; inside or within:

"The larynx is internal to the neck."

External

Toward the outer surface:

"The ribs are external to the lungs."

Lateral

Pertaining to or situated at the side:

"The arms are lateral to the body."

Medial

Pertaining to or situated toward the midline:

"The tongue is in the medial aspect of the oral cavity structures."

Proximal

Nearest to the body point of reference:

"The shoulder is proximal to the elbow."

Distal

Farthest from the body point of reference:

"The fingers are distal to the wrist."

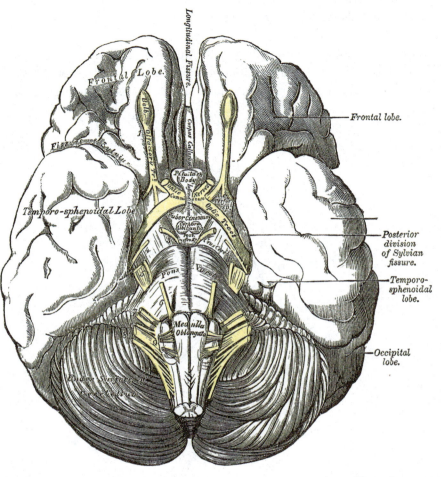

Ventral	**Dorsal**
Toward the belly:	Toward the backbone:
"The navel is on the ventral aspect of the abdomen."	"The spine is on the dorsal aspect of the abdomen."
Cephalad/Cranial	**Caudal**
Referring to the head	Referring to the tail
"The head is cephalad to the neck."	"The neck is caudal to the head."

Planes of Reference

Planes of reference refer to the various real or imaginary views, sections, or slices of the body. They are surfaces formed by slices through two points of reference.

A **frontal plane**, sometimes called the **coronal plane**, is a vertical or longitudinal line, and all parallel ones, extending from the medial anterior to the posterior point of the specimen. It cuts the specimen into front-to-back (anterior and posterior) sections.

A **transverse plane** is a horizontal cross-section line, and all parallel ones, extending from the inferior caudal to superior cephalic regions. It cuts the specimen into superior and inferior sections, that is, from top-to-bottom.

A **sagittal plane,** a vertical longitudinal section, and all parallel ones, divides the body into left and right sections. It extends from the ventral or anterior region to the dorsal or posterior region and divides the specimen into right and left anterior and posterior sections.

The "midsagittal plane," roughly corresponding to the middle of the sagittal suture of the skull is a special sagittal plane, because it is so often used in anatomic illustration. Since the body is roughly symmetrical on its right to left sides (the heart and lungs notwithstanding) the midsagittal plane is of interest because it demonstrates the body's core structures.

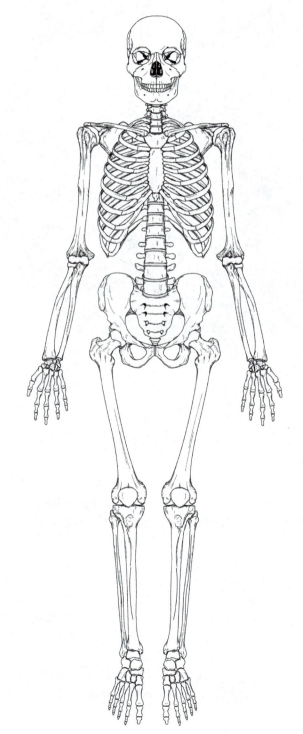

Topographical Regions

Throughout this text, certain anatomical regions will be cited as the locations of structures of interest. The names of these regions are often mentioned in research literature, scholarly discussions, conferences, and clinical reports.

The **cranial region** begins at the vertex or most superior portion of the skull and extends inferiorly to the first cervical vertebra. In an adult, the mandible's angle is sharp enough that the body of the mandible extends to a horizontal plane slightly lower than the plane of the first vertebra. Structures of the cranial region include the oral and nasal cavities, orbits and the brain. It includes the entrances to the airway and digestive system.

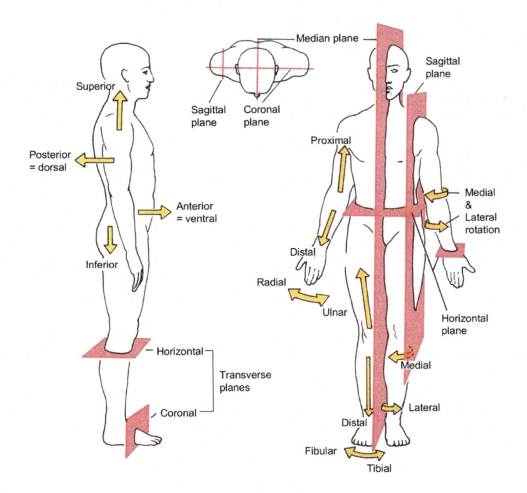

The **cervical region** begins at the horizontal plane of the first cervical vertebra and extends inferiorly to the horizontal plane of the first thoracic vertebra. The cervical region contains the great vessels and nerves that connect the cranial region to the rest of the body. It also contains the larynx.

The **thoracic region, or thorax**, begins at the first thoracic vertebra and extends inferiorly to the diaphragm. The diaphragm is a dome-shaped muscle of respiration, with its most superior central portion located at the plane of the first and second lumbar vertebrae, and its more distal portions located at the levels of the lower ribs and sternum. Thus, the inferior border of the thoracic region defies a simple horizontal reference. The thoracic region contains the lungs, heart, and great vessels as well as important muscles of respiration.

The **abdominal region, or abdomen**, begins at the inferior border of the thorax and extends to the pelvic floor. This region contains the viscera of the digestive and urogenital systems, great vessels, as well as some accessory respiratory muscles.

The **upper extremities (limbs)** are what we call colloquially, the "arms." Important in communication as organs for writing, gesturing and signing, as well as for environmental exploration, balance and orientation, the upper extremities are composed of several parts. The most proximal part is called the "arm," and it articulates with the "forearm," distally. The distal part of the forearm articulates with the hand at the wrist.

The **lower extremities (limbs)** are known popularly as "the legs." These extremities are important in a communicative sense for normal ambulation and developing and maintaining a cognitive awareness of the environment. The proximal part of the lower extremity that articulates with the pelvis is called the "thigh." The distal thigh articulates with the "leg," the distal end of which articulates with the foot at the ankle.

Speech and Hearing Systems

Anatomical and physiological **systems** are networks of correlated and semi-independent functionally related structures. There is no consensus regarding the number of systems in the human body and definitions of what organs constitute them also vary. Various anatomists have described as few as seven (Romanes, 1972) and as many as eleven (Tortora and Grabowski, 1990) body systems including: skeletal, articular (joints and ligaments), muscular, digestive, vascular, nervous, respiratory, urinary, reproductive, endocrine (ductless glands), and integumentary (skin, nails, hair). In addition, some systems can be grouped together to form system complexes, and systems can exist within systems.

Speech and hearing scientists and practitioners identify systems involving communication. These systems are composed of parts of other systems. "With just a moment of thought, it becomes apparent that no one of these systems is independent of the others. The speech mechanism draws heavily on some systems and less heavily on others, but either directly or indirectly it is dependent upon all the systems in the body" (Zemlin, 1998, p. 30).

The muscular, vascular, nervous, digestive and respiratory systems are fundamentally involved in human communication and play primary roles, while the reproductive or urinary systems, for example, are indirectly involved in communication and have ancillary roles. Ancillary systems are addressed in the context of their capacities in supporting the fundamental systems.

Separation of the structures involved in human communication into central and peripheral groups may be convenient from practical and learning standpoints. This in no way implies a complete separation, for each division is dependent on the other. Still, such separation simplifies and compartmentalizes study of the structures.

In clinical practice, most evaluation protocols call for examinations of the *peripheral speech mechanisms*, with subsequent diagnosis often phrased as involving a *central language* process. Audiologists frequently classify hearing disorders in terms of the relative effects of *central auditory disorder* or *peripheral hearing loss*.

For clinical purposes, the peripheral hearing system begins at the pinna (external ear) and extends proximally to the middle of the brainstem, at the junction of the pons and cerebellum. Essentially, it is composed of structures of the skull and central nervous system and includes the cochlear division of the vestibulocochlear nerve. At its proximal extent, nerve fibers of the peripheral hearing system synapse with neurons of the central auditory system at the dorsal and ventral cochlear nuclei. It is at this point that the central auditory system begins. The central auditory system extends from the cochlear nuclei to the association areas of the cerebral cortex.

The peripheral speech system includes the respiratory system and parts of the digestive system. This includes structures of the skull and neck: the mouth, nose, pharynx and upper respiratory passages. It also includes structures of the thorax: the lower respiratory passages and lungs, as well as the peripheral nervous system, including the appropriate cranial and spinal nerves. The central speech system is the central nervous system and consists of the brain and the spinal cord.

Adding gestural and graphic language to the human communication picture, it is clinically useful to appropriate the muscles, integumentary and skeletal structures of the upper extremities and face, including the eyes and optic nerves, as well as certain cranial and spinal nerves, to the peripheral system. Central gestural and graphic systems include the brain and the spinal cord.

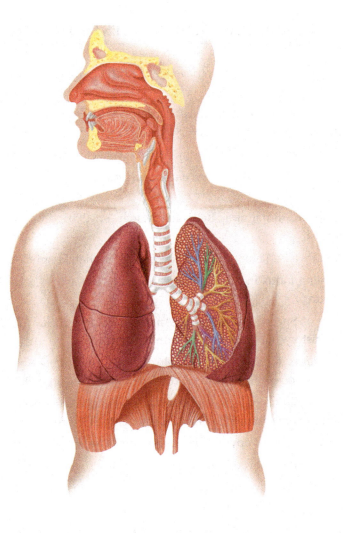

CHAPTER SUMMARY

The language of speech and hearing anatomy and physiology reflects, and in many ways directs, humankind's arduous road to knowing itself. It has evolved and continues to evolve, change, and adapt to technological, scientific, and cultural changes.

Human anatomy, the study of the structure, and human physiology, the study of the function of *homo sapiens*, are interdependent disciplines, especially where it concerns acts of communication. Dissections of cadavers and modern imaging technology are two primary methods for studying speech and hearing anatomy and physiology, and they are augmented by texts, models and online resources. Conventional anatomical and physiological terms are based on the *Basle Nomina Anatomica* system developed in the 1800s. This system of nomenclature began with terms having Latin bases, reflecting the universality of that language in the nineteenth century. In the twentieth century, English joined Latin as a universal language in the modern *(Paris) Terminologica Anatomica*.

Scholars in anatomy and physiology have developed several subdivisions of the sciences, and proprieties for representing the positions and planes of the body. Primary and ancillary anatomical and physiological systems work in concert to enable human communication.

REFERENCES AND SUGGESTED READING

Anatomy. (2011). In *Encyclopædia Britannica*. Retrieved May 18, 2011, from http://www.britannica.com/EBchecked/topic/22980/anatomy

Anatomical nomenclature. Encyclopædia Britannica. Retrieved May 20, 2011, 2005, from Encyclopædia Britannica Online. http://search.eb.com/eb/article?tocId=284

Aristotle 384–322 BCE. Retrieved May 11, 2011, from http://www.ucmp.berkeley.edu/history/aristotle.html

Bois, J.S. (1966). *The art of awareness: A textbook on general semantics*. Dubuque, IA: WM. C. Brown Company Publishers.

Casserius, Julius. Encyclopædia Britannica. Retrieved March 14, 2005, from <http://www.britannica.com/eb/article?tocId=9020648>

Collice, M., Collice, R., and Riva, A. (2008). Who discovered the Sylvian Fissure? *Neurosurgery 63*, 623–628.

Daly, L.W. (1968). Combining forms in medical terminology. In *Dorland's Pocket Medical Dictionary* (21st ed.). Philadelphia: W. B. Saunders.

Duchan, J. (2004). History of Speech Pathology in America. Retrieved May 20, 2011, from <http://www.acsu.buffalo.edu/~duchan/history_subpages/leeedwardtravis.html>

Ebers, G. (1875). *Papyros Ebers* (Ebers Papyrus). Leipzig, 1875. Retrieved April 19, 2011, from <http://www.britannica.com/eb/article?tocId=9031856> and <http://en.wikipedia.org/wiki/Ebers_papyrus>

Eustachi, B. (1562). *Epistola de auditus organis.* Cited in Whonamedit? A dictionary of medical eponyms. Rretrieved May 11, 2011, from http://www.whonamedit.com/doctor.cfm/1433.html

Farag, T.I. (2005). The Unveiled Ebers Papyrus. *The Ambassadors On-line Magazine.* Vol. 8, Issue 1, January.

Hippocrates (c.450—c.380 BCE). Retrieved May 12, 2011, from http://www.iep.utm.edu/hippocra/

History of Anatomy. Wikipedia. Retrieved May 18, 2011, from <http://en.wikipedia.org/wiki/History_of_anatomy>

Kent, R.D. (1997). *The Speech Sciences.* San Diego: Singular.

Korzybski, A. (1933). *Science and sanity: An introduction to non-Aristotelian systems and general semantics.* Lakeville, CT: International Non-Aristotelian Library.

Lockhart, R.D., Hamilton, G.F., and Fyfe, F.W. (1959). *Anatomy of the Human Body.* Philadelphia: J. B. Lippincott Company.

Lu, W. (2000). Zhongwai yixue fazhan shi bijiao. [Comparison between the developmental history of Chinese medicine and foreign medicine]. *Zhonghua Yishi Zazhi (Chinese Journal of Medical History) 30*, 35–39.

Islamic Culture and the Medical Arts. National Library of Medicine Home Page. (1998). Retrieved May 20, 2011, from <http://www.nlm.nih.gov/exhibition/islamic_medical/islamic_10.html>

Longrigg, J. (1988). Anatomy in Alexandria in the Third Century B.C. *British Journal for the History of Science 21*, 455–488. Retrieved May 20, 2011, from http://www.jstor.org/stable/pdfplus/4026964.pdf?acceptTC=true

Nomina Anatomica (1956). Revised by the International Anatomical Nomenclature Committee, Baltimore, MD: Williams and Wilkins Co.

Pick, T.P. and Howden, R. (Eds.). (1977). Gray, H. *Anatomy, Descriptive and Surgical: A Revised American, from the Fifteenth English Edition* (Original work published in 1901). New York: Bounty Books.

Robinson, A. (1927). Introduction. In A. Robinson (Ed.) *Cunningham's Text-book of Anatomy* (5th ed). New York: William Wood and Company.

Romanes, G.J. (Ed.). (1972). *Cunningham's Textbook of Anatomy* (11th ed.). London, Oxford University Press.

Travis, L.E. (1957). *Handbook of Speech Pathology*. New York: Appleton-Century-Crofts.

Travis, L.E. (1978). The cerebral dominance theory of stuttering, 1931–1978. *Journal of Speech and Hearing Disorders, 43*, 278–281.

Tshanz, D.W. (2003). Hunayn bin Ishaq: The great Translator. *Journal of the International Society for the History of Islamic Medicine. 1*, 29–40.

Wood, Kenneth Scott (1971). Terminology and nomenclature. In Lee Edward Travis (Ed.) *Handbook of Speech Pathology and Audiology.* Englewood Cliffs, NJ: Prentice-Hall.

Wilkins, R.H. (1964). Neurosurgical classic-XVII Edwin Smith Surgical Papyrus. *Journal of Neurosurgery, March, 1964*, 240–244. Reprinted in *Cyber Museum of Neurosurgery*. Retrieved March 2, 2005, from <http://www.neurosurgery.org/cybermuseum/pre20th/epapyrus.html>

Zemlin, W. (1998). *Speech and Hearing Science* (4th ed.). Boston: Allyn & Bacon.

CHAPTER 2
Cytology and Histology

CHAPTER PREVIEW

This chapter is an examination of the basic building blocks of the human body: cells and tissues. These essential components are not only the constituents of larger structures, but their study makes understanding of the larger structures more accessible. Material on cytology and histology is intended to cover the topics for students who either did not have a general human anatomy course or who need a review source. First, basic components of cells and sub cellular structures are covered. The text then moves systematically to cover progressively larger structures, including four basic tissue types and their developmental anatomical origins. This chapter provides a foundation for material readers will see again and again in subsequent chapters describing the gross anatomy of structures used for human communication.

CHAPTER OUTLINE

Cell Anatomy and Physiology
  Plasma Membrane
  Cytosol
  Cytoskeleton
  Organelles
  Nucleus
  Ribosomes
  Endoplasmic Reticulum
  Golgi Apparatus
  Mitochondria
  Lysosomes and Peroxisomes
Tissues
Epithelial Tissue
  Characteristics of Epithelial Tissue
  Functions of Epithelial Tissue
  Types of Epithelial Tissue
  Descriptive Terms for Epithelial Tissues
Connective Tissue
  Characteristics of Connective Tissue
  Functions of Connective Tissue
  Types of Connective Tissue
  Loose Connective Tissue

Dense Connective Tissue
Skeletal Connective Tissue
Cartilage
Bone
Articulations of the Skeleton
Fluid Connective Tissue
Muscle Tissue
  Characteristics of Muscle Tissue
  Functions of Muscle Tissue
  Types of Muscle Tissue
  Skeletal Muscle Nomenclature
  Muscle Power
  Movement Terms
Nervous Tissue
  Characteristics of Nervous Tissue
  Functions of Nervous Tissue
  Types of Nervous Tissue
  Neurons
  Glial Tissue
Embryology of Cells and Tissues
Chapter Summary
References and Suggested Reading

Cell Anatomy and Physiology

The study of cells is *cytology*. The word stems from the Greek *cyt* meaning "cell" and any word containing that stem is related to some aspect of cell anatomy or physiology. Cells are the basic physiological constituents of the human body and take on many different forms, depending upon their specific functions.

About 100 trillion cells comprise the human body. Their size is usually measured in microns (1/1000 of an inch). A red blood cell is about 7 to 8 microns in diameter. Other cells are relatively large, including some in the nervous system (neurons) having fibers that are visible to the naked eye and are measured in meters.

Cells can be placed into one of two broad categories. The most simple cells are *prokaryotic*. Prokaryotic cells have no organelles, not even nuclei. Bacteria, for example, are prokaryotic. The more complex and familiar cells are *eukaryotic*. They have separate organelles and a much more complex genome.

The study of tissues is *histology*. *Tissues* are collections of cells that serve a specific function and that typically have a common embryonic origin. The entomological root is *histio* which means "tissue."

The cells of the human body begin in their embryonic stage as identical *daughter cells*. Following their genetic predisposition, after a series of identical divisions, the identical cells differentiate into the component cells of the various tissues. These differentiated cells are so created in their various forms to perform their vital functions.

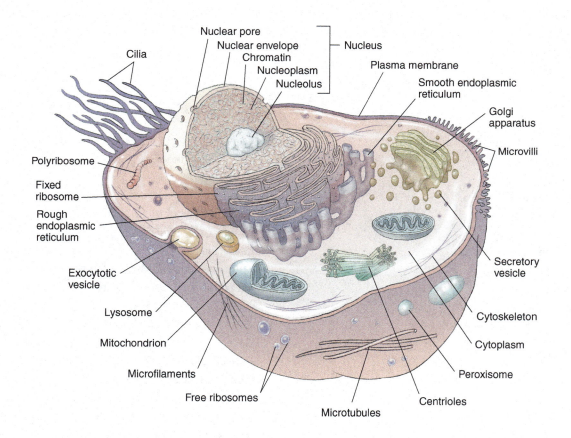

Regardless of differences in form or functions, all cells have certain common characteristics. These include an outer membrane, called the *plasma membrane*, a substance called *cytosol*, a *cytoskeleton*, and several types of subcellular entities called *organelles*.

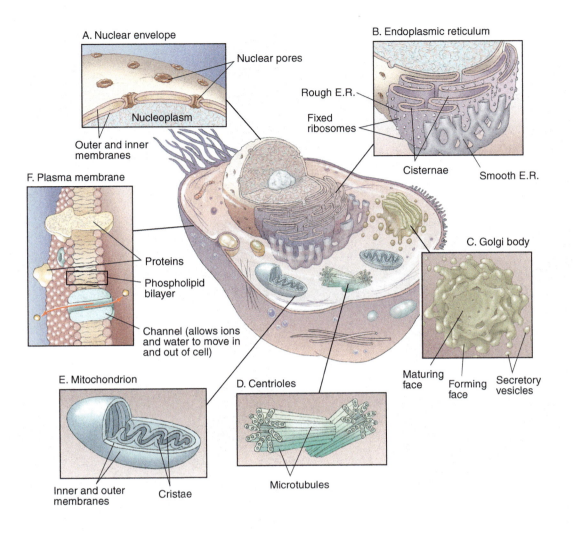

A. Nuclear envelope

Nuclear pores

Nucleoplasm

Outer and inner membranes

B. Endoplasmic reticulum

Rough E.R.

Fixed ribosomes

Cisternae

Smooth E.R.

F. Plasma membrane

Proteins

Phospholipid bilayer

Channel (allows ions and water to move in and out of cell)

C. Golgi body

Maturing face

Forming face

Secretory vesicles

E. Mitochondrion

Inner and outer membranes

Cristae

D. Centrioles

Microtubules

The Plasma Membrane

The plasma membrane of a cell is its outer coating. It is formed from combinations of proteins and lipids in a mosaic arrangement. The specific arrangement and chemistry of plasma membrane molecules allow the plasma membrane to separate the inner cell from its outer environment and to permit passage of certain substances in and out of the cell. It is this special characteristic that allows the cell to maintain its functional integrity while providing a conduit for nourishment and waste evacuation.

The ability to allow transit of various substances through the plasma membrane not only permits cell maintenance and survival, but also allows each type of cell to perform its specific bodily functions. It accomplishes this feat because each cell has special *proteins* embedded among the lipids that comprise its structure. These proteins, called *integral proteins*, form conduits through the membrane. Proteins specific to the various types of cells permit transit of specific substances, including hormones, nutrients, or neurotransmitters which allow each cell type to perform its specific bodily functions. Passage of substances intermediated by integral proteins are referred to as *passive transport* processes. *Active transport* processes involves mutation of the adenosine triphosphate (ATP) molecule to allow large particles to enter or exit a cell. ATP is a complex molecule which provides energy for cell functions, including active transport.

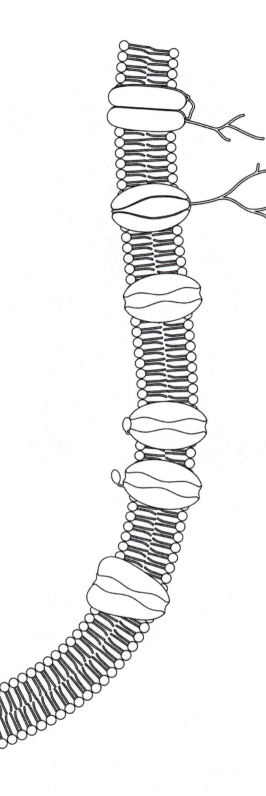

Cytosol

Cytosol is a gelatinous liquid confined within the plasma membrane in which the organelles and cytoskeleton are suspended. It is composed of about 75% to 90% water and contains several solid components, including solute amino acids and more complex amino acid combinations called *proteins*, simple and complex *carbohydrates*, *fatty lipids*, and *inorganic substances*. The cytosol receives this material from extracellular fluids by transport across the plasma membrane. Once they are within the plasma membrane, cytosol acts as a medium within which these solid substances support the cell's activity by carrying out chemical reactions.

Cytoskeleton

Some integral proteins of the plasma membrane form attachment points for tiny filaments that form the cell's skeleton. This skeleton is called the *cytoskeleton*, and it is the primary structural basis for the shape and integrity of the cell. The cytoskeleton also underlies cellular movements, including those of substances within the cell's inner environment and those of the cell itself in cases where a cell is mobile.

Cytoskeletal structures are formed of protein fibers. These microscopic structural components are formed of molecular combinations called polymers and come in various sizes.

The smallest of the cytoskeletal proteins are *actin filaments*. These are also called *microfilaments*. Actin filaments are protein rods of varying length that provide structure and support for the plasma membrane in the same manner that poles support a tent. As their name suggests, microfilaments are formed of the protein *actin*. Varying polarity of the actin molecules can create movement within the cell for functions such as secretion. It can also create movement of the entire cell. In muscle cells, actin molecules interact with larger myosin molecules to shorten the entire cell. Actin and myosin molecules form the smallest muscle contractile unit, the sarcomere. Embryonic cells that migrate from one region to another also rely on the variable polarity of actin-based cytoskeletons to make their embryonic journeys.

Next in size are the *intermediate filaments*. These cytoskeletal structures are formed of several types of proteins most of which are bound together to form polymers. Intermediate filaments provide most of the structural integrity of the cell. In cells that have specific structural purposes, such as connective tissue cells, intermediate filaments bind with one another for additional strength. Certain intermediate filaments called *keratins* are produced by specialized cells to form hair and nails.

The largest cytoskeletal components are *microtubules*. Microtubules not only provide structure to the cytoskeleton, but they form polymer pathways upon which certain organelles move within the cell. Microtubules are formed of the protein *tubulin* and are polarized, like tiny magnets, with "positive" and "negative" ends. The negative ends of microtubules are embedded in structures called *centrosomes*. The centrosomes are located near the cell nucleus, and the "positive" ends of microtubules are free to create a weak attraction for loose tubulin proteins within the cytosol. Thus, microtubules are in a constant state of growth and disintegration.

MUSCULAR DYSTROPHY AS A DISEASE OF THE CYTOSKELETON

Muscular dystrophy is a group of progressive diseases of the muscular system, characterized by loss of muscle cells, weakening and wasting the skeletal muscles. Nine types of muscular dystrophy have been identified, the most two most common of which are the Duchenne and Becker types (Richman & Schub, 2011).

The Duchenne and Becker types of muscular dystrophy are familial genetic conditions resulting from the failure of the muscle cell's cytoplasm to properly anchor to the extracellular matrix, undermining its ability to withstand the stress of contraction. It is believed to be related to a mutation of the "DMD" (dystrophin, muscular dystrophy) gene. This gene signals cells to create dystrophin, a protein essential to the cytoskeleton's attachment to the muscle cell's extracellular matrix (Bellayou, Hamzi, Rafai, Karkouri, Slassi, Azeddoug, & Nadifi, 2009).

Organelles

Organelles are subcellular entities that serve as tiny crucibles for chemical reactions within the cytoplasm. The structure and arrangement of organelles keep the many chemical reactions taking place within the cytoplasm from interfering with one another. Several types of organelles exist, each of which serve a particular purpose. These include *the nucleus, ribosomes,* two types of *endoplasmic reticulum, the Golgi apparatus, mitochondria, lysosomes,* and *peroxysomes*.

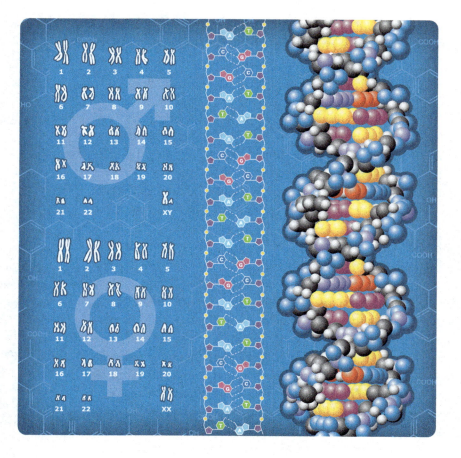

GENETIC DISORDERS AND CHROMOSOMAL DISORDERS

Inherited disorders fall into four main categories. Three of these are genetic disorders and the fourth is chromosomal. They are categorized according to the fundamental manner in which they affect the reproduction of the cells. Genetic disorders arise from the mutation of one or more gene. Chromosomal disorders result from abnormalities of entire chromosome chains.

Genetic Disorders may be of three types: monogenic disorders, polygenic disorders, and mitochondrial disorders. Some of these are apparent at birth, and are manifested by an unusual physical appearance or by metabolic disturbances. Others do not appear until later life. Many genetic disorders cause death in infancy or early childhood.

Nucleus

The nucleus of a cell is usually spherical or oval in shape and contains the cell's reproductive center. Most of the cell's hereditary information is contained within the nucleus on combined DNA and protein strands called *chromosomes*. This information is encoded on the chromosomes in series of *genes*. Genes are formed of *deoxyribonucleic acid* or *DNA*. Their presence and sequential arrangement in a chromosome provide the biological foundation the organism's development and function.

Cells in human beings are *diploid* cells, meaning that each cell receives half its genetic information from each parent. Each cell contains two copies of each chromosome, one copy coming from each parent. In the human being and other "higher" animals, each parent contributes twenty-three chromosomes. It is estimated that twenty to twenty-five thousand genes are embedded on these chromosomes (Stein, 2004). Human beings use these genes as design patterns to create all the proteins that will ultimately form every structure in the body.

The nucleus is surrounded by a nuclear envelope. This separates the nuclear material from the surrounding cytosol. In structure it is a double walled version of the plasma membrane, formed of phospholipid molecules. The pores in the nuclear envelope are larger than those in the plasma membrane allowing ready communication and transport of substances between the nucleus and cytosol.

Several *nucleoli* are also contained within the nucleus. These are aggregations of RNA, DNA, and proteins, and produce *ribosomes*. During cell division, the nucleoli disappear, to reappear in the daughter cells when division is complete.

Ribosomes

Ribosomes are granular organelles formed, as stated above, by nucleoli. They are composed of proteins, and *ribonucleic acid (RNA)*, and have no membranes. Ribosomes are found in the cytoplasm in several places, attached either to rough endoplasmic reticulum, contained in mitochondria, or as free bodies.

Ribosomes can synthesize proteins by "reading" the codes presented in the form of a particular type of RNA called "messenger" RNA. It then combines the structures of individual amino acids and combines them to form particular proteins as needed.

Endoplasmic Reticulum

Two types of endoplasmic reticulums are attached to the nuclear envelope and extend throughout the cytosol. Both types contain tubes and spaces called *cisterns* for protein synthesis and are covered by membranes. The two types of endoplasmic reticulums are called *rough* and *smooth*.

Rough endoplasmic reticulum is so-called because it contains ribosomes on its outer membrane. The ribosomes contained on the rough endoplasmic reticulum synthesize most of the proteins needed by the cell. These proteins are stored temporarily within the cisterns of the rough endoplasmic reticulum and are ultimately secreted from the cell.

The membranes of smooth endoplasmic reticulum contain no ribosomes. The tubules and cisterns of smooth endoplasmic reticulum are sites of lipid synthesis and of synthesis of certain detoxifying enzymes.

Golgi Apparatus

The Golgi apparatus, discovered by Camillo Golgi in 1898, and the endoplasmic reticulum have a functional relationship for refinement of cellular proteins. As created by the endoplasmic reticulum, proteins are not ready to be used by the cell or to be exported. The Golgi apparatus modifies the raw proteins and prepares them for their intended purposes.

Each Golgi apparatus has two distinct faces and a middle section. The face oriented toward the endoplasmic reticulum is called the *cis* face and receives the newly created basic proteins. These proteins are captured by the Golgi apparatus and transported through its interior chambers. In the middle section of the Golgi apparatus, the proteins are passed from chamber to chamber, steadily becoming more refined and prepared for function. When the proteins reach the other side of the Golgi apparatus, they exit through the *trans* face and are released into the cytosol. At this stage, proteins are ready to become structural parts of the cell, or are exported in such forms as hormones or enzymes for use in other bodily locations. The exportation of proteins through the cell membranes is called *exocytosis*.

Single gene (monogenic) disorders are related to a single gene. They are inherited in recognizable patterns or follow Mendelian laws of inheritance. Monogenic disorders include many rare diseases, syndromes, or morphological traits. These arise from defects in protein synthesis caused by the presence or absence of a particular gene.

Single gene disorders can be *autosomal* or *sex-linked* depending upon the nature of the chromosome on which the gene is located. If a single dominant gene located on an autosome can cause the disorder, the condition is said to be *autosomal dominant*. If a pair of recessive autosomal genes is required, the condition is *autosomal recessive*. If the condition the result of a gene on the X or Y chromosome, it is sex-linked. X chromosome linked conditions can be dominant or recessive.

Polygenic (-Multifactorial or Complex) disorders result when multiple genes combine with environmental influences. Polygenic disorders are sometimes difficult to recognize, but they include relatively common developmental defects including maxillofacial clefts. In many individuals, polygenic phenotypes are expressed in behavior, such as alcoholism or schizophrenia. Environmental influences may precipitate demonstration of the trait in some individuals, while other individuals will appear normal. Some postulate a genetic threshold for manifestation of a trait (Falconer, 1965). These individuals will not manifest the trait unless environment conditions are right.

Environmental influences in polygenic multifactorial inheritance may take the form of *teratogens*. Teratology is the study of diseases or conditions that cause congenital malformations, and substances in the environment that can cause defects in embryonic or fetal development are *teratogens*. Examples of teratogens include ethyl (beverage) alcohol, lead, uranium, and tobacco.

Susceptibility to many teratogenic agents depends on the genotype of the fetus and the manner in which it interacts with the environmental factors.

Susceptibility to teratogenic agents also depends upon the developmental stage at the time of exposure. Depending upon the cellular makeup of the individual at a given stage of gestation, a teratogen may cause varying malformations. Further, different teratogens affect different types of cells in different ways.

The effects of exposure to teratogens may range from functional disorders to dysmorphic changes, delayed development, and death. The extent of teratogenic effects is directly related to the amount of exposure or dosage. Naturally, the total dosage consists not only of the quantity of the teratogen presented during a single exposure, but also by the total number of exposures.

Mitochondrial Transmission. Mitochondrial transmitted conditions are carried in the DNA of mutated mitochondria. Readers will recall that mitochondria are extranuclear organelles found in each cell. Diseases transmitted through mitochondria are progressive and do not have a uniform effect on cells. This is because cells can have widely variable numbers of mutated mitochondria, and the extent of the disability appears to depend upon the proportion of the cells' mitochondria involved. The manifestations of mitochondrial transmitted conditions are not apparent at birth, but eventually cause disability.

It is widely believed that, since individuals inherit their cell mitochondria from the female parent, that transmission of mitochondrial traits are through mothers. However, there appears to be some evidence of rare paternal transmission. (Schwartz and Vissing, 2002).

Mitochondria

Mitochondria are the locations at which most of a cell's energy production takes place. In the mitochondria, simple carbohydrates are combined with enzymes to produce *adenosine triphosphate (ATP)*, a chemical essential for most of the cell's activities, such as protein synthesis, movement, transportation of substances, and the like.

It is in the mitochondria that cellular respiration takes place. This process, called the *Krebs cycle*, combines oxygen with carbohydrates in the presence of special enzymes to produce ATP, water and hydrogen ionic energy. It is the chemical basis for life. While some ATP production occurs in the cytoplasm, it is a relatively inefficient process and takes place without oxygen. Inside the mitochondria, though, the presence of oxygen greatly increases the amount of ATP that can be created in chemical reactions.

Mitochondria have double membranes. The inner membrane is folded into *cristae*. The folds of the cristae increase its inner surface area and allows for greater volume of ATP production. The fluids contained within the inner mitochondrial membrane are called its *matrix* and consist of proteins and enzymes that catalyze the Krebs cycle.

Deoxyribonucleic acid, DNA, is found in mitochondria, a fact that leads some biologists to believe that mitochondria are the descendants of prokaryotic cells that have evolved in symbiotic relationship with the eukaryotes in which they dwell (Cooper, 1997). Mitochondrial DNA forms the genetic infrastructure by which mitochondria replicate independently of the cell's replication.

Lysosomes and Peroxysomes

Lysosomes are the "digesters" of the cell. Their function is to break down or dissolve molecules in the cytosol. They also break down and recycle worn out organelles in a progress called *autophagy*. Lysosomes are generally spherical in shape and are enclosed by membranes. They contain a variety of enzymes by which they perform their roles. In addition, the interior pH of a lysosome is about 100 times more acidic than that of the cytosol. The membranes of lysosomes are particularly resistant to the chemical properties of their contents, and allow the lysosome to perform its function without interfering with the normal function of the cell.

Peroxysomes neutralize toxic substances in the cell by creating chemical reactions with them that produce hydrogen peroxide. Substances that may be toxic in the bloodstream, such as alcohols and aldehydes, react with peroxysomal enzymes and are neutralized. In addition to toxic substances, fatty acids are broken down in the peroxysomes.

Since both break down unwanted substances within the cell, peroxysomes are similar to lysosomes in function. However, the two organelles are different in two major ways. First, peroxysomes are self-replicating, like mitochondria, although they do not contain DNA. The biogenesis of lysosomes is still being investigated (Duclos, Corsini, and Desjardins, 2003; Kloer, *et al.*, 2010), but they appear to be synthesized in the cytosol by transient interactions during endocytosis. *Endocytosis* is an active transport process by which the plasma membrane

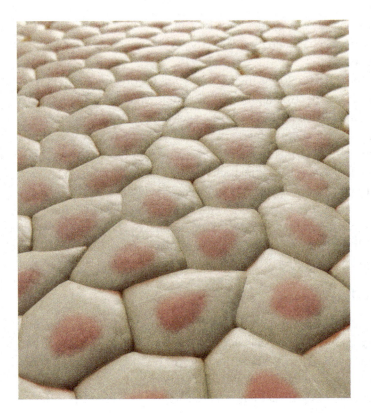

forms a vesicle around an outside particle and that outside material enters the cell passing through the transport proteins in the plasma membrane. Another difference between lysosomes and peroxysomes is structural. The peroxysomes have a crystaline structure, contained in a single membrane, while the lysosome is formed as a two-layer phospholipid structure.

This brief summary of cell structure is general. Cells vary widely in size, shape, and function, and a detailed analysis of cytological science is beyond the scope of this text. It must be emphasized that, depending upon the functions of various cells, the structures and organelles described above present varying preeminence. For example, the cytoskeleton is preeminent in muscle cells and connective tissue cells, secretory bodies play important roles in some epithelial cells, and the plasma membrane of nerve cells is unique in form and function.

Tissues

An aggregation of cells organized to perform a specific bodily function is called a *tissue*. Discernable differences in forms and functions of tissues in the body have resulted in differentiation of several *tissue types*.

Anatomists have differing philosophies regarding how many tissue types there are, and several schemes of tissue classification exist. These include classifications according to embryological origin, structural characteristics, or functional similarities. Embryological classification groups tissues according to the *germ layer* from which it developed. A structural classification looks at the tissue according to its form or shape. The functional schema groups tissues according to their roles in body physiology.

Chromosomal disorders are alterations in either the *number* of chromosomes or in the *structure* of a single chromosome. They may involve any of the twenty-three pairs of chromosomes. The viability of the fetus depends upon which chromosome is affected and/or to what extent a chromosome is affected.

The incidence of chromosomal errors in live births is about one in 200 (Jung, 1989). It has been estimated that as many as 5% of live births have some kind of chromosomal aberration (Carter, 1969). They are produced by microscopically detectable cytogenetic aberrations that occur during fertilization or shortly thereafter. Thus, they are most frequently not inherited (familial).

Certain chromosomal anomalies are aberrations in the *number of chromosomes*. The normal human somatic cell has a karyotype with forty-six chromosomes, arranged in twenty-three pairs. Aberrations can occur in the somatic cells, rendering a karyotype with an abnormal number of chromosomes.

If the total number of chromosomes is a multiple of twenty-three, the condition is a *polyploidy*. For example, if there are three times as many chromosomes in a somatic cell as normal, the condition is a *triploidy*. Such a condition yields 69 chromosomes, with the sex chromosomes being XXY if the individual phenotype is male or XXX if she is phenotypically female. Triploidy occurs during meiosis, where one gamete has an extra set of chromosomes. Polyploid individuals have poor survival statistics.

Aneuploidies are chromosome numbers that are not multiples of twenty-three. An aneuploidy can arise during meiosis or mitosis. This is the state during which the structure or number of chromosomes is most unstable. The odd number of chromosomes comes about by the development of unpaired chromosomes during the division process. Aneuploidies involve the grouping of certain chromosomes in singles or in triplets.

When a karyotype has only forty-five (out of forty-six) chromosomes, there is one chromosome that is not paired with another from the parent cell. This singleton effect has only been observed with sex chromosomes.

When a chromosome pair in a germ cell fails to split during meiosis, the event is called *meiotic nondisjunction*. The resulting gamete has forty-seven chromosomes, rather than the normal forty-six. Because the chromosome complement is abnormal, the resulting fetus will form abnormally.

The best known aneuploidies are *trisomies*. In a trisomy, a triplet of chromosomes occurs where there is normally a pair. This gives the individual an extra chromosome in the karyotype.

Trisomies can involve any of the chromosome pairs, but usually occur in the shorter (higher numbered) pairs. They are most often on chromosomes numbers 3, 18, 21, or sex chromosomes. Fetuses having trisomies of other chromosomes usually do not survive.

Down syndrome, also called *trisomy 21*, is a commonly encountered aneuploidy. The syndrome begins before fertilization, during the formation of the gametes of one parent. Chromosome pair number twenty-one remains intact, that is, does not split up like normal gamete chromosomes are supposed to do during meiosis. The resulting gamete has two twenty-first chromosomes from one parent and the regular one from the other. This leaves any zygote formed by the mating of the two gametes with three chromosomes of the twenty-first number. At that stage, the pattern is set. Any further cell division, now mitotic in nature, in that zygote, embryo, fetus, or individual, will have the trisomy of chromosome group number twenty-one. The causes of the original nondisjunction are unknown, but there is a positive correlation between frequency of occurrence and parental age (Fisch, Golden, Hensle, Olsson and Liberson, 2003).

This text groups tissues into four broad types, distinguished by structural and functional characteristics. These types are epithelial, connective, muscle, and nervous tissues.

Epithelial tissue forms the linings of hollow organs and the protective coverings of body surfaces, membranes of body cavities, and the secretory masses of glands. Connective tissues form the skeletal structure of the body and mediate maintenance functions. Muscle tissue mediates gross movement inside and outside the body. Nervous tissue provides communication between body structure and coordination of body function.

A notion of cell collections implies consideration of the space between the plasma membranes of adjacent cells. This space is called *extracellular*, and may contain a variety of material, depending upon what type of tissue is under consideration. Indeed, the nature of the intercellular space is one characteristic by which tissue type is determined.

Epithelial Tissue

Epithelial tissues form the protective coverings of the body glands and sensory end organs. In particular, it covers the face and hands, and also the linings of the oral, nasal, and pharyngeal cavities. For this reason, the clinician should be familiar with some generalities of epithelial tissue.

The cells of epithelial tissue are arranged in a broad and flat pattern. This arrangement is often termed mosaic, owing to the side-by side cellular situation, and reminiscent of the tiles in a mosaic art form. The arrangement of cells in epithelial tissue suits it well for its functions of forming a protective outer layer or functional inner lining of organs and of secreting certain bodily fluids as glandular tissue. Besides its cellular arrangement, epithelial tissue is distinguished by its sparse blood supply.

Characteristics of Epithelial Tissue

Epithelial tissue is distinguished from other tissues by its intercellular structure and its cellular arrangement. It usually has a close relationship to other tissue types and obtains its blood supply through diffusion. Epithelial tissue has its own nerve supply, and where it forms sensory end organs, it is closely related to nervous tissue. Epithelial tissue has very little intercellular material. Its cells are closely packed together, with very little material in the intercellular spaces.

The cells of epithelial tissue are arranged in a flat or mosaic pattern. They also have a free surface, meaning that one end that is not covered with another tissue type. The free surface faces either the outside of the body or the inside of a body cavity, such as the stomach.

Epithelial cells are usually situated over a layer of connective tissue, from which they are separated by a *basement membrane*. Basement membrane is formed of structural substances secreted by the epithelial cells.

Functions of Epithelial Tissue

Epithelial tissues are found throughout the inside and outside of the body. Epithelial tissue functions are related to the locations of the body structures they cover. These functions are protective, absorptive, secretory, and sensory.

Protective Function: Epithelial coverings of the outside of the body (the outer skin or epidermis) must prevent drying, protect from abrasion, and inhibit the infiltration of micro-organisms.

Absorptive Function: Epithelial tissues line the intestines, absorbing material from the lumen of the gut for nutrition.

Secretory Function: Epithelial cells secrete mucous in the linings of various parts of the body, such as the respiratory tract and stomach.

Sensory Function: The epithelia of the ears, eyes, tongue and nose (to name a few) contain special cells that respond to certain environmental changes by sending an action potential along the nervous tissues to which they are connected. This specialized type of epithelial tissue is *neuroepithelium*.

Types of Epithelial Tissue

Collectively, epithelial tissue is referred to as *epithelium*. Although there are instances in which the distinctions are not clear, epithelium may be classified into two broad groups: *covering and lining* epithelium and *glandular* epithelium.

Within the covering and lining groups are three types of epithelial tissue are: *proper epithelium, endothelium,* and *mesothelium.*

Proper Epithelium

Proper epithelium forms the protective inner and outer coverings of the body. This includes the skin, as well as the mucous membranes of the digestive tract and airway.

Mitotic chromosome nondisjunction occurs during a mitotic division. Somatic cells divide by mitosis. In mitotic nondisjunction, the duplicate chromosomes do not properly separate. Mitosis involves the distribution of four chromosomes, in two pairs, to each of two daughter cells. Each daughter cell forms with a set of duplicate chromosomes, in pairs, just like the parent cell. In mitotic nondisjunction, one new cell gets only one copy of the chromosome. This cell will not develop further. The other daughter cell will have three copies of the chromosome, or a *trisomy*. Once, again, the die is cast, and all further cells having trisomy in the karyotype will have abnormal development.

Mitotic aneuploidy can result in an individual having cell pedigrees with more than one compliment of chromosomes. Such a condition is a *mosaic*. In mosaic individuals, the cells have chromosomes from the same zygotes, but mitotic anomalies give some of them different genetic composition. Mosaicism may be more severe if the mitotic anomaly begins early in gestation. The mosaic karyotype may vary from cell to cell or tissue to tissue. Most mosaic conditions have cells of normal and trisomic composition.

Chromosomal disorders that involve a defect in the *structure of the chromosome* include anomalies in the entire protein string or in a segment of the string. Structural aberrations result from the breakage of chromosomes during cell division. Broken chromosomes usually repair themselves with no net loss of genetic information in the daughter cells. The total amount of genetic material is the same, even if the actual genes are moved from one broken chromosome site to another. Such an event is a *dislocation*. Thus preserved, the phenotype of the resulting individual is unchanged, but the individual is at increased risk for abnormal gamete production (Robbins, Kumar, and Cotran,

Structure of the Epidermis

Dead keratinocytes, those on the surface flake off

Stratum corneum

OLD

Stratum lucidum

Stratum granulosum

Living keratinocytes

Stratum spinosum

Dendritic cell

Melanocyte

YOUNG

Stratum basale

Dividing keratinocyte (stem cell)

Tactile cell

Dermis

Sensory nerve ending

1994). Sometimes, the chromosome does not break, but is seriously weakened. One "arm" of the chromosome is loosely connected to the rest of the body. One such instance is "Fragile-X" syndrome, said to be the most common of inherited mental retardation (Anderson, 1994).

Sometimes the genetic code is seriously changed. This occurs when genetic material is lost or a deletion. Most deletions result in death of the zygote or embryo. Survivors include individuals born with *Cri du Chat* syndrome, a developmental disorder that manifests in intellectual disability, morphogenic anomalies, and often a high pitched cry reminiscent of that of a cat. This syndrome results when the short arm of the fifth chromosome pair is lost.

Mesothelium

Mesothelium forms the protective sack-like walls of the four main body cavities. These four membranes are the *peritoneum,* which lines the abdomen; two *pleural membranes (pleura),* which line the inside of the thorax and cover each lung; and the *pericardium,* which surrounds the heart.

Endothelium

Endothelium forms the very smooth linings of the lumen of blood and tlymph vessels.

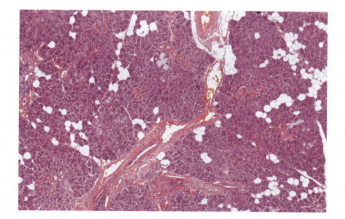

Glandular Epithelium

Glandular epithelium forms the secretory structure of the endocrine and exocrine glands. These organs produce and secrete special fluids essential for normal body maintenance. Such substances include hormones, for physiological regulation, saliva for digestion, and mucin to aid respiratory system maintenance.

Glands may be further classified as *endocrine* or *exocrine*. Endocrine glands secrete their special fluids, called hormones, into the intercellular space, and, by diffusion, into the bloodstream. Examples of endocrine glands are the thyroid gland and pituitary glands. Hormones are powerful chemical regulators of biological balance. Exocrine glands secrete their fluids through ducts or directly onto linings and surfaces, including the internal surfaces of hollow organs. Most exocrine glands are multicellular and secrete fluids through ducts. Exocrine fluids include saliva, a powerful digestive chemical, cerumen, the waxy substance found in the ear canal, and mother's milk. In some cases, such as in the lining of the upper respiratory tract, exocrine glands are unicellular. Unicellular glands called *goblet cells* are found among clusters of lining ciliated columnar epithelial cells, and secrete their fluids directly onto the inner surface.

LYSOSOMES AND TAY SACHS DISEASE

The enzymes that allow the lysosome to perform its digestive functions are produced in the endoplasmic reticulum and transferred through the Golgi apparatus to be specifically targeted to the lysosome (Schulze-Lohoff, Hasilik, and von Figura 1985). Synthesis and targeting of these enzymes are believed to be genetically determined, and mutations of the genome are believed to result in disorders of protein synthesis leading to mental retardation and related disorders (Vellodi, 2005).

Descriptive Terms for Epithelial Tissues

Epithelial tissues are described in terms of their shapes and the number of layers composing the tissues. A characteristic of epithelial tissue is that its superficial cells have a free end. The free ends may be turned toward the inner cavity of hollow organs, forming their linings, or toward the exposed surface of a body part. At the other end of an epithelial cell is attached to a *basement membrane*, which, in turn, is attached to a deeper layer of connective tissue. As mentioned above, the basement membrane tissue provides support for the blood and nerve supply to and from the epithelial tissue.

Cell Shape

One way in which epithelial tissues are described is in terms of the shapes of the composite cells. The shape of an epithelial cell depends, in part, upon the mechanical forces acting upon the cell, and in part, upon the functions of the cells. A cytoskeleton also provides structure for cell shape.

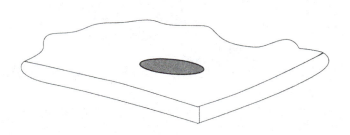

Squamous epithelial cells have a flat cell shape, and their nuclei take on a similar form. These cells form the inner and outer tissues of parts of the respiratory and digestive tracts, and are arranged like the tiles on a floor. Because they are flat, substances can slide over them easily. Squamous cell carcinoma is particularly dangerous, and speech-language pathologists treating patients who have had their larynges removed surgically (laryngectomy) will be aware of the devastating results of that cancer.

Cuboidal epithelial cells are shaped like cubes because of the way they are packed together. These cells can have secretory or absorptive functions.

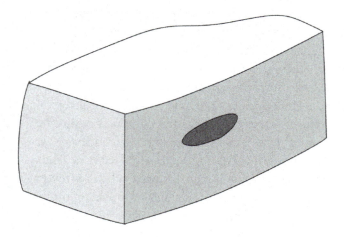

Columnar epithelial cells are cylindrical, long and narrow, and have a functional end. Columnar cells often have cilia at their free ends. One example of a ciliated columnar cell is found in the basilar membrane of the cochlea. These "hair cells" turn hydraulic energy into auditory impulses, to be interpreted by the brain as sound. Other examples are found in the nose and activated by chemical changes in the air to be interpreted as smell. Other columnar epithelial cells line the insides of the respiratory tract. They have steadily waving cilia at their free ends, which wave steadily and help remove debris from the respiratory passages.

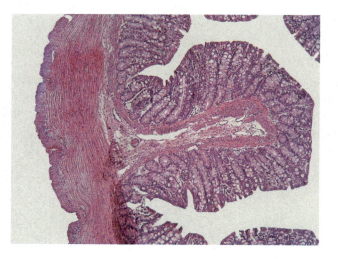

Goblet epithelial tissue cells are shaped as their name implies. These special cells are a type of columnar cell, and have secretory functions in the respiratory, digestive, urinary, and reproductive tracts, where they form unicellular glands. In these locations, they secrete *mucous*, a viscous and sticky liquid that lubricates and helps clean the passages and may be called *glandular epithelium*. Goblet cells become distended at their free ends when the mucus accumulates.

In some cases, epithelial cells aggregate and form *multicellular glands*. These clusters of epithelial cells secrete substances in two ways. *Exocrine* glands secrete through ducts. The parotid salivary gland, which is located at the mandibular ramus, is an example of a multicellular exocrine gland. It secretes saliva, a digestive fluid, into the oral cavity through a duct opening at approximately the level of the second upper molar. *Endocrine* glands secrete substances directly into the bloodstream by diffusion. An example is the pituitary gland, a multicellular endocrine gland located at the base of the brain that secretes growth and metabolism regulating hormones and is closely related to the hypothalamus.

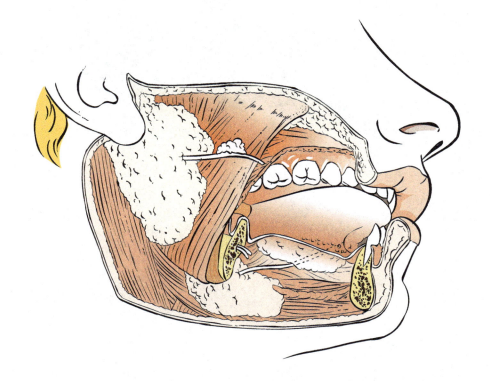

Depending upon their functional status or the position of the body parts in which they are situated, epithelial cells may change shape from cuboidal to columnar, or even appear pyramidal. For example, the epithelial cells lining the insides of curved surfaces may be shaped like pyramids. When distended with secretory substances, columnar cells may appear cuboidal.

Layers of Epithelial Cells

Epithelial cells are also described in terms of how many layers of cells are present. Cells of epithelial tissues may be arranged in a single layer or in two or more layers.

When arranged in a single layer, the epithelial tissue is said to be *simple*. The simple epithelial layer can be squamous, columnar or cuboidal. An example of simple epithelium is in the walls of arteries.

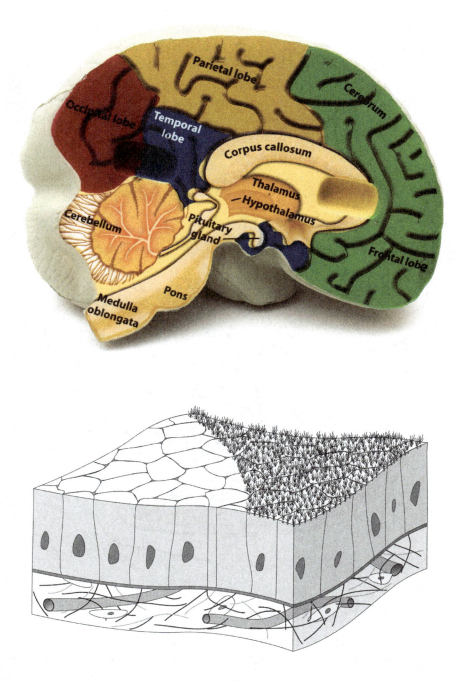

Stratified epithelial tissue consists of two or more layers. Stratified epithelial tissue lines the oral cavity, including the tongue. Cells in stratified epithelial tissue having free ends at the surface are referred to as *apical cells*. Cells situated between apical cells and the basement membrane have no free surfaces.

Stratified epithelial cells can be squamous, columnar, or transitional, but the shape description only applies to the outer layer. They exist in layers because the tissues are located in places where the outer call layer is exposed to wear. As the outer cells wear off, the lower cells take their places. The surface of the tongue is partly composed of stratified squamous epithelium, as are the surfaces of the vocal folds and ventricular folds in the areas where they contact on adduction.

Pseudostratified epithelium appears stratified because the cell nuclei are arranged at various levels along the columnar cells. This arrangement makes the cells appear to be arranged in several layers, where it is actually composed of several types

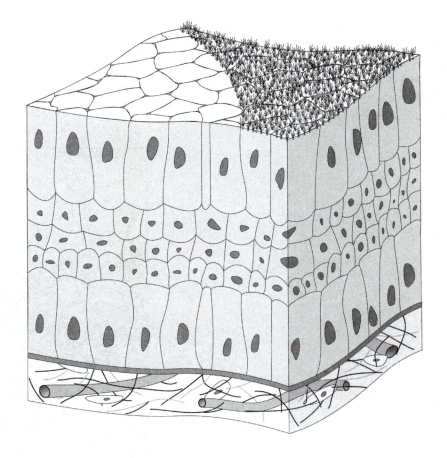

of epithelial cells arranged in a single layer. The lining of the laryngeal cavity, except at the medial surfaces of the vocal and ventricular folds, is lined with simple pseudostratified columnar epithelium.

Connective Tissue

Connective tissue forms the structural framework of the body and conducts essential transport, protective and storage maintenance functions. It is the most widely distributed tissue type and can be said to form the structure of the body, itself. For that reason, familiarity with the nature and distribution of connective tissue is essential to the study of anatomy and physiology of human communication.

Characteristics of Connective Tissue

Connective tissue is distinguished by its great variability in composition and form. Cells of connective tissues also have an exceptional ability to change form. Thus, cells of one type can assume the roles of other types of connective tissue cells. This characteristic makes them particularly suited for repair of structural damage.

Most connective tissue has an abundant blood supply, delivered through arteries and veins. Exceptions to this are cartilage, with no vascular system, and tendons, with poor circulation.

In composition, connective tissue has few cells and a great deal of material between these cells. This intercellular material is called matrix, and can take on many forms, from being liquid, as with blood, to being like sand, as with bone. Matrix is nonliving, and is composed of fibers and amorphous ground substances. Except in the case of blood or lymph, in which the matrix is liquid, matrix is usually created by the connective tissue cell, in response to chemical and, possibly, physical influences. The archetypical connective tissue cell is the *fibroblast*, and it is felt that these basic cells can, under the right circumstances, develop into some, if not all, other connective tissue cell types (Alberts, Johnson, Lewis, Raff, Roberts, and Walter, 2002).

Connective tissue is categorized according to the nature of its matrix. This makes it vary widely in form function. Matrix makes the connective tissue liquid, rigid or plastic.

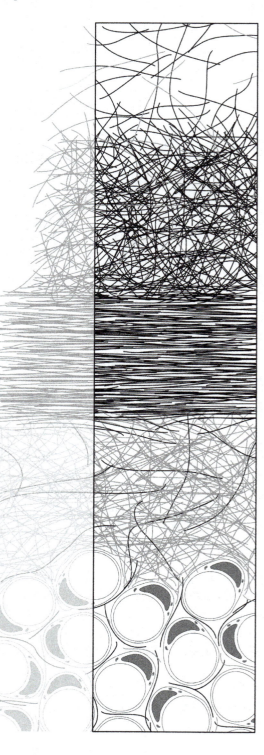

In form, connective tissue can take on a wide array of shapes. As bone, it can be long and thin, or flat. As fat or part of the skin, it can assume the shapes of the underlying tissues it surrounds. As a liquid, connective tissue assumes the shape of whatever vessel contains it.

Functions of Connective Tissue

Connective tissues variability in form suits them for a similar variability in function. Functions of connective tissues include the following:

Structural Binding

Ligaments bind skeletal members, bone and cartilage. Tendons bind musclesto bone or cartilage.

Structural Separation

Cartilage and fascial tissue separate body structures so they may function without binding against other structures.

Adipose Connective Tissue Functions

Adipose (fat) tissue is a kind of connective tissue that stores nutrients and insulates against heat loss.

Fluid Connective Tissue Functions

Blood and lymph are fluid connective tissues. These tissues are formed in the bones, which are connective tissue. Fluid connective tissue functions include transportation of gasses, nutrients, wastes, and hormones, regulation of temperature, hydration and pH, defense against microorganisms and toxins, and coagulation to stop leakage from the circulatory system.

Types of Connective Tissue

Connective tissues may be grouped into four general types: *loose, dense, skeletal* and *fluid* connective tissues. These are distinguished by their matrices. Although the mechanism is not clear, the four types of connective tissue have the abilities to replace one another, one type assuming the form of another. This is especially obvious with skeletal connective tissue, as the conversion of cartilage into bone.

Loose Connective Tissue

Loose connective tissue forms a base layer to help bind other cell types together and to perform certain functions associated with fat storage. As its name would suggest, loose connective tissue matrix fibers are loosely arranged among a liquid extracellular ground substance.

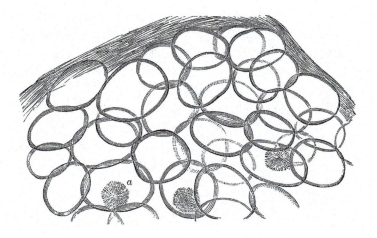

Three types of loose connective tissue are: *areolar, adipose,* and *reticular.* Areolar connective tissue is a binding type of connective tissue, and it is found all over the body. It surrounds nerves, organs, arteries, and veins, and binds other tissues together. Areolar connective tissue provides the blood supply for avascular epithelium. The connective tissue situated just deep to the skin epithelial tissue cells, and serving as their basement membrane, is areolar connective tissue. Adipose tissue has special cells that can store fats and oils (triglycerides). Reticular tissue is very loosely constructed and, like areolar tissue, serves a binding function in some organs.

Dense Connective Tissue

Because of its tightly packed fibers and flexible matrix, dense connective tissue is very strong and elastic. Dense connective tissue forms the tendons and ligaments of the body. Tendons connect muscles to bones and to cartilages, and ligaments connect bones to bones, cartilages to cartilages, cartilages to bones, and teeth to bones. Dense connective tissue also forms some thinner membranes called fascia and aponeuroses, which are broad, flat tendons.

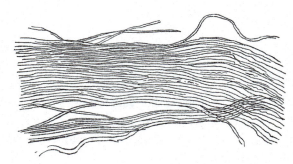

Skeletal Connective Tissue

Skeletal connective tissues give rigid structure to the body, creating a framework that supports the other tissue types. Skeletal connective tissues include *cartilage* and *bone*.

Cartilage

Cartilage is a flexible skeletal tissue. It gives structure to organs of the respiratory tract and the auditory system. It provides cushions between bones. Cartilage has no blood vessels. Cartilage cells, called *chondrocytes,* obtain their circulatory needs through diffusion of substances through the matrix. Because of this very slow process, cartilage injuries take much longer to heal than do bone injuries.

Types of Cartilage

Cartilage may be grouped into three types: *hyaline cartilage, elastic cartilage,* and *fibrous cartilage.*

Hyaline cartilage is bluish-white and translucent. Most of the body's cartilage is hyaline. Hyaline cartilage forms the skeleton of the larynx and the trachea. Its flexibility allows hyaline cartilage to maintain the patency of the airway while being moved about by changes in thoracic dimensions. This flexibility diminishes with age, in a process called *calcification,* or *ossification,* during which changes in the matrix ground substance occur. With increased laryngeal hyaline cartilage rigidity may come changes in phonatory and respiratory function.

Early in life, immediately following birth, hyaline cartilage matrix may lack sufficient rigidity to keep the airway patent. During this stage, a benign *inhalatory stridor* may result from lack of structural support for the vocal folds.

Elastic cartilage is yellowish and like rubber. Elastic cartilage is found in distal portion of the ear canal, the epiglottis, some smaller cartilages of the larynx, and the Eustachian tubes.

Fibrous cartilage is the most durable cartilage in the body. It is most notable in the intervertebral discs of the spinal column. Its toughness and density allow it to be an effective cushion against shocks transmitted through the spinal column during daily activities.

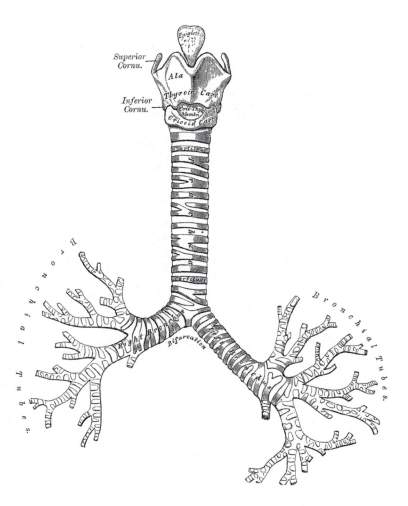

Bone

Bone is the most rigid connective tissue. It also has more intercellular substance than any other tissue type. This intercellular substance, or matrix, is composed of a collagenous substance and "ground salts." Ground salts are mostly calcium phosphate (Saunders, 2007). Bone cells, called *osteocytes, osteoblasts*, and *osteoclasts*, are sparsely distributed within the matrix. Bone tissue has a rich system of blood vessels and canals. This allows bone to develop fast and heal readily. Bone development is dynamic and ongoing throughout life.

Bones are covered inside and out with a very thin connective tissue membrane. Outside, the membrane is called *periosteum*, and supplies vascularity and innervation. It also helps healing in case of fracture (Beniker, McQuillan, Livesey, Urban, Turner, Blum, Hughes, and Haggard, 2003). The membrane inside the bone is called *endosteum*, and functions in the genesis of bone cells, as its composite fibroblasts change into osteoblasts. Inside the bone is a spongy interior, which contains *marrow*, either red or yellow. Red marrow manufactures red blood cells. Yellow marrow is adipose tissue. The spongy space between flat bones (like those of the skull) is called *diploe*.

Types of Bones

Bones come in a wide variety of shapes. Most of them are classified according to their shapes.

Long Bone

Long bones are longer than they are wide. They form the skeletons of the extremities. Long bones have several parts. The shaft forms the long body of the bone. Either end of a long bone is called its ephysis or its head. The rounded, smooth surface at which bones connect is called an *articular facet*, a *condyle*, or an *epiphysis*.

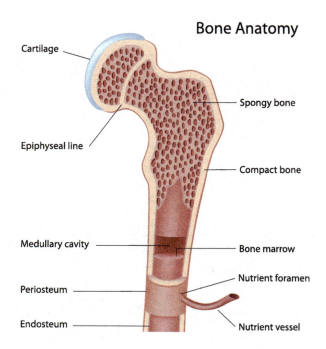

Bone Anatomy

Cartilage

Epiphyseal line

Spongy bone

Compact bone

Medullary cavity

Bone marrow

Periosteum

Nutrient foramen

Endosteum

Nutrient vessel

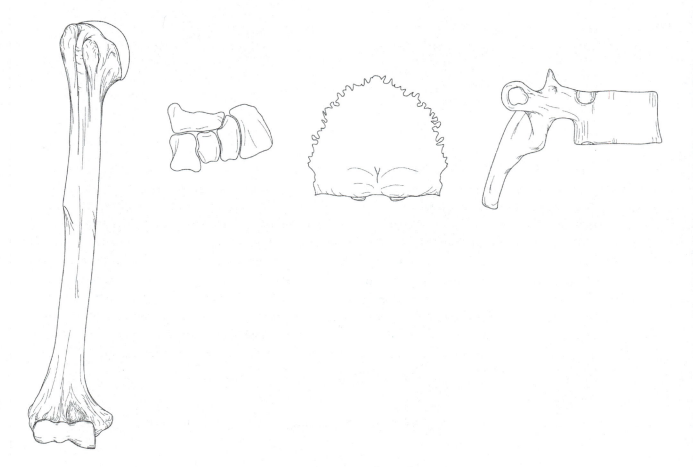

Short Bone
Short bones are formed like long bones, only they are shorter. They are found in the fingers and toes (phalanges).

Flat Bone

Flat bones have plates of bone sandwiching a layer of marrow (diploe). They form the bones of the upper skull (calvaria).

Irregular Bone

Irregular bones have no clear classification, since they have no consistent shape. An obvious example is a vertebra.

Some bones are classified by their locations, rather than their shapes.

Sutural or Wormian

Named after the sixteenth-century Danish anatomist, Ole Worm, these are bones that are located between the bones of some cranial sutures, most commonly the lambdoid suture. The presence of sutural bones is common in normal children, but when they occur in great numbers, has been associated with several congenital conditions, including *osteogenesis imperfecta, otopalatodigital syndrome,* and *Down syndrome* (Cremin, Goodman, Spranger, and Beighton, 1982).

Sesamoid bones, so named because their appearance is similar to that of a sesame seed, are contained in the substance of some tendons. The patella bones, commonly known as "kneecaps," are sesamoid bones, and the several sesamoid bones located in the feet allow vertical movement of the toes.

Articulations of the Skeleton

Bones meet each other at *articulations*. Articulation as a noun refers to the functional union between parts of the body. As a verb it means movement of one part of the body with respect to another. The ends of the bones that meet at moveable articulations are covered with hyaline cartilage. This cushions the articulation with the other bone, which, itself, has a similar hyaline covering.

Articulation Types

Articulations may be broadly classified either according to *functional* or *anatomical* categories. The functional classification sorts articulations according to the *degree of movement inherent in the joint*, and the anatomical classification sorts articulations according the *structures found in the joints interfaces*.

Functional Classifications

Functional articulation classifications include synarthrodial, amphiarthrodial, and diarthrodial joints.

Synarthrodial articulations are immoveable but may give slightly under pressure. The sutures of the skull are synarthrodial, as are those of the pelvis. During birth, some of the cranial and pelvic articulations are moveable, easing the newborn transfer through the birth canal. After birth, these articulations begin a slow fusion process.

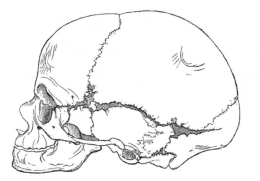

Amphiarthrodial articulations are slightly moveable. The articulations of vertebrae are examples of amphiar-throdial articulations.

Diarthrodial articulations are freely moveable. The articulations of the extremities or of the mandible with the temporal bone (temporomandibular joint) are good examples.

Anatomical Classification

Articulations may also be classified in terms of the anatomical structures situated at the points of articulation.

Fibrous articulations are immoveable and correspond to the synarthrodial functional classification. The material between the bones of a fibrous joint is dense connective tissue. The articulations of the teeth with their alveolar sockets in the maxilla and mandible are examples of fibrous articulations.

Cartilaginous articulations correspond to the amphiarthrodial functional classification. They are distinguished by the presence of a binding hyaline or fibrous cartilage between the bones. The articulations of the ribs with the spine or sternum are examples.

Synovial articulations are freely moveable and correspond to the diarthrodial functional classification. The synovial joint is so named because of the presence of a *synovial cavity* separating the articulating bones. This cavity contains a hyaline *articular cartilage* covering the epiphyses of the articulating bones. Surrounding the synovial space and binding the bones together is an *articular capsule*, formed on the outside of dense connective tissue and on the inside, of areolar connective tissue. The outer layer of the articular capsule serves as an articular ligament in some joints. The inner layer of the articular capsule secretes *synovial fluid*.

Synovial fluid has several functions that aid in the joint's function. The viscosity of synovial fluid is provided by *hyaluronic acid* or mucin. As a viscous fluid, it lubricates and cushions the joint. This fluid becomes thinner with increased movement allowing freer movement of the bones. It also contains *phagocytic cells* that destroy microbes and remove nonliving matter, nutrients to nourish the articular cartilage, and enzymes to reduce wastes from its chondrocytes.

Fluid Connective Tissue

Fluid connective tissue forms the blood and lymph. This type of connective tissue is distinguished by the loose molecular bonds of its intercellular matrix.

Blood

Blood consists of *formed elements* including *red blood cells (erythrocytes)*, several types of *white blood cells (leukocytes)*, and *platelets (thrombocytes)*. In general, erythrocytes are gas transporters, leukocytes are infection fighters, and thrombocytes are clotting agents. They are formed in the bone marrow. The formed elements are suspended in a yellowish transparent fluid, *plasma*, which is blood's matrix.

Blood flows through the *cardiovascular* system, propelled by the heart. During its flow, blood exchanges gases in the lungs, receives nutrients in the digestive system, is filtered in the liver and kidneys, and receives hormones from the endocrine system.

Blood flows from the heart to the organs through *arteries* and returns to the heart through *veins*. Because they are carrying blood from the beating heart, a rhythmic swelling and contraction known as a *pulse* can be observed by palpating the site where an artery runs close to the surface. Veins have no pulse.

Arteries are thick vessels with three layers or *tunics*. The inner tunic is composed of simple squamous epithelium and a basement of collagen fibers. The middle tunic is smooth muscle fiber and elastic fibers. The outer tunic is composed of elastic fibers and collagen. The space inside an artery, through which the blood flows, is called its *lumen*.

Veins have the same general construction as arteries, but their tunics are much thinner. The middle tunic of a vein also has fewer smooth muscle fibers. The lumens of veins contain epithelial tissue valves to prevent blood from flowing backward.

Blood has several functions important functions in body maintenance.

Blood Functions:

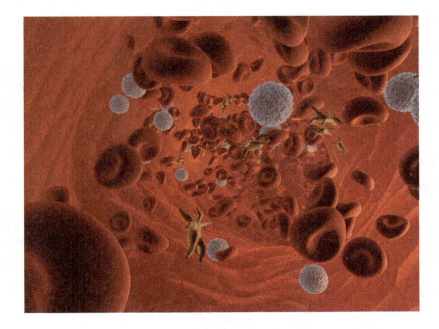

Oxygenation, Nutrition, Evacuation

Blood transports oxygen and carbon dioxide to and from the cells. It also transports nutrients to the cells and waste material from them.

Heat Distribution and Chemical Regulation

Blood regulates body temperature, regulates acid/alkali balance (pH), regulates water composition of the cells, and transports hormones to the cells.

Healing

Blood coagulates to prevent blood loss following injury.

Immunity

Blood defends against destructive microorganisms and destroys certain poisons.

Lymph

Lymph is a clear and viscous fluid connective tissue that circulates in the *lymphatic system*. In composition, lymph is much like blood, with cellular components suspended in a plasma matrix. The lymph cell is called a *lymphocyte* and is a type of white blood cell. Lymph receives its liquid plasma matrix from blood plasma fluids that pass through capillary walls. Lymph contains no red blood cells or platelets.

The lymphatic system is closely related to the cardiovascular system. It is composed of lymphatic vessels, along which are situated *lymph nodes*, the *spleen*, and several glands, including the *thymus* and *tonsils*. Lymph nodes are lymphocyte-producing organs situated in the neck, armpits, chest, abdomen, pelvis, and groin. Lymph is propelled through the lymphatic system by skeletal muscle contraction and by the give and take of thoracic/abdominal movements during breathing.

Lymph plasma flows into the lymphatic system from the spaces between cells (*interstitial spaces*). It then infuses into the capillaries of the lymphatic system and circulates throughout the body. Lymph returns to the venous blood stream, just before it reaches the heart.

Lymph Functions:

Lymph *transports* nutrients and waste materials to and from the cells and *fights infections*.

Muscle Tissue

Muscle tissue mediates the body's external and internal movements and creates heat. These specialized tissues allow observable movements, and equally as important, maintenance of posture, in structures such as the peripheral speech and hearing mechanisms, trunk and extremities. Muscle tissue also provides less observable movement, such as those internal movements that propel blood and lymph through their circulatory systems or the food through the digestive system.

Characteristics of Muscle Tissue

The principal characteristic of muscle tissue is its ability to contract and, when acted upon by opposing muscles, to relax. Muscle tissue accomplishes contraction through propagation of electrochemical changes at the molecular level. Thus, another distinguishing characteristic of muscle tissue lies in its special sensitivity to electric charges.

Muscle tissue is organized into groups called *muscles*. Muscles provide movement force only through their contractions, and this force is generated in a vector parallel to the lines of their fibers. In other words, a muscle can only *pull*.

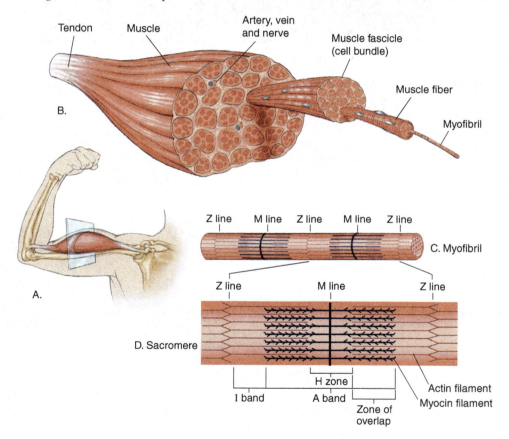

The cells that form muscle tissues are quite complex. Muscle cells, or *muscle fibers*, are very large and are formed by the fusion of many embryonic muscle cells called *myoblasts*. The result is an elongated adult muscle cell with many nuclei and a specialized cytoplasm called *sarcoplasm*. Muscle fiber contractile properties result from interaction of two proteins that compose the muscle cell cytoskeleton: *actin* and *myosin*. Actin and myosin, along with several other proteins contained within the sarcoplasm, are arranged roughly parallel to one another in elongated chains called *sarcomeres*.

The cytoskeletal proteins of a muscle cell are sufficiently compressed that they crowd other essential support organelles out toward their cell membranes. In muscle cells, the cell membranes are called *sarcolemmas*. The sarcolemma is specially constructed to allow chemical reactants to penetrate deep into the cell's center for contraction or relaxation. Muscle cells also contain many mitochondria to support their special need for adenosine triphosphate (ATP). The contracted status of a muscle is maintained by glycolysis and can continue as long as the chemical substrates, ATP and creatine phosphate (CP) are present and the accumulation of resultant *lactic acid* does not interfere.

Muscle contraction and the movement of body parts to which they are attached results from the shortening of successively larger and more complex *contractile units*. The smallest of these contractile units is the sarcomere, formed largely of actin and myosin, and is part of the individual muscle cell's cytoskeleton. When a stimulus is applied, a change in the relative polarity of the actin and myosin molecules causes them to slide along one another, shortening the sarcomere. An aggregation of sarcomeres, arranged end-to-end, is a *myofibril*. As the sarcomeres that form a myofibril become shorter, so the myofibril becomes shorter, as well. Thus, the chain of contraction continues. Several myofibrils bundle into groups to form a muscle fiber. A muscle fiber is, as mentioned above, the multinucleated muscle cell. A group of muscle fibers is called a *fascicle*, and several fascicles form the muscle.

The length of a muscle depends upon how many myofibrils are connected, and the thickness of a muscle is the result of the number of myofibril bundles organized in parallel arrangements. A muscle can contract to about half its relaxed length. The magnitude of its potential force depends upon its thickness, with the thicker muscle having the potential to generate greater force than the thinner one.

Contraction stimuli normally arrive from the discharge of motor neurons. When the stimulus ceases, or when the muscle becomes fatigued, external force, such as that produced by an opposing muscle, can pull the actin and myosin molecules away from one another with very little effort. This successively lengthens the progressively contractile units.

Types of Muscle Tissue

Three types of muscle tissue include *striated,* also called *skeletal* or *voluntary*; *smooth*, also called *visceral* or *involuntary*; and *cardiac*. The first two types are named for the appearance of their cells, while the last is named for its location.

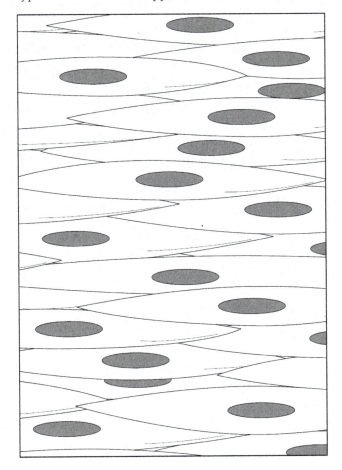

Striated muscle tissue is so named for its parallel arrangement of muscle fascicles. These bundles are easily visible, and their parallel arrangements result from the similar arrangements of their composite fibers. Striated muscle fibers are also known as skeletal fibers, because they provide motor function for the skeleton, including the trunk and extremities.

Striated muscles may be contracted through conscious, volitional programmed effort. This fact does not imply that they are always contracted through such voluntary effort. In fact, many striated muscle functions are conducted unconsciously and involuntarily, including postural maintenance, vegetative respiration, some aspects of walking, and protective responses of the upper extremities.

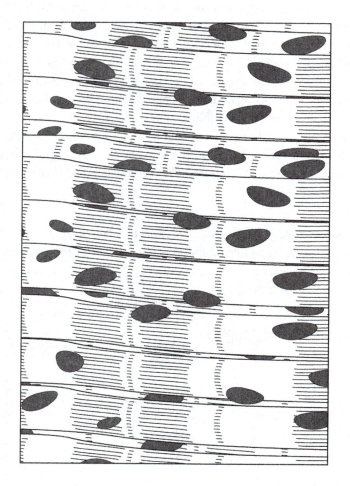

The fact that striated muscles may be easily activated through volition makes their function uniquely susceptible to learned programming through *operant conditioning*. Operant conditioning of voluntary muscle functions establishes patterns of movements based on the conditions that follow. This is of particular importance to the speech-language pathologist since striated muscle tissue predominates in the distal vocal tract and in the respiratory muscles. Most of the efforts put forth in speech therapy may be reduced to one or another operant conditioning paradigms.

Striated muscle tissue may be distinguished according to the speed of its contractile ability. *Fast* muscle fibers contract quickly and powerfully, but they cannot maintain their contractile status for prolonged periods. *Slow* muscle fibers react much slower to stimulation than their fast counterparts, but they are specialized to sustain their contractile status much longer. Most muscles in the human body are composed of both fast and slow fibers, but fast muscle tissue predominates in some muscles, such as the extraocular muscles, while slow muscle tissue is more dominant in the lower leg.

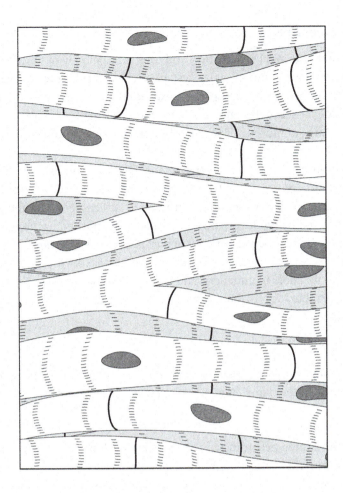

Smooth muscle tissue lacks the parallel arrangements of striated fibers, and the striated appearance. These muscles are found in the digestive and genitourinary passages. Smooth muscles are ideally suited for visceral functions, particularly peristalsis, because of their annular and longitudinal laminar arrangements.

Smooth muscle fiber contractions are much slower than those of even the slowest striated muscle fibers, and they differ further from striated muscle in that stimulation of one fiber may cause stimulation of adjacent fibers without intervention of the nervous system. The slow twitching of smooth muscle fibers allows them to maintain contractile status for long periods.

Smooth muscle tissue is not normally under voluntary control. Input from the autonomic nervous system can trigger the smooth muscles to maintain status or to relax and become quiescent. This phenomenon makes smooth muscles more susceptible to *classical conditioning*, by which contraction of smooth (visceral) muscle may be stimulated by manipulating environmental conditions before the muscle contracts. Classical conditioning may be useful in some speech-language therapy paradigms, such as relaxation therapy to help control speech fluency.

Cardiac muscle tissue is found only in the heart. Cardiac muscle fibers appear to have characteristics of both striated and smooth muscle fibers. They have one function, which is to provide force for the pumping of blood. Cardiac muscle has striations, although these are not arranged in parallel patterns, as are striated muscle fibers. Instead, the myofibrils of cardiac muscle branch out and communicate with adjacent myofibrils.

As is the case with smooth muscle fibers, cardiac muscle tissue is self-stimulating, although nervous system input can effect changes in the rates of its contractions. Since it is not normally under voluntary control, manipulation of antecedent conditions, as in classical conditioning, is the usual clinical approach to effecting changes in cardiac muscle contractions.

Skeletal Muscle Nomenclature

Movement is fundamental to normal life activities, and thus, a large body of terminology is associated with muscles and with form and function of the muscular system. These terms apply to the system of striated or skeletal muscles and their associated connective tissues, and exclude the smooth and cardiac muscles. Muscular system terms apply to the parts of muscles, in general and to the movements they effect, as well as to the identity of individual muscles and groups of muscles. They are of interest to any professional who is concerned with movement or its disorders.

Muscles effect movement by virtue of their attachments. They accomplish their movement functions by exerting a force against a resistance, and most usually in the form of a lever and fulcrum system. In such systems, the muscle applies a force to a moveable body part by virtue of its attachment and anchor to a relatively stable body part. Attachments are provided by tendons. The part of a muscle attached to the relatively stable anchor is called its *origin*. The part attached to the relatively moveable part is called its *insertion*.

Muscles names reflect one or more characteristics of the muscle, including its size, shape, location, number of origins, the locations of its ends, its function, and the direction of its fibers. These characteristics are usually expressed in Latin, or a combination of Latin and English, according to the conventions of the *Terminologica Anatomica*.

Muscle Names According to Size

Muscles that are relatively large are given the name *maximus,* and those that are relatively small are called *minumus.* Muscles that are long are given the name *longus,* while those that are short are called *brevis.*

Muscle Names According to Shape

The most familiar muscles named according to their shapes are found in the thorax, shoulders, and abdomen. These include the *deltoid* muscles, which are named according to their triangular shapes, the *splenius* muscles in the neck, which look like bandages, the *serratus* muscles, so named because of their jagged appearance,

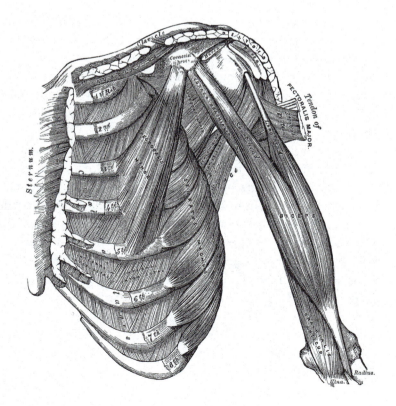

the *trapezius* muscles, which have shapes reminiscent of an irregular quadrilateral, and the diamond-shaped *rhomboid* muscles. The *diaphragm* is so named because it represents a fence or partition between the thorax and abdomen.

Other, less familiar, muscles named according to their shapes include the *gracilis* muscle, located in the lower extremity, named for its long, slender appearance, the *piriformis*, a muscle of the gluteal region, and *pyramidalis*, in the pubic region, both shaped like pyramids, and the *uvulae* muscle, a very small muscle, located at the distal end of the velum, in the posterior oral cavity, which is vaguely reminiscent of a grape.

Muscles Names According to Location

Many muscles are named according to body regions or structures nearby. For example, the *external intercostal* muscles are named for the ribs, to which they attach.

Muscles Names According Number of Origins

The suffix, "*-ceps*," in a muscle name derives from the Latin "*caput*," meaning "head," and refers to the number of a muscle's origins. Thus, "biceps" indicates a muscle has two heads, or origins, "triceps" indicates three origins, and "quadriceps" means four such attachments.

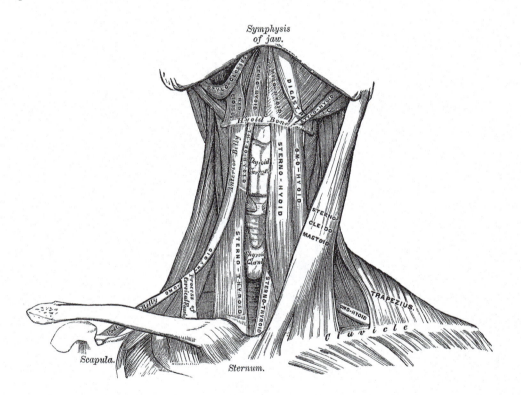

Muscles named according to their number of origins are located in the extremities. Following the Latin indicator of their origin numbers, they also have names indicating their locations. Thus, the muscle of the upper extremity having two origins is named *biceps brachii*, and the one in the lower extremity having two origins, *biceps femoris*.

Muscles Names According to the Locations of Their Ends

Muscles names given according to the locations of their origins and insertions begin with the origin. Since muscles can only contract, determining muscle functions from names that indicate a muscles origin and insertion is easy. The muscle that draws the thyroid cartilage inferiorly toward the sternum is, thus, named *sternothyroid*.

Muscles Names According to Function

Muscle function names reflect the results of that muscle's contraction. For example, the name *levator labii superiorus* indicates that contraction of that muscle lifts the upper lip, while *depressor labii inferiorus* lowers the lower lip. The following names also apply to movement terminology.

Latin terms used for muscle functions include (Adapted from Anderson, 1994):

Abductor: one that moves away from the midline

Adductor: one that moves toward the midline

Arrector: one that makes stands on end

Articularis: one that moves in relationship with another structure

Buccinator: one that blows a trumpet

Compressor: one that reduces in size

Constrictor: one that tightens

Corrugator: one that wrinkles

Cremaster: one that hangs

Depressor: one that moves inferiorly

Detrusor: one that thrusts

Dilator: one that increases in circumference

Erector: one that stands upright

Extensor: one that stretches out

Flexor: one that pulls in

Levator: to lift

Masseter: one who chews

Obturator: one that blocks or closes

Opponens: to oppose

Procerus: stretched

Pronator: one that bends forward

Risorius: one that laughs

Rotatores: one that turns about

Supinator: one that bends backward

Tensor: one that stretches

Muscles Named According to the Directions of Their Fibers

In the cases of muscles named according to the directions their fibers take relative to the midline of the body, *rectus* indicates the fibers run parallel to the midline, *oblique(us)* indicates they run at angles to the midline, and *transvers(us)* indicates the fibers course perpendicularly to the midline.

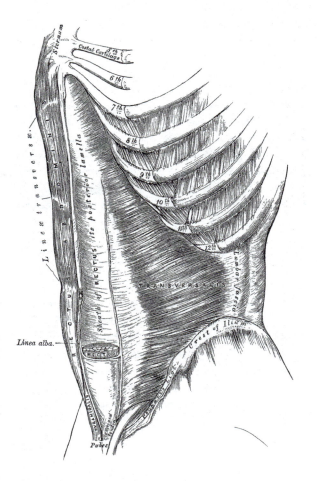

Muscle Power

A muscle can contract to about half its length. Longer muscles can, thus, cause longer movements. However, since longer muscles usually contain fewer fascicles, there is a relationship between muscle length and muscle power.

Muscle force is directly related to the number of fibers recruited to accomplish a contraction, and that force, applied over a given distance, represents the quantity of muscle power. The greatest number of fibers produces the greatest force. Therefore, thicker muscles have the potential to produce greater force. Fibers are gathered into bundles called fasciculi, or fascicles, and the orientation of these fasciculi relative to the tendinous attachments is one determinant of a muscle's power.

Muscles having fascicles in parallel arrangement with the tendons are called *parallel* (and colloquially, *strap*) muscles. An example of a parallel muscles is the *zygomaticus major* muscle of the face. Those having parallel fascicles, but with a pronounced *belly* in the middle are called *fusiform*. The bellies of the *digastric* neck muscle are examples. Muscles with concentric, circular fibers are called *circular*. The *oribicularis oris* muscle at the oral opening is such a muscle.

Pennate muscles have short fascicles arranged around their tendons. These are among the most powerful skeletal muscles. Some are *unipennate,* with fascicles on one side of a long tendon, itself about equal in length with the fascicles. The *masseter* muscles, which elevate the mandible, are unipennate. Some pennate muscles are *bipennate* with short powerful fibers, gathered into fascicles on two sides of the central tendon. *Rectus femoris,* in the anterior thigh, is such a muscle. *Multipennate* muscles have short fascicles, running at oblique angles to central tendons, which, themselves, are multiple in number and course in several directions. The *deltoid* muscles of shoulder and thorax are multipennate.

Movement Terms

Since muscles can only contract, another muscle or group of muscles must act to stretch the original muscle. Thus, many skeletal muscles and muscle groups are arranged in opposition.

For every movement, one muscle or group of muscles accomplishes the movement, and one opposes the movement. Muscles that accomplish a particular movement of interest are called *agonists* or *prime movers*. Opposing muscles are called *antagonists* or *reciprocal movers*. More movement terms are in the list of muscle names according to function.

Paralysis is the loss of muscle function secondary to interruption or compromise of the nerve supply. Causes can be vascular, infectious disease, degenerative disease, congenital malformation, trauma, or poison. Although the term *paralysis* usually implies motor or movement disorder, it also applies to loss of sensation as do the terms *anesthesia* and *paresthesia*.

The term *paresis* applies to incomplete loss of motor function, in cases where such distinction is required, and most often as a combined suffix, as in *hemiparesis*, indicating the partial loss of muscle function in the upper and lower extremities of one side, while *hemiplegia* applies to the complete loss of movement in both extremities of a single side. *Diparesis* or *diplegia* refers to the partial or complete loss of muscle function in both upper or both lower extremities.

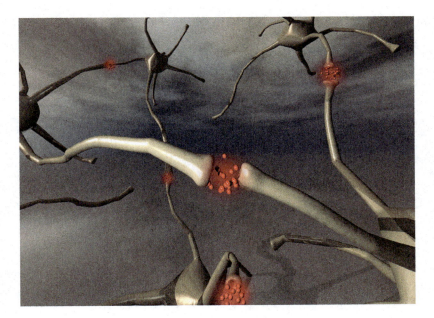

At rest, all skeletal muscles have, a certain degree of contractile status which is called *tonus*. Tonus resists stretch by antagonistic muscles and maintains posture. Paralysis terms may be extended to specify the type of movement dysfunction, that is, whether there is too much or too little tonus and whether muscle contractions are under voluntary control. *Hypertonic* or *spastic* muscles have too much resistance to stretch, while *hypotonic* or *flaccid* muscles have too little. *Clonic* muscles resist stretch in a quivering, rhythmic series of contractions. Muscle *tremors* are phasic, uncontrolled contractions of opposing muscle groups. Tremors may be *intentional*, when they worsen or occur only during or voluntary movement, and *unintentional* or *essential* when they occur without the volition for movement.

Nervous Tissue

Nervous tissue allows communication among all the various tissue types. In this way, changes in environmental conditions can be sensed and adaptive responses can be initiated, maintained and terminated appropriately. Communication is accomplished when an internal or external signal or *stimulus* is presented peripherally. Stimuli may take form of any change in homoeostatic conditions, including chemical, mechanical, photic, or acoustic energy variations. A stimulus of sufficient magnitude initiates a series of cellular changes called *afferent impulses*. Afferent impulses are relayed centrally, processed, and as appropriate instigate a series of cellular changes called *efferent impulses* that are propagated distally.

Characteristics of Nervous Tissue

The chief characteristics of neural tissue are its specialized excitability and its ability to transmit the excitement in a specific direction. This excitability takes the form of changes in chemical polarity across the neural plasma membrane. Polarity changes of the cell membrane are called *action potentials*, and are propagated along the cell membrane in a single direction and from cell to cell. However, only neural tissues, that is, collections of neurons, perform this specific function.

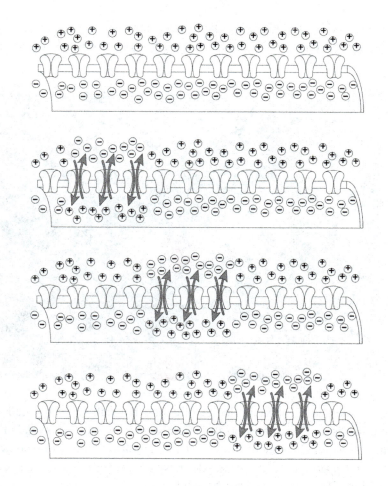

Other nervous system tissues, called *glial cells* or *neuroglia*, or *glia*, provide support, both literally and figuratively, for neural tissue. Such support includes suspension of the neural cells in position, maintenance of a chemical barrier between the circulatory system and the brain, amplification of action potential propagation, enhancement of neurite regeneration, and insulation of the neural cells.

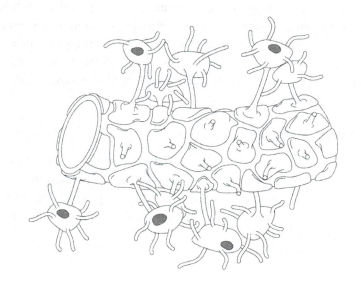

Types of Nervous Tissue

Two general types of tissue compose the nervous system: neurons and glia. Neurons perform the functions of the nervous system, whereas glial tissue provides various types of support for the neurons.

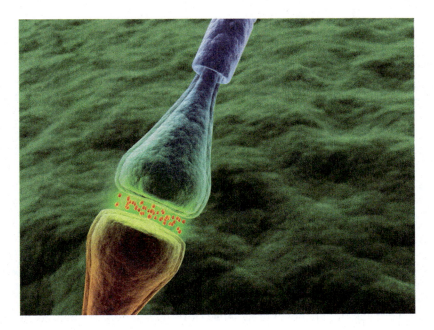

Neurons

Neurons are the functional cells of the nervous system. To meet that end, they are the conveyors of afferent and efferent action potentials to and from one another and the other body systems.

Two or more neurons communicate at a *synapse*. The neuron from which the action potential enters the synapse is the *presynaptic neuron*, and the neuron receiving the action potential is the *postsynaptic neuron*. Between the two neurons is a *synaptic cleft*.

A neuron consists of a cell body, also called the *perikaryon*, or *soma*, and one or more processes called *neurites*. Neurons take on a variety of shapes, depending largely upon the number and structure of neurites.

The cell body of the neuron is similar to the cell bodies of other tissue types. It contains the organelles of the cell and is the locus of cellular respiration and for chemical synthesis to support most of the cell's functional activities.

An important function of the neuron's organelles is the synthesis of *neurotransmitters*. These are chemical messengers that enter a synapse and facilitate the propagation of action potentials from one neuron to another. Neurotransmitters are of multiple types and are stored in vesicles along the microtubular cytoskeleton from the cell body distally to the terminal ends of neurites. When a presynaptic neuron is stimulated, the ionic change releases the neurotransmitter from its vesicle through exocytosis, eliciting a corresponding response in the postsynaptic neuron.

More than fifty neurotransmitters have been identified, and these fall into two broad categories: small-molecule transmitters and neuroactive peptides. Depending upon its chemical structure, neurotransmitter molecules may facilitate or modulate action potential transmission across a synapse by reacting with large post-synaptic receptor proteins in the postsynaptic neuron's plasma membrane. *Acetylcholine* is a small-molecule fa-cilitating neurotransmitter. It is found at the myoneural junction where it initiates muscular contraction. Neuron organelles also synthesize enzymes to break down the neurotransmitters and end synaptic transmission. One such enzyme is *acteylcholenesterase (AchE)* which breaks acetylcholine down (hydrolyzes) into non-facilitating molecules, acetate, and choline.

Neurites are of two general types. *Dendrites* convey action potentials toward the cell body and are afferent, whereas *axons* convey action potential away from the cell body and are efferent. Dendrites, then, receive information from sensory receptors, or from other neurons, and axons transmit information.

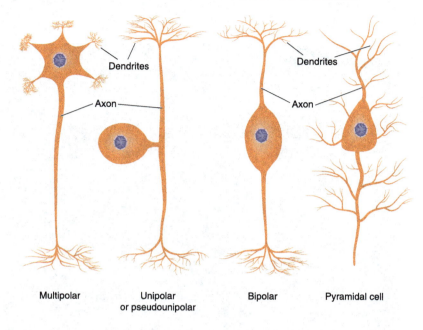

Multipolar   Unipolar   Bipolar   Pyramidal cell
or pseudounipolar

Unfortunately, such broad terms are not entirely satisfactory in distinguishing neurites. Morphological distinctions between dendrites and axons are more accurate. Thus, dendrites may be distinguished from axons by the presence of receptor proteins their cell membrane structures. Similarly, axons may be distinguished by the presence of an *end brush* and *terminal boutons* in which vesicles await the signal to release their stored neurotransmitters into a synaptic cleft.

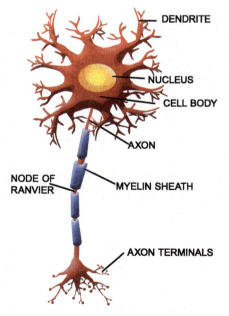

In many examples, axons are distinguishable by the presence of a lipid sheath called a *myelin sheath*. This substance is the product of glial cells that grow around an axon in a concentric arrangement, forming a sheath along the length of the axon. In the central nervous system, myelin is produced by oligodendrocytes, whereas in the peripheral nervous system it is produced by Schwann cells. A myelin sheath accelerates propagation of action potentials along the length of an axon, and is present on axons which must extend over relatively great distances. Examples of such long axons are those that form the *white columns* of the spinal cord, carrying action potentials rostrally and caudally along its length.

Neurons vary widely in form and function, and schemata of classification are elusive. Variation in the morphology of neurons is so great that a universally accepted way to classify them has yet to appear.

Perhaps the most widely disseminated neuron classification system is that which groups neurons according to the number of neurites present. By this system, neurons may be *unipolar*, having a single neurite, *bipolar*, having two neurites, and *multipolar*, having more than two neurites.

Unfortunately, classification according to the number of neurites has its shortcomings. *Unipolar* neurons are rare in the postpartum human nervous system. True unipolar neurons have one axon and no dendrites, and are anomalies in the adult human being (Leeson and Leeson, 1973). Instead, unipolar neurons exist in the embryonic human nervous system in invertebrate nervous systems.

Those neurons that are frequently called unipolar neurons are actually a morphogenesis of *bipolar* neurons and are more properly called *pseudounipolar* neurons. These have a single process that extends a short distance from the cell body and bifurcates. One limb of the bifurcation is afferent whereas the other is efferent. Both limbs have myelin sheaths.

The most commonly described pseudounipolar neurons are those afferent nerves having cell bodies in the trigeminal ganglia and in the dorsal root ganglia of the spinal nerves. These neurons convey action potentials associated with touch, *somesthesis*, from the periphery to the central nervous system.

Bipolar neurons are found in organs of the special senses. Bipolar neurons form the spiral ganglia in the inner ears, and are cell bodies of the cochlear division of the eighth cranial nerve. These neurons convey action potentials associated with audition to the central nervous system.

Most of the neurons in the human nervous system are multipolar. The motor neurons in the ventral horns for the spinal cord are good examples. These neurons receive impulses from diverse central nervous system tracts and convey them to the muscles via the spinal nerves.

Other classification systems for neurons have attempted to cover examples not covered by the "number of neurites" system. Such systems are commonly interchanged, and include those that classify according to the shape of the cell body, the configuration of the dendritic tree, the length of the axon, the type of neurotransmitter involved, and the rate of action potential generation.

Multipolar neurons having cell bodies shaped roughly like pyramids are called *pyramidal cells*. These are found in the cerebral cortex which is just deep to the brain's surface. *Stellate* cells have multiple dendrites, extending in multiple directions for maximum communication between adjacent brain neurons, and have the appearance of stars. Motor neurons use acetylcholene for their voluntary motor functions, and are called *cholenergic*. *Golgi* neurons are of two types, *type I* having a long axon, and *type II* having short axons. *Alpha motor neurons* generate fast, rapidly decaying action potentials and *gamma motor neurons* generate slow, sustained action potentials.

Glial Tissue

Glial tissue is far more predominant in the nervous system than is nervous tissue. Most of its widely varied cells develop from the same ectodermal embryonic cells as do neurons, but through genetic programming, they assume their various forms and functions. The general term *glia* denotes all the various forms of glial tissue.

Until recently, most anatomical texts characterized glia as not being involved directly in propagation of neuronal action potentials (Zemlin, 1998). Indeed, the word glia comes from the Latin word for "glue," and was applied to this tissue in 1846 by German physician Rudolph Virchow suggesting its role as a binder of neurons (García-Marín, García-López, & Freire, (2007). Among those functions generally attributed to glia are creation and maintenance of the optimum chemical environment for neuronal function, insulation and facilitation of action potential transmission along the axon, and lining of the central nervous system cavity. When nervous tissue damage occurs in the central nervous system, and most neurons do not regenerate, glial tissue may replace nervous tissue by a process called *gliosis*.

More recently, neuroscientists are recognizing more active roles for glia. Although glial cells do not appear to generate action potentials, they do have the same types of protein neurotransmitter receptors embedded in their plasma membranes as are found in postsynaptic neuron plasma membranes. There is growing evidence that they are intimately involved in information exchange between and among networked neurons (Nedergaard, 1994; Parpura, Basarsky, Jeftinija, and Haydon, 1994; Pasti, Volterra, Pozzan, and Carmignoto, 1997).

Four Types of Glial Tissues

The most numerous of the glial cells are the star-shaped *astrocytes*. These cells are interspersed among the neurons of the central nervous system and were once thought to serve merely as space fillers. When neural tissue is destroyed, astrocytes proliferate and form a fiber network to support the remaining healthy tissue through a process called *astrocytosis*. However, astrocytes are the targets of much current research, and are being recognized as having other important roles, including regulation of neuronal extracellular environment and direct involvement with action potential propagation.

The myelinating glia are ologodendroglia and Schwann cells. These cells form sheaths around axonal membranes by growing in concentric circles around the axons, Ologodendroglia are found in the central nervous system, and Schwann cells in the peripheral nervous system.

As mentioned previously, myelin is a fatty layer that surrounds an axon. It is interrupted in regular intervals at spaces called *nodes of Ranvier*. The function of the myelin layer is to increase the velocity of action potential propagation in axons that have to extend great distances or where fast processing is important. The presence of a myelin sheath on a neurite is sufficient to distinguish it from a dendrite despite the course taken by its action potential. Thus, myelinated neurites that convey action potentials toward a neuron cell body are axons.

In the peripheral nervous system, the myelin sheath and the Schwann cells that form it are collectively called *neurilemma* or *epineurium*. They appear to support axon regeneration following injury.

The presence of a myelin sheath around axons gives the tissue they form an off-white color. This color gives rise to the term *white matter* of the nervous system. The white matter of the nervous system forms most of its bulk. It forms the major ascending and descending columns on the external surface of the spinal cord, the major inter- and intra-hemispheric connections of the cerebral hemispheres and the lateral connections of the spinal cord. It also forms the axons of peripheral nerves. Unmyelinated neurons have a gray appearance and form the *gray matter* of the nervous system, especially the cerebral cortex.

Ependymal cells form the lining of the nervous system. In the embryo, this lining takes the form of a tube. As development occurs, however, the tube expands at its rostral end, in several places, ultimately forming the ventricles of the brain and the central spinal cord canal. Ependymal cells are ciliated and help circulate cerebrospinal fluid.

Microglia support nervous system function by protecting it from disease, in the same way as do microphages in the circulatory system. These are the smallest of the glial cells, and due to their phagocytotic abilities, are thought to derive from blood microphages. Microglia are also reactive to various acute and chronic neuroinflammatory substances, including the amyloid plaque that accompanies *Alzheimer's disease* (Streit, Mrak, and Griffin, 2004).

Embryology of Cells and Tissues

An understanding of the developmental anatomy or *embryology* of tissues is essential to the understanding of the form and function of the larger structures they compose. Knowledge gained from histological embryology clarifies not only the understanding of gross structural interaction, but also the understanding of the cause and progress of certain diseases. This section is concerned with the developmental anatomy of cells and the tissues they form. Later sections will cover the embryology of specific systems of interest.

Since tissues are basic components of all the systems to be covered in later sections, a few observations about basic histological developmental anatomy are appropriate here. These observations will provide a foundation for the developmental anatomy of other systems.

Three basic stages in developmental anatomy are: *preembryonic, embryonic,* and *fetal*. Each is marked by distinct morphogenic changes.

The preembryonic stage begins at conception when sperm and ovum unite to form a *zygote*. It extends to the appearance of the first somite, about two weeks later. During this stage, the cells divide in a process called *cleavage*. The first cell divides into two cells. These two cells divide, in turn, into four cells, and so on. Cells thus formed are identical. Through repetitive cleavage, the mass of the developing individual increases to form a cluster of cells called a *morula*.

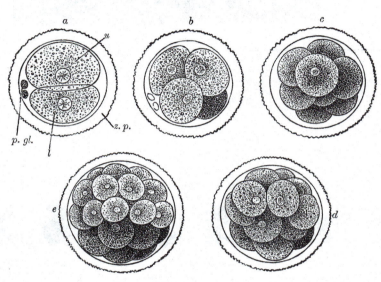

The embryonic stage occurs from the end of the second week until about eight weeks postconception. During the embryonic stage, cells *differentiate*, that is, they take on specific structural characteristics that will enable them to serve their functions, both before and after birth. The first differentiation occurs when cells of the morula arrange themselves into an inner cell mass, or *embryoblast*, which will become the fetus, and a *trophoblast*, which will become the placenta. At this point, the morula is called a *blastula*.

Embryonic Germ Layers

During the third gestational week, differentiation becomes more complex as the embryoblast cells differentiate into two laminae called *germ layers*. This is accomplished through invagination of the hollow blastula, creating inner and outer layers of tissue. The developing embryo is now called a *gastrula*. Cells forming its layers undergo their own morphogenesis. The outer lamina is called *ectoderm* and the inner layer called *endoderm*. In this manner, the embryo begins differentiation of its internal and external structures including the digestive system on the inner side and the superficial layer of skin on the outer side.

Endoderm will form the embryonic gut, ultimately forming the inner lining of the proximal digestive tract. Endoderm will also form the inner lining of proximal respiratory tract, liver, pancreas, and other glands.

The folding of external tissue, ectoderm, into the interior of the gastrula results in some ectodermal presence in the interior of the advanced embryo, the fetus and, ultimately, the adult. Thus, ectoderm will form the epidermis, nails, hair, teeth, and some exocrine glands, as well as the distal linings of respiratory and digestive tracts including the anterior of the oral cavity. It also forms the nervous system, which begins as a longitudinal folding of ectoderm along the dorsal axis.

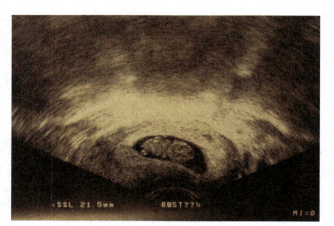

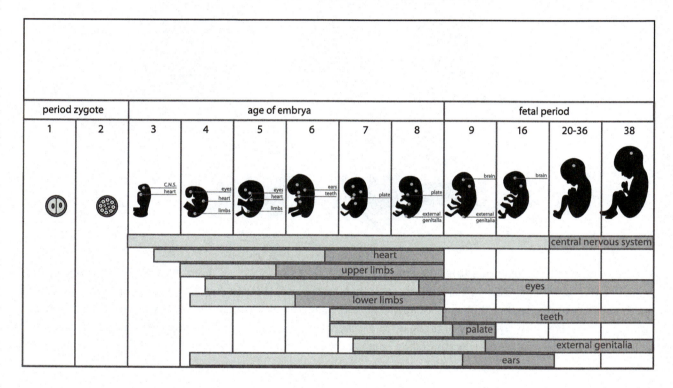

period zygote		age of embrya						fetal period			
1	2	3	4	5	6	7	8	9	16	20-36	38

Through continued *gastrulation,* the period during which cells differentiate and migrate, a middle lamina germ layer forms, between the endoderm and ectoderm. This tissue is called *mesoderm*. The morphogenesis of mesoderm is complex. Initially, mesoderm differentiates into three subtypes of tissue: *mesenchyme, somites,* and the *notochord*.

Mesenchyme cells are loosely organized and migratory. They will migrate to various locations in the embryo and ultimately form the connective tissues of the body, including cartilages and bones, the circulatory system, and the dermis.

Somites are swellings, situated in pairs along the dorsal axis of the embryo. They develop laterally to what will eventually become the nervous system, appearing first at the rostral end of the embryo, but caudal to what will become the skull. Up to forty-four pairs of somites may appear along the dorsal axis of the human embryo, but several will disappear. The final number of somites in the human embryo is thirty-one. The somites form most of the body wall, including the axial skeleton, muscles, gonads, excretory system, and outer coverings of the internal organs. The word *somatic* refers, then, to those structures of the body wall.

The *notochord* is an axial rod of mesoderm that is destined to become the spinal column. It will become the main axial support for the entire body.

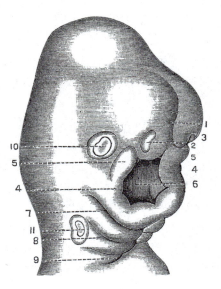

Each of the three gastrular laminae, germ layers, gives rise to a specific group of organs, including examples containing epithelial, connective, muscle, and nervous tissues. Association of these laminae with their derivative organs can be useful in understanding the scope and extent of some diseases, for anomalies of a particular germ layer may result in malformation or malfunction of some or all other organs derived from that germ layer. For example, ectoderm gives rise to both the integumentary and nervous systems, and certain nervous system disorders may also be associated with skin anomalies.

Distinction of germ layer origins of the four basic tissue types, epithelial, connective, muscle and nervous tissues is complex since some tissue types originate from more than one germ layer. Ectoderm gives rise to both nervous and epithelial tissues. Mesoderm is the origin of some connective and all muscle tissues, and endoderm gives rise to various examples of internal epithelial tissues. The smooth muscle tissue of internal organs are derivatives of migratory mesenchymal mesoderm (Gravesa and Yablonka-Reuvenia, 2001).

CHAPTER SUMMARY

This overview of the elemental aspects of cytology and histology provides a foundation for the further study of speech and hearing anatomy and physiology. Such a foundation will provide a basis for a more complete understanding of the structures described in later sections. This chapter will also provide a basic reference for future questions that may arise in clinical work.

Although there are smaller living entities, the cell is the basic physiological unit in animals. After birth, cells vary widely in form and function, but all share some basic constituents, including a plasma membrane, which contains cytosol, a cytoskeleton, and several specialized organelles.

Four basic tissue types form all organs of the body. Epithelial tissue form protective coverings for the organs, connective tissue mediates maintenance and provides structure, muscle tissue mediates movement and creates heat, and nervous tissue coordinates functions among all the tissue types.

Originally identical following fertilization, cells begin to differentiate during the first two weeks of gestation, and this differentiation becomes more complex as gestation continues. During the embryonic stage, usually considered from two to eight weeks following conception, the primary activity is the differentiation of cells and refinement of the four basic tissues types into multiple forms with special functions to carry out life's activities.

REFERENCES AND SUGGESTED READING

Alberts, B., Johnson, A., Lewis, J., Raff, M., Roberts, K., and Walter, P. (2002). *Molecular Biology of the Cell. 4th ed.* New York: Garland. Retrieved May 26, 2011, from: <http://www.ncbi.nlm.nih.gov/entrez/query.fcgi?db=Books&cmd=search&term=connective+tissue>

Anderson, K. N. (Ed.) (1994). *Mosby's Medical, Nursing, and Allied Health Dictionary* (4th Edition). St. Louis: Mosby.

Bear, M.F., Conners, B.W., and Paradiso. M.A. (2001). *Neuroscience: Exploring the Brain.* Baltimore: Lippincott.

Bellayou, H., Hamzi, K., Rafai, M., Karkouri, M., Slassi, I., Azeddoug, H., & Nadifi, S. (2009). Duchenne and Becker muscular dystrophy: contribution of a molecular and immunohistochemical analysis in diagnosis in Morocco. *Journal of Biomedicine & Biotechnology*, doi:10.1155/2009/325210

bone matrix. (n.d.) *Saunders Comprehensive Veterinary Dictionary, 3 ed.* (2007). Retrieved May 26 2011 from http://medical-dictionary.thefreedictionary.com/bone+matrix

Cooper, G.M. (1997). *The Cell: A Molecular Approach.* Sunderland, MA: Sinauer Associates.

Carter, C.O. (1969). *An ABC of Medical Genetics.* Boston: Little, Brown and Co. Cited in Northern, J., and Downs, M. (1974). *Hearing in Children.* Baltimore: Williams and Wilkins.

Falconer, D. (1965). The inheritance of liability to certain diseases estimated from the occurrence among relatives. *Annals of Human Genetics, 29*, 51–76.

Fisch, H., Hyun, G., Golden, R., Hensle, T., Olsson, T. and Liberson, G. (2003). The influence of paternal age on Down syndrome. *The Journal of Urology, 169*, 2279-2278. Retrieved May 25, 2011 from http://www.harryfisch.com/pdf/Paternal%20age%20and%20Down%20Syndrome.pdf

Fuller, G.M., and Shields, D. (1998). *Molecular Basis of Cell Biology.* Stamford, CT: Appleton and Lange. García-Marín, V., García-López, P., & Freire, M. (2007). Cajal's contributions to glia research. *Trends in Neurosciences, 30*, 479–487. doi:10.1016/j.tins.2007.06.008

Darras, B.T., Korf, B.R., and Urion, D.K. (2005). Dystrophinopathies. *Gene Reviews*, Retrieved May 26, 2011, from: <http://www.ncbi.nlm.nih.gov/books/NBK1119/#dbmd>

Fitzhenry, R.I. (1993). *The Harper Book of Quotations* (3rd ed). New York: HarperCollins.

Gravesa, D. C., and Yablonka-Reuvenia, Z. (2001). Vascular Smooth Muscle Cells Spontaneously Adopt a Skeletal Muscle Phenotype: A Unique Myf5-/MyoD+ Myogenic Program. *Journal of Histochemistry and Cytochemistry, 48*, 1173–1194. Retrieved May 26, 2011, from: <http://www.jhc.org/cgi/content/full/48/9/1173>

Hogenboom, S., Wanders, R.J., Waterham, H.R. (2003). Cholesterol biosynthesis is not defective in peroxisome biogenesis defective fibroblasts. *Molecular Genetic Metabolism. 80*, 290–295.

Jung, J. (1989). *Genetic Syndromes in Communicative Disorders.* Austin, TX: Pro-Ed.

Kloer, D., Rojas, R., Ivan, V., Moriyama, K., van Vlijmen, T., Murthy, N., &... Bonifacino, J. (2010). Assembly of the biogenesis of lysosome-related organelles complex-3 (BLOC-3) and its interaction with Rab9. *The Journal of Biological Chemistry*, 285, 7794–7804. Retrieved from EBSCOhost.

Leeson, T.S., and Leeson, C.R. (1973). *Descriptions and Explanations: Practical Histology a Self-Instructional Laboratory Manual in Filmstrip.* London: W.B. Saunders.

Lewis, B. (1990). Familial Phonological Disorders: Four Pedigrees. *Journal of Speech and Hearing Disorders, 55*, 160–170.

Nedergaard M. (1994). Direct signaling from astrocytes to neurons in cultures of mammalian brain cells. *Science, 263*, 1768–1771.

Oxford Dictionaries (2011) http://oxforddictionaries.com/

Parpura, V., Basarsky, T.A., Liu, F., Jeftinija, K., Jeftinija, S., Haydon, P.G. (1994). Glutamate-mediated astrocyte-neuron signaling. *Nature, 369*, 744–747.

Pasti, L., Volterra, A., Pozzan, T., Carmignoto, G. (1997). Intracellular calcium oscillations in astrocytes: a highly plastic bidirectional form of communication between neurons and astrocytes in situ. *Journal of Neuroscience, 17,* 7817–7830.

Richman, S., & Schub, T. (2011). Muscular Dystrophy, Duchenne's. Retrieved May 25, 2011, from http://ehis .ebscohost.com/ehost/pdfviewer/pdfviewer?sid=64657603-ba07-4d2a-b594-e3b708520c08%40sessionmgr114& vid=5&hid=124.

Robbins, Kumar, and Cotran (1994). *Pathologic Basis of Disease.* Philadelphia: W.B. Saunders.

Stein, L. D. (2004). Human genome: End of the beginning. *Nature, 431,* 915–916.

Streit, W. J., Mrak, R.E., and Griffin, W. T. (2004). Microglia and neuroinflammation: a pathological perspective. *Journal of Neuroinflammation, 1,* 14, 1742–2094. Retrieved May 26, 2011, from: <http://www.jneuroinflammation .com/content/1/1/14>

Schulze-Lohoff, E., Hasilik, A., and von Figura, K. (1985). Cathepsin D precursors in clathrin-coated organelles from human fibroblasts. *The Journal of Cell Biology, 101,* 824–829.

Schwartz, M., and Vissing, J. (2002). Paternal inheritance of mitochondrial DNA. *New England Journal of Medicine, 347,* 576–580.

Tortora, G. J., and Grabowski, S.R. (1996). *Principles of Anatomy and Physiology.* New York: Harper-Collins.

Ullian, E. M., Christopherson, K. S., and Barres, B. A. (2004). Role for glia in synaptogenesis. *Glia, 47,* 3, 209–216. Retrieved April 7, 2005, from: <http://www3.interscience.wiley.com/cgi-bin/fulltext/109088516/HTMLSTART>

Van Blerkom, J. (2004). Mitochondria in human oogenesis and preimplantation embryogenesis: engines of metabolism, ionic regulation and developmental competence. *Reproduction, 128,* 269–280.

Vellodi A. (2005). Lysosomal storage disorders. *British Journal of Haematology. 128,* 413–431.

Zemlin, W. R. (1998). *Speech and Hearing Science: Anatomy and Physiology (4th. Ed.).* Englewood Cliffs, N.J.: Prentice-Hall.

IMPORTANT TERMS

Acetylcholine: The most common excitatory neurotransmitter.

Acetylcholinesterase: An enzyme secreted into the synaptic cleft that breaks down (hydrolyzes) acetylcholine.

Actin: Protein rods or microfilaments of varying length that provide structure and support for the cell plasma membrane, in the same manner that poles support a tent. In muscle cells actin molecules interact with larger myosin molecules to shorten the entire cell. Actin and myosin molecules form the smallest muscle contractile unit, the sarcomere.

Action Potential: Change in polarity across the neuron cell membrane.

Active Transport: Metabolic transport processes that involve ATP mutation and allow large particles to enter or exit a cell.

Aldehyde: Toxic organic substances neutralized by intracellular processes.

Articular Cartilage: Hyaline cartilage within the synovial cavity of a freely moveable joint, covering the ephises of the articulating bones.

Articulation: In anatomy and physiology: Noun: The joining of bones or cartilages. Verb: Movement of a body structure in relation to another body structure.

Axon: One type of protoplasmic processes of neurons.

Basement Membrane: A layer of tissue secreted by epithelial cells separating them from supporting connective tissue.

Carbohydrate: Organic combination of carbon and water, literally, "watered carbon."

Centrosome: Organelle situated at the negative end of the cytoskeletal tubules.

Chondrocyte: Cartilage cell.

Chromosome: Protein strands of genetic material within the cell nucleus.

Cilium: A strand of actin extending from the end of a ciliated cell to create movement of extracellular substances or of the whole cell.

Classical Conditioning: Behavioral establishment technique whereby the reinforcement or punishment is administered before the behavior is done.

Cleavage: Separation of a single cell into two identical cells.

Clonic: Muscular resistance to stretch in a quivering, rhythmic series of contractions.

Columnar Cell: Cylindrical, long and narrow epithelial cells, usually with a functional, ciliated end.

Contractile Unit: A muscular tissue component that shortens in a hierarchical series, beginning at the sarcomere (molecular) level and ultimately resulting in shortening of the entire muscle.

Cubiodal Cell: Epithelial cells shaped like cubes because of the way they are packed in together. These cells can have secretory or absorptive functions.

Cytology: The scientific study of cells.

Cytoskeleton: Protein support structure of cells.

Cytosol: Thin protoplasm outside the cell nucleus.

Daughter Cell: One of the pair of cells resulting from cleavage.

Dendrite: Protoplasmic projection of a neuron distinguished from axons by the presence of receptor proteins their cell membrane structures. Action potentials propagate toward the cell body via dendrites.

Diploe: Spongy marrow-like layer between the outer and inner laminae of calvaria bones.

DNA: Deoxyribonucleic acid; the macromolecular basis for genes.

Endocrine Gland: Glands that secrete their special fluids, called hormones, into the intercellular space, and by diffusion, into the bloodstream.

Endocytosis: Means of cellular material transport by which the plasma membrane forms a vesicle around a particle and outside material enters the cell passing through the transport proteins in the plasma membrane.

Endoplasmic Reticulum: Cellular organelle containing tubes and spaces called *cisterns* for protein synthesis and are covered by membranes.

Endosteum: Membrane within the bone *endosteum* which functions in the genesis of bone cells, as its composite fibroblasts change into osteoblasts.

Ependymal cells: Neuroglial cells that form the lining of the nervous system.

Epiphysis: Articular and of a long bone, usually with a convex, rounded surface.

Epithelium (simple, stratified, pseudostratified): Tissue type characterized by a mosaic arrangement of cells, forming the outer or inner lining of a body part.

Eukaryote: Cells having separate organelles within the plasma membrane.

Exocrine Gland: Glandular epithelial tissue formation that secretes fluids through ducts or directly onto linings and surfaces, including the internal surfaces of hollow organs.

Exocytosis: Cell transport means whereby materials are transported from inside to the outside of the cell.

Extracellular: Material outside the cell plasma membrane.

Flagellum: Protein cell projection with a whip-like mechanical action intended to propel the cell through a medium.

Formed Element: One of several blood components, including red blood cells (erythrocytes), several types of white blood cells (leukocytes), and platelets (thrombocytes).

Gastrula: Embryonic stage marked by invagination of the hollow blastula creating inner and outer layers of tissue.

Gene: Formations of deoxyribonucleic acid or DNA. Their presence and sequential arrangement in a chromosome provide the biological foundation the organism's development and function.

Germ Layer: One of three laminae of embryoblast cells begun through gastrulation or invagination of the hollow blastula creating inner and outer layers of tissue. There are ultimately three germ layers, ectoderm, mesoderm, and endoderm from which all body tissues derive.

Glia: Connective support tissue of the nervous system.

Goblet Cell: Type of epithelial cell with a swollen free end containing secretory fluids.

Golgi Apparatus: Cellular organelle that modifies the raw proteins and prepares them for their intended purposes.

Histology: The scientific study of tissues.

Hypertonia: Muscular condition wherein there is too much resistance to stretch.

Hypotonia: Muscular condition wherein there is too little resistance to stretch.

Inhalatory Stridor: Noise produced on inhalation caused lack of structural support for the vocal folds.

Integral Protein: Protein within the plasma membrane that allows material to cross through it.

Intermediate Filament: Mid-sized cytoskeletal components that provide most of the structural integrity of the cell.

Keratin: A type of intermediate filament with a durable structure found on wear resistance body parts such as the hair and nails. Also found on the free surfaces of the tongue's filiform papillae.

Lactic Acid: A chemical compound that forms during muscular activity.

Lumen: Interior of a tubular organ, such as a artery or intestine.

Lymphocyte: Lymphatic cell.

Lysosome: Cellular organelle whose function is to break down or dissolve molecules in the cytosol.

Marrow: Spongy material within the bone.

Mesenchyme: Loosely organized and migratory embryonic cells. They will migrate to various locations and ultimately form the connective tissues of the body, including cartilages and bones, the circulatory system and the dermis.

Microfilament: Protein rods of varying length that provides structure and support for the plasma membrane.

Microglia: Neuroglial cells that support nervous system function by protecting it from disease.

Micron: 1/1000th of an inch.

Microtubule: The largest cytoskeletal components. Microtubules not only provide structure to the cytoskeleton, but they form polymer pathways upon which certain organelles move within the cytoplasm.

Mitochondrion: Organelles at which most of a cell's energy production takes place.

Morula: Embryonic stage at which the dividing cells form a cluster; from Latin for mulberry.

Notochord: Mesodermal derivative which, in human beings, forms the spinal column.

Muscle Fiber: A muscle cell.

Myelin: Lipid sheath surrounding some neural axons. Myelin increases the rate of action potential propagation.

Myosin: In muscle cells, actin molecules interact with larger myosin molecules to shorten the entire cell. Actin and myosin molecules form the smallest muscle contractile unit, the sarcomere.

Neurite: Protoplasmic projection of neurons; neurites are either axons or dendrites.

Neurotransmitter: A chemical compound that enter a neural synapse and facilitates the propagation of action potentials from one neuron to another.

Nucleus: Largest organelle within a cell containing the genetic material.

Oligodendrocyte: A type of neuroglia that produces myelin in the central nervous system.

Operant Conditioning: Behavior establishment paradigm wherein the reinforcement of punishment is presented after a behavior is done.

Organelle: A subcellular structure.

Ossification: The changing of cartilage into bone.

Osteocyte: A bone cell.

Passive Transport: Passage of substances through cell plasma membranes intermediated by integral proteins and requiring no energy expenditure on the part of the cell.

Perikaryon: A neuronal cell body.

Periosteum: Thin connective tissue membrane covering the outer surface of bones.

Peroxysome: A type of cellular organelle that neutralizes toxic substances in the cell by creating chemical reactions with them that produce hydrogen peroxide.

Plasma: A yellowish transparent fluid that forms the matrix of blood and lymph.

Plasma Membrane: The outer structure of a cell.

Postsynaptic: That part of a synapse receiving the action potential input.

Presynaptic: That part of a synapse transmitting the action potential.

Prokaryote: Primitive cell having no organelles.

Protein: An organic amino acid chain which forms the structural basis for all body parts. Protein structures are determined by genes, and are widely variable, depending upon their specific functions.

Pulse: Rhythmic swelling of an artery.

Ribosome: A granular organelle that can synthesize proteins by "reading" the codes presented in the form of a particular type of RNA called "messenger" RNA.

RNA: Ribonucleic acid. A type of nucleic acid onto which genetic information is transcribed from DNA.

Schwann Cell: Myelin producing glial cell in the peripheral nervous system.

Soma: Neuronal cell body.

Somite: A swelling along the dorsal surface of the developing embryo. The somites form most of the body wall, including the axial skeleton, muscles, gonads, nervous system, excretory system, and outer coverings of the internal organs.

Squamous Cell: A flat type of epithelial cell; squamous cells are usually located in areas susceptible to mechanical wear.

Synapse: The connection between two communicating neurons across which action potentials are propagated.

Synaptic Cleft: The space within a synapse into which neurotransmitters are secreted.

Synovial Fluid: The fluid within the capsule of a freely moveable articulation.

Tissue: A collection of cells having similar form and function.

Tremor : Uncontrollable muscular contractions; can be intentional or unintentional.

Tubulin: A protein that forms the microtubules of the cytoskeleton.

Tunic: One of three layers of an artery.

CHAPTER 3

General Developmental Anatomy and Human Growth

CHAPTER PREVIEW

This chapter focuses on human developmental anatomy in general, beginning with a treatment of basic genetics and genetic terms. Next is a chronological study of development of mechanisms of human communication from fertilization, through the embryonic and fetal stages and on to postnatal development and old age. This chapter on developmental anatomy covers general changes in the human embryo and fetus. An important facet of education in human communication and its disorders is the study of communication processes across the life span. Chapter sidebars describe genetic and chromosomal disorders commonly involved in disorders of communication.

CHAPTER OUTLINE

Developmental anatomy is the study of individual growth from fertilization to birth. During this phase in life, huge changes take place that will serve as the starting place for the remainder of that individual's life. An understanding of developmental changes occurring before birth is essential for a proper understanding of postnatal anatomy and functional relationships of body parts. It is also a strong foundation upon which to build an understanding of the scope of problems and anomalies that may occur during the prenatal period.

Genetics

The study of developmental anatomy begins with the study of genetics, the science of heredity and the mechanics by which parents pass traits to their offspring. Genetic science explores the manners in which specific traits are transmitted from one or both parents.

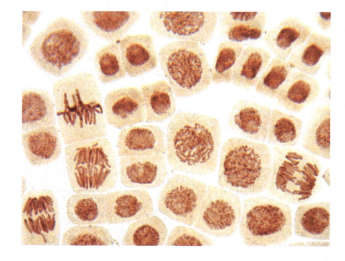

Individual development may be said to begin with the creation of a *zygote* following fertilization. To create a zygote, each parent contributes twenty-three chromosomes, double strands of DNA containing the *genome*, or array of genes unique to an individual, via their *gametes*, or *germ cells*. Twenty-three chromosomes represent half of the forty-six total number of chromosomes to be inherited by the individual. The number of chromosomes contributed by one parent is called the *haploid* number of chromosomes, and forty-six the *diploid* number.

The mother's twenty-three chromosomes are delivered in her gamete, an *ovum*, and the father's in his gamete, the *spermatozoon* or *sperm*. Twenty-two of a parent's chromosomes are called *autosomes*, numbered one through twenty-two, and members of the remaining pair are called *sex-linked chromosomes*. The sex-liked chromosomes are labeled "X" and "Y." Females carry two X chromosomes and males carry an X and a Y.

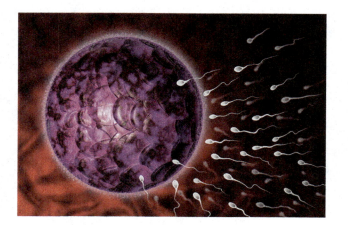

The genetic array of each gamete is the foundation of the individual's potential structure and function. This potential is called the individual's *genotype* and includes the genetic potential of each parent and his or her ancestors. Thus, traits such as height, eye color, body build, and certain physiological or mental functions, to name a few, are transmitted through the genome of each parent, passed down through the generations. Genetic possibilities are known as *genetic traits*, and not all are readily observable. Of course, an individual can only manifest one set of traits. These manifested, observable traits are the individual's *phenotype*.

Whether a genetic characteristic is expressed phenotypically depends upon certain *rules of inheritance*, which were first published in the mid-nineteenth century by Gregor Mendel (1865) and based on his plant research. Inheritance rules rely on laws of probability to predict the chances of phenotypic expression. Such chances are contingent upon whether a trait is located on an autosome or on a sex-linked chromosome, and whether the trait is *dominant* or *recessive*.

Dominant traits for a characteristic express themselves when in pairs or when combined with a recessive trait. Recessive traits must paired in order to express themselves. A gene that transmits a specific trait of interest in called an *allele*. Individuals whose genotypes include two dominant or two recessive alleles are *homozygous*, and individuals whose genotypes include one dominant and one recessive allele are *heterozygous*.

Four modes of genetic inheritance or transmission include *autosomal dominant*, *autosomal recessive*, *X-linked*, and *multigenic/polyfactorial*. These modes apply only to chromosomal inheritance.

Autosomal dominant transmission occurs when at least one parent has a dominant trait. In this case, the trait will be expressed if the dominant gene is combined with another dominant gene or if the dominant gene is combined with a recessive gene. It is possible for a given autosomal dominant trait to appear in each generation, and every affected person has at least one affected parent.

When a person having an autosomal dominant trait mates with a person who does not have that trait, there will be, on the average, an equal number of children who have the trait and who do not. Children who manifest the trait can be assumed to be heterozygous. Children of parents having the trait who do not manifest that trait phenotypically will have unaffected offspring. In autosomal dominant transmission, males and females are affected in equal numbers.

Autosomal recessive traits are expressed when two recessive alleles are paired. In this mode of transmission, if a child manifests a trait, and the parents do not, both parents are presumed to be heterozygous. Mendelian probabilities predict, in the pairing of heterozygous traits, a 25% chance that the normal parents' offspring will be affected, a 50% chance these offspring will be carriers, and a 25% chance children will be homozygous for the dominant allele.

If a person with a recessive allele mates with a dominant version of the allele, all the children will be phenotypically normal heterozygotes. If a person whose phenotype expresses a recessive trait mates with a heterozygote, an average of one half of their children will demonstrate that trait. The other half will not demonstrate the trait, but will be carriers. Parents who both manifest a recessive trait, meaning that they are homozygous, will transmit that allele to all their children. Males and females are equally likely to be affected.

WHY STUDY GENETICS?

The study of genetics is of interest to practitioners in communication sciences because genetic factors can affect development of peripheral or central structures involved in communication. Many genetic anomalies or abnormalities occur as clusters of symptoms. These clusters ore encountered often enough in clinical practice that they become known as **genetic syndromes.** Syndromes are named after their physical manifestations or after the person who first wrote about it or, sometimes, after an individual who was affected by the condition.

Some genetically based communicative disorders are secondary to **anatomic anomalies**. The speech-language pathologist can see differences in the shape of the skull or extremities. **Abnormal skull dimensions** can alter the manner or place of phoneme production. Maxillary fistulae can prevent the build up of intraoral air pressure required to produce plosive of fricative phonemes distal to the site.

The appendicular skeleton may be affected as well as the axial skeleton. Malformations of the upper extremities may affect writing ability. They may also affect the manner in which an augmentative device is actuated.

Some genetic anomalies are much less visible. The patient appears to be normal in appearance. However, malformations of nervous system structures can affect mental or motor functions involved in communication. The presence of cognitive or communicative dysfunction may be the primary symptom.

In any case, visible or invisible, the responsible habilitation team must do a little fact finding to complete the treatment regime of the child. Since each individual has unique potentials and characteristics, practitioners encounter wide variations among children seen in clinical programs. Few of the children who have recognized genetic syndromes fit all the criteria described in textbooks. Clinician's must assess cognitive, linguistic, motoric, and anatomical abilities and possibilities of each child with the goal of developing a care plan that meet the challenges of the conditions that are present and ultimately helps bring out the potential of each child.

Management of cases in which genetic anomalies play a role is most often conducted by a *team* of health care professionals. Members of the team can include, in addition to the speech-language pathologist and the audiologist, physicians with specializations in plastic surgery, pediatrics, and internal medicine, a dentist and orthodontist, a physical therapist, occupational therapist and educational specialist, a genetic counselor, and a psychologist. Team members usually meet periodically and discuss the status and disposition of each case. Basic knowledge of genetic fundamentals adds to the speech-language pathologist's ability to act effectively as a team member.

If the anomaly is genetic, the genetic counselor instructs the parents about certain genetic principles. Sensitivity to personal feelings is important when dealing with a person's pedigree. However, parents are usually glad to know the presence or absence of familial disorders.

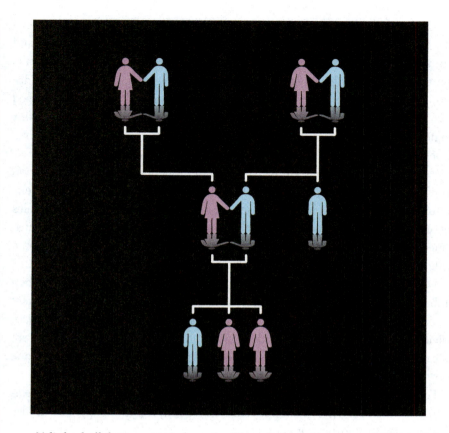

X-linked alleles are carried on the X sex-linked chromosome. These inheritance traits are transmitted by the female parent to the male child, since the male must inherit a Y chromosome from his father in order to be male. Father-to-son inheritance is never observed. X-linked recessive traits are usually expressed in males and rarely expressed in females. A father carrying a recessive X-linked trait and a homozygous normal mother will have all normal children. Any daughters they have will be carriers.

X-linked dominant traits are manifested in males who are hemizygous, having only one gene for the trait, since males have a single X chromosome, and in females who are heterozygous. Mendelian probability predicts that female with a recessive X-linked trait will transmit the trait to half of her sons and that half of her daughters become carriers. More males than females can be expected to express the recessive X-linked trait. The recessive allele appears to skip generations. It may be expressed in fathers and pass unexpressed through carrier daughters and granddaughters to reappear in grandsons and great grandsons.

Polygenic/multifactorial inheritance depends upon the combination of multiple alleles in interaction with environmental factors. This mode of transmission is generally applied to anomalies which manifest as clinical syndromes (Lewis, 1990).

Since there are few known specific genes in polygenic/multifactorial form of transmission, existence of inheritance must be inferred according to several criteria. First, the trait must occur more often in relatives than in the general population, and the risk to relatives must decline with increasingly remote degrees of relationship. Further, studies of twins show a higher frequency of *concordance*, where both twins have the trait, for monozygotic than dizygotic twins. Monozygotic twins reared apart have greater concordance than expected by chance. In cases where incidence of the effect is lower in one sex, a sex-linked condition may be present. Other evidence indicates a genetic rather than an environmental factor operating in the onset of a condition including the existence of adopted children who resemble their biological parents rather than their adoptive parents in disease frequency and biological relatives who have increased disease frequency but spouses who do not, even though living in the same environment.

Inheritance through chromosomal transmission is not the only mode by which offspring may inherit traits. Mitochondria in the ovum contain sufficient DNA to transmit traits to the zygote and subsequent cells. Mitochondrial DNA is only transmitted through the ovum, since the sperm contains few mitochondria.

The Embryonic Stage

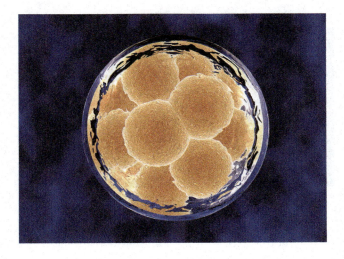

The embryonic stage is that stage of development during which cells increase in number and differentiate in to the stem cells for the various tissue types. This stage lasts from fertilization to the end of the eighth week. During this stage, the developing individual is called an *embryo*, a word which derives from the Greek *"to grow in."* Systems develop and differentiate during the embryonic stage, and external features, including the face, become recognizable. Tissue differentiation is covered in Chapter Two.

Some scholars (e.g., Brooks and Zeitman, 1998) consider the period between fertilization and the end of the third gestational week to be the "preembryonic period," while others (e.g., O'Rahilly and Muller, 1996) fix the beginning of the embryonic period at fertilization. In either scheme, the embryo is called a *zygote* at the instant of fertilization.

Fertilization occurs when a single spermatozoon penetrates the outer membrane of the ovum and the zygote is formed. When this occurs, the plasma membrane of the ovum changes in a manner to disallow entrance of any more spermatozoons. The twenty-three chromosomes of sperm and ovum arrange into homologous pairs, according to the particular genes they contain, and begin to create identical reproductions of themselves through mitosis for somatic cells and meiosis for germ cells. Thus, all cells in the human body are descendants of a single cell.

GENETIC TERMS

GENETICS: the scientific study of heredity.

CYTOGENETICS: the study of chromosomes

CHROMOSOMES: Thread-like structures in the nucleus of the cell that contain the genetic information encoded by functional units of DNA called genes.

Autosome: Any of the twenty-two pairs of chromosomes that are not sex chromosomes. Autosomes are responsible for transmitting any trait that is not sex-linked.

GENES: Hereditary polypeptide DNA units transmitted from parents to their offspring through meiosis or mitosis.

Meiosis: Division of a cell into four haploid (with half as many chromosomes as the parent cell) daughter cells. The type of cell division associated with the formation of either the sex cells (sperm or eggs).

Mitosis: Division of a cell into two identical cells containing a complete compliment of chromosomes. The common mechanism of cell replication growth and development. Autosomes replicate through mitosis.

FAMILIAL OR HEREDITARY: Terms applied to traits that are common among members of a person's family.

GENETIC: Caused by the presences of or changes in genes. Genetic conditions may not necessarily present difficulties. Some genes enable individuals to excel in mental or physical ways. Genetic conditions are not always familial. Mutations of genes can be secondary to exogenous factors, such as gamma radiation.

CONGENITAL: Conditions are those associated with or are present at birth, and may or may not be familial.

GENETIC COUNSELOR: An expert in genetics.

KARYOTYPE The genetic inventory of an individual or species.

PEDIGREE: The family history of genetic traits.

INHERITANCE PATTERNS: GENETIC, CHROMOSOMAL, AND MITOCHONDRIAL.

GENETIC INHERITANCE: Transmitted by one or more genes.

   Autosomal: Transmitted by one or more genes located in the twenty-two autosome pairs, or non sex chromosomes: Not sex-linked or mitochondrial.

   Single Gene (Monogenic) Inheritance: Transmission by a one gene.

   Autosomal Recessive: Transmission requires a recessive gene from each parent.

   Autosomal Dominant: Transmission requires a single dominant gene from one parent.

   Sex-Linked: Transmission by genes contained in the sex chromosomes.

   Polygenic Inheritance: Transmission by two or more genes.

CHROMOSOMAL INHERITANCE: Inheritance of traits through numerical or structural anomalies of the chromosomes.

   Numerical Anomalies:

   Euploidy: having the normal number of chromosomes.

   Aneuploidy: having an abnormal number of chromosomes.

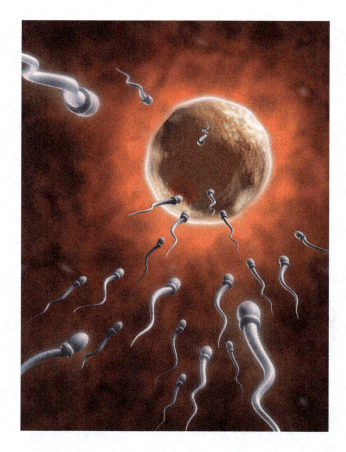

The human zygote exists until its first division, termed a *cleavage*. Cleavage may occur between one and four days post-fertilization, while the zygote is still in the fallopian tube (Fenwick, Platteau, Murdoch, and Herbert, 2002; Fisch, Rodriguez, Ross, Overby, and Sher, 2001). Further duplication of cells occurs as the dividing embryo, now called a *morula*, migrates along the fallopian tube to implant itself in the uterine wall. At the implantation stage, the zygote becomes a *blastocyst*. The blastocyst develops to form an embryo and, later, becomes a fetus.

At the very beginning of development in the human body, as well as those of all mammals, cells belong to one of two main groups: *somatic* cells and *germ* cells. Within their two groups, the cells are initially identical, but during the course of embryonic development, they begin to differentiate.

Somatic cells form the larger group of cells. These cells form the body tissues, organs, and systems. Each somatic cell contains forty-six chromosomes, and divides by mitosis. In mitosis, the somatic cell splits into two identical *daughter* cells, each having an identical chromosomal complement.

Germ cells are specialized for the continuation of species. The mature male or female germ cell is called a gamete. As mentioned earlier, the male gamete is a spermatozoon, and the female gamete is an ovum. Each gamete contains the haploid number of chromosomes from the parent cell and divides by meiosis. In meiosis, the haploid number of chromosomes is maintained, preserving the normal number of chromosomes for a species.

Branchial Arches

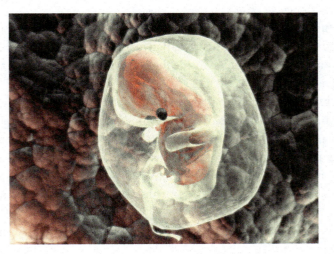

Monosomy: having a missing member of a chromosome pair.

Trisomy: having a triplet of chromosomes in the place of a pair.

Structural Anomalies:

Inversions: reversal of chromosome DNA sequence.

Ring: chromosome pairs form a ring.

Translocation: part of the chromosome sequence is out of sequence.

MITOCHONDRIAL INHERITANCE: transmitted by the female parent through DNA contained in the ovum. All female of spring will inherit the trait; no male offspring will inherit the trait.

During the embryonic stage, an important development occurs with the formation of the *branchial arches*, also called *gill* or *pharyngeal* arches. These are swellings of the ventral surface, with the closed aspect of the arch on the ventral surface, and the open ends projecting dorsally. The first arch appears at the rostral end of the embryo, in an open area called the *primitive foregut*, on approximately day twenty-two of gestation. The primitive foregut, an endodermal tube, begins at the rostral end and extending caudally, will develop to become the respiratory system and the rostral part of the alimentary system. Other divisions of the tube, a midgut and hindgut, will extend all the way to the anal plate forming the remainder of the alimentary system. Subsequent arches develop in a craniocaudal sequence, with the last arch appearing about day twenty-nine. They will become the face and neck, within which will be the anterior alimentary and respiratory tracts. Understanding derivative structures of the branchial arches enhances the understanding of functional and anatomical head and neck relationships.

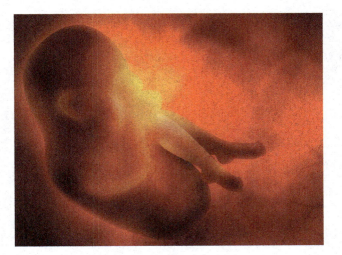

Branchial arches form in all vertebrate animal species, and their study is a focus of comparative anatomists. To comparative anatomists, arches are also known as *gill* arches, deriving, as they do, from the gill structure of ancient jawless fishes. In the zebra fish (*brachydanio rerio*), widely studied because of its transparency and ability to reproduce in great numbers, seven branchial arches form and are the foundations of the mandible and breathing apparatus. In the human being, six branchial arches form, but only five appear. The fifth arch is very small and disappears early in development. Remaining arches are numbered *one* through *six*, with arch number *five* omitted in deference to comparative anatomical consistency.

Four of the six branchial arches appear first as swellings on the sides of the embryo. They grow ventrally to meet and fuse with their opposites on ventral midline surface of the embryo. Between each arch is a branchial *groove* or *cleft*, on the outside of the embryo, and a branchial *pouch*, formed inside the embryo. Arches, grooves, and pouches serve as the foundations for laryngeal, oral, and nasal development.

From the ectodermal layer of the branchial arches comes the external skin of the maxilla, mandible, pinna, and external ear, as well as the internal epithelium of the oral and pharyngeal cavities. The endodermal, deep layer gives rise to the glands, blood vessels, and internal epithelium.

The mesoderm of the branchial arches forms from migratory *mesenchyme* having two points of origin. That which forms the nerves and connective tissues of the head and neck derives from the primitive mesencephalon of the embryonic neural crest, whereas that which is destined to form the muscles originates in the ordinary embryonic mesoderm.

Each branchial arch is composed of an outer layer of ectoderm and inner layer of endoderm and a middle layer of mesoderm. From these germ layers develop a cartilaginous bar, a muscular component, an artery, and a nerve for each arch.

Branchial arch cartilages will develop into the skeleton of the distal vocal tract, parts of the orbits, and the tiny ossicles of the middle ear. The first arch develops two cartilaginous swellings, destined to become the jaws. The more rostral cartilage of the first arch is called the *palatopterygoquadrate plate*. It will develop in to the greater wing of the sphenoid bone, also known as the *alisphenoid bone* and the incus of the middle ear. The caudal cartilage of the first arch is known as *Meckel's cartilage* (after Johan Friedrich Meckel, "The Younger," 1781–1833). It gives rise to the malleus, maxilla, zygomatic bone, squamous portion of the temporal bone, and the mandible. The cartilage of the second arch is known as *Reichert's Cartilage* (after Karl Bogislaus Reichert, 1811–1883). It gives rise to the lesser hyoid cornua and that part of the hyoid corpus between them, the stylohyoid ligament, the styloid process

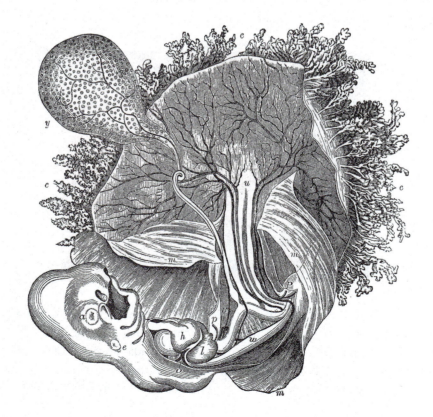

of the temporal bone, and the stapes. The third branchial cartilage is the source of the lower part of the hyoid corpus and the greater hyoid cornua. From the fourth branchial arch cartilage comes the epiglottis, thyroid, and cuniform laryngeal cartilages, while the sixth arch cartilage becomes the cricoid, arytenoid, and corniculate laryngeal cartilages.

Each branchial arch is associated with an artery derived from the multiple primitive aortic arches of the embryonic circulatory system. These arteries will change greatly in form, as will the fetal circulatory system, either to disappear or to become parts of the postnatal circulatory system.

The artery of the first arch is destined to become the maxillary artery, supplying blood to the superior oral region. The second arch artery will become the tympanic branch of the maxillary artery and will supply blood to the middle ear. The artery of the third arch will become the common carotid artery and the proximal part of the internal carotid artery, supplying blood to the face and cranium. The fourth and sixth arches develop asymmetrically. The artery of the fourth arch, on the left, will become the aortic arch, after birth, and contribute to the left subclavian artery. On the right, the artery of the fourth arch will become the brachiocephalic artery and contribute to the right subclavian artery, whereas the artery of the sixth arch will become the pulmonary arteries.

Each branchial arch is also associated with a pair of cranial nerves. The nerve of the first arch is the trigeminal nerve (cranial nerve V), that of the second is the facial nerve (cranial nerve VII), while the glossopharyngeal nerve (cranial nerve IX) develops in the third arch. The fourth and sixth branchial arches share innervation from the vagus nerve, with the fourth arch receiving the superior branch of the laryngeal nerve and the sixth arch receiving the recurrent laryngeal branch.

The first branchial arch (Meckel's Arch) divides into two sections corresponding to the jaws and opening of the alimentary tract: a maxillary one and a mandibular one. The maxillary sections will become those parts of the face associated with the superior oral cavity, including small parts of the sphenoid bone that contribute to the orbits, while the mandibular section will become the mandible and anterior two-thirds of the tongue. The first arch also gives rise to the head of the malleus and part of the incus. From its ectoderm will come the skin beginning over the ventral part of the pinna and ranging ventrally to cover the cheeks, lips, and oral vestibule. Its mesenchyme will develop into the facial dermis, including the trigeminal nerve, via which touch sensation will be transmitted to the central nervous system. Muscles developing from first arch mesenchyme migrate to their destinations, pulling rami of the trigeminal nerve (cranial nerve V) with them. These muscles include the muscles of mastication, the mylohyoid, anterior digastric belly beneath the mandible, the tensor veli palatini, and tensor tympani.

From the second branchial arch will develop support structures for the mandible, tongue, and larynx, as well as parts of the middle ear. Oral and laryngeal support derivatives include the upper part of the hyoid corpus and its lesser cornua, the styloid process of the temporal bone and the ligament connecting it to the lesser hyoid cornua, the stylohyoid ligament. Second arch middle ear derivatives include the manubrium of the malleus, long crus of the incus, and head and crura of the stapes. Second arch ectoderm gives rise to the epidermis of the dorsal pinna and the rostral neck. Mesenchyme of the second arch gives rise to muscles of facial expression, posterior digastric belly, stylohyoid and stapedius muscles, all receiving their motor innervation from cranial nerve VII, the facial nerve. The second branchial arch is also known as the *hyoid* arch.

The third branchial arch is the embryonic source of the remainder of the hyoid bone, including its lower corpus and greater cornua. The posterior one-third or pharyngeal part of the tongue is also a derivative of the third branchial arch, as is stapedial footplate. From third arch mesenchyme comes the stylopharyngeus muscle.

The fourth branchial arch is the embryonic source for more caudal respiratory structures, including parts of the larynx. The laryngeal thyroid cartilage derives from the fourth arch. Most authorities hold that the epiglottis comes from the fourth arch, as well (Brookes and Zietman, 1998), although some evidence exists that the epiglottis forms from mesenchyme not originally part of that arch (Larsen, 2001). Musculature developing from the fourth arc includes the three pharyngeal constrictors, levator palatini, and cricothyroid muscles.

The remainder of laryngeal cartilages and intrinsic laryngeal muscles are derived form the sixth branchial arch. These cartilages include the cricoid, arytenoid, and corniculate cartilages. Muscles derived from the sixth branchial arch originate from mesenchyme of the occipital somites, and include the lateral and posterior cricoarytenoid muscles, the thyroarytenoid, and vocalis muscles.

The first branchial cleft develops into the external auditory meatus. The second through fourth branchial clefts are covered by the proliferation of the second branchial arch tissues.

Branchial pouches, formed on the inner embryonic wall, also give rise to adult structures. These pouches first differentiate in to ventral and dorsal sections before developing into their postnatal structures. The dorsal first branchial pouch section becomes the tympanic cavity and eustachian tube. From the second branchial pouch come the palatine tonsils. The lingual tonsil derives from the third pouch.

TABLE 3.1: Branchial Arch Derivatives

	Arch I	Arch II	Arch III	Arch IV	Arch VI
Cartilage	A. (Palatopterygo-quadrate) Greater Wing: Sphenoid; Incus B. (Meckel's) Malleus; Maxilla; Zygomaytic; Squamous Portion of Temporal; Mandible	Lesser Cornua: Hyoid; Part of Hyoid Corpus; Stylohyoid Ligament; Styloid Process: Temporal Bone Stapes	Lower Hyoid Corpus; Greater Hyoid Cornua	Epiglottis; Thyroid; Cuneiform	Cricoid; Arytenoid; Corniculate
Artery	Maxillary	Tympanic	Common Carotid Proximal Internal Carotid	Left: Aortic Arch, Part of Left Subclavian Right: Brachiocephalic Part of Right Subclavian	Pulmonary
Nerve	Trigeminal (V)	Facial (VII)	Glossopharyngeal (IX)	Vagus (X)	Vagus (X)
Adult Structures	Incus; Malleus; Upper and Lower Jaws, inc m.m. Mastication; Anterior 2/3 of Tongue	Stapes; Support for Mandible, Tongue and Larynx; m.m. Facial Expression	Hyoid; Stylopharyngeus m.m. Posterior 1/3 of Tongue	Larynx; Cricothyroid m.m. Pharyngeal Constrictors; Levator Veli Palatini	Larynx; Intrinsic Laryngeal m.m. except Cricothyroid

Somites

Another important embryonic development is the appearance of *somites* along the dorsal surface of the embryo. Somites are roughly cubical paired swellings that begin to appear at the rostral end of the embryo during the third and fourth gestational week.

The first somites appear in the occipital cranial region. More somites appear along the dorsal embryonic surface, growing in a cranial to caudal direction, until about forty-four pairs are present. Some of these disappear, and the final number of somites in the human embryo is thirty-one. The number of somites is sometimes used to represent the age of an embryo.

Somites will develop into the skin, bones, and musculature of the body wall. Hence, functions and structures of the body wall are termed *somatic*. Each paired somite is associated with one of the thirty-one paired spinal nerves.

The Fetal Stage

The fetal stage begins at start of the ninth gestational week and continues until birth, at thirty-eight to forty weeks. This stage is marked by very rapid overall growth of the entire fetus as well as by continued differentiation and development of organs and systems. The rate and amount of growth during the fetal stage is determined by genetic and environmental factors.

Due to the adducted and flexed postures of the fetus' extremities, overall fetal length is measured from the crown of the skull to the end of the sacrum. This measurement is called, "Crown-Rump Length (CRL)." In overall length, from crown to rump (CRL), the typical fetus grows tenfold, from approximately 3 to 4 cm to 36 to 50 cm. In an eight-week-old fetus, the head may occupy 43% of the CRL, while, at birth, the head typically occupies 31%. Fetal weight increases vastly, as well, from between 1 and 4.5 grams to between 3,250 and 3,500 grams (Brooks and Zietman, 1998; Larsen, 2001).

Systems and organs continue their developmental course during and after the fetal period. Skeletal elements are all present early in the fetal period, but they continue development through ossification of existing cartilaginous structures. Some of this ossification continues after birth.

The nervous system proliferates during and after the fetal period, with brain maturation continuing at a rapid rate up to two years after birth. The brain develops quite slowly, contributing greatly to an extended period of dependence by the infant upon others for years after birth.

The fetal circulatory system begins to circulate blood by the end of the fourth week, but the circulatory system is vastly different from that which will circulate blood after birth. This is due to the fact that the fetus continues to circulate the mother's blood. In fact, the fetal circulatory system will continue to evolve after birth.

Respiratory development represents a change from a system capable of breathing only in a liquid environment to one capable of gas transfer on dry land. At approximately twenty weeks, pulmonary alveoli are developed to the extent that natural respiration is possible.

PRINCIPLES OF TERATOLOGY

Teratology is the study of diseases or environmental substances that cause congenital malformations. Environmental agents that may precipitate polygenic expression are *teratogens*. These agents are chemical, such as lead, or energy such as radiation, but are not behavior, such as poor nurturing technique.

Teratogens may cause congenital anomalies with or without the influence of genetic predisposition. Susceptibility depends on the genotype of the fetus and the manner in which it interacts with the environmental factors.

Susceptibility to teratogenic agents (substances that cause congenital developmental anomalies) depends upon the developmental stage at the time of exposure.

The teratogen may cause different malformations at different times in gestation.

Different teratogenic agents affect different cells in different ways.

The final results due to teratogens are death, malformation, growth retardation, and functional disorders.

As exposure increases, the extent of deviant development increases from no effect to lethal effect.

Postnatal Growth and Development

Growth represents one of the essential characteristics of living organisms. In human beings, there are several manifestations of this phenomenon (Sinclair, 1976). *Multiplicative* growth involves the increase in the number of cells in a given tissue or organ. *Auxetic* growth is the increase in cell size. *Accretionary* growth is an increase in the amount of non-living material, such as intercellular matrix, in a tissue. Human growth involves all three of these growth types.

Human growth in so-called "developed" cultures in the twenty-first century may be separated into four main stages: infancy, childhood, adolescence, and adulthood. Researchers specializing in these general areas further divide them into smaller groups. Bogin (1999), for example, identified nine stages in human growth. These included the prenatal stage, which has been discussed above, the neonatal period, infancy, childhood, a juvenile stage, puberty, adolescence, adulthood, and old age. Others descriptions of human growth stages are similar (e.g., Timiras, 1972; Freiberg, 1983), but all follow a general trend in development of human beings and other species emerges. This is the *sigmoidal developmental curve*.

The graphic sigmoidal curve plots time on the abscissa and the growth parameter on the ordinate. The result is a roughly S-shaped curve, having slow acceleration at first, followed by a rapid growth rate over time, and succeeded by a return to a slower rate. A sigmoidal growth curve can describe various parameters of growth, including those of physical as well as psychometric measures.

The first month after birth represents the *neonatal period*. During this period, the individual has undergone an extreme change in environment, leaving a liquid one of relatively consistent temperature and arriving at a gaseous one with potentially extreme temperature variations. The neonate must now rely also on his or her own respiratory system for metabolic gas exchange and on the digestive system for food intake and waste elimination. Bogin (1999) identified this period as distinct because the rate of growth is at its greatest during the entire postnatal period. The special branch of medicine that deals with human beings from prenatal stages to adulthood is called *pediatrics*.

The *infancy* period follows the neonatal period, and lasts to the third year, the typical time for cessation of maternal lactation. Thirty percent of the infant's body's size is occupied by the head and neck. Another important milestone in infancy is the eruption of the deciduous teeth. This is an important phenomenon for nutrition, for the infant no longer depends upon milk alone for all nutritional needs. It is also an important stage in speech development, since tooth eruption provides new articulatory possibilities and changes the shape of the oral cavity. Great leaps in cognitive and psychomotor development also occur, accompanying proliferation of neural cells in the brain during the infancy period, although the infant is still dependent upon older caretakers for safe, productive decision making.

The period of *childhood* lasts from the third to the seventh year. The childhood period is characterized by a deceleration in the growth rate, with changes still occurring at a steady but slowed rate (Bogin, 1999). Children are usually weaned from the breast, but still depend upon others for nutrition, with nutrition being an important factor in maintenance of the overall growth rate. During the childhood period, the deciduous teeth are shed and permanent ones erupt. The brain also reaches its adult weight.

Researchers have identified several factors that seem to precipitate normal development in children. These include heredity, nutrition, illness, socioeconomic status, and psychological environment. With the exception of heredity, deficiencies these factors may precipitate slowing of the growth curve. However, amelioration of these deficiencies, present in the short term, may be followed by a period of *catch-up growth*, during which the child will grow at an accelerated rate until the normal developmental trend is recapitulated (Bryson, Theriot, Ryan and Pope, 1997).

Following the childhood stage, the individual enters *adolescence*, a period marked acceleration in overall growth and by the onset of puberty, or, more accurately, *gonadarche*, during which hormones are produced by interaction of pituitary and gonadal functions (Bogin, 1999). Adolescence typically begins earlier in females than in males, but in both genders, the period is marked by another accelerated phase in skeletal tissue growth. Adolescent development also includes appearances of *secondary sexual characteristics*, including development of external genitalia, distinctions in male and female body types (sexual dimorphism), and a decrease in vocal fundamental frequency (pitch), most marked in males (Peterson and Barney, 1952).

Adulthood is reached when multiplicative skeletal cell growth ceases and reproductive maturity is reached. This stage is marked by stability in growth and resistance to factors that inhibit growth, such as disease. In skeletal growth, continued increases in tissue mass occurs as a result of auxetic growth. Reproductive maturity refers not only to the ability to produce viable sperm and ova, but to the ability to father a child in males and to the ability to bring an infant to full-term (Bogin, 1999).

The end of adulthood is called *old-age* or *senescence*. During this period, the individual's resistance to environmental stresses declines and the ability of cells for mitotic division diminishes.

Several theories have been advanced for the decline in tissue adaptive abilities. One is the interaction of genetic and environmental factors. This theory posits that a genetically encoded trait is the reaction to certain environmental stimuli to produce a halt to some forms of cell duplication. This reaction is called *apoptosis*, and is beneficial in formation of the structures of the brain or skeleton, for example, but limiting in later life. Another theory arises from the observation that certain cells of the body, such as heart and skeletal muscle cells and neural tissue cells, do not multiply enough to replace themselves (Tortora and Grabowski, 1996). Environmental factors, such as the presence of *free radicals* (electrically imbalanced molecules), and auto-immunological factors have also been suspects in the ageing process.

Old age is characterized by superficial signs, such as graying and thinning of the hair, wrinkling of the skin, loss of dentition, and decrease in muscle mass. There are internal signs, as well, including decreased elasticity of epithelial and connective tissues and decreases in the number of central nervous system neurons.

Such external and internal changes can result in disturbances in speech, language and hearing processes. Decreased cochlear function, for example, results in an age-related decrease in hearing sensitivity known as *presbycusis*. Loss of teeth and associated changes in alveolar processes of the maxilla and mandible can be associated with changes in speech articulatory functions. Tissue changes in the eye can cause difficulty with visually based communicative functions such as reading. More pervasive are the changes in brain function that may accompany decreases in the population of cortical or subcortical neurons. *Organic brain syndrome* is a general loss of cognitive functions accompanying cerebral cortical neuron loss. Various forms of movement disorders accompany decreases in the populations of cells in subcortical structures, such as the basal nuclei or brainstem.

CHAPTER SUMMARY

This chapter covering developmental anatomy and human growth presented a general look at the individual from conception to senility. Where possible, we offered some focus on form and function of structures and foundations specific to human communication. More detailed examinations of development of those structures are contained in their specific sections.

REFERENCES AND SUGGESTED READING

Bogin, B. (1999). *Paterns of Human Growth (2nd Edition)*. Cambridge, UK: Cambridge University Press.

Brookes, M. and Zietman, A. (1998). *Clinical Embryology: A Color Atlas and Text*. Washington, DC: CRC Press.

Bryson, S.R., Theriot, L., Ryan, N.J., and Pope, J. (1997). Primary follow-up care in a multidisciplinary setting enhances catch-up growth of very-low-birth-weight infants. *Journal of the American Dietetic Association, 97,* 386–390. Rertrieved May 30, 2011, from: <http://www.sciencedirect.com/science?_ob=ArticleURL&_ udi=B758G-48CDY8X-3F&_user=109269&_coverDate=04%2F30%2F1997&_alid=358679725 &_rdoc=8&_fmt=full&_orig=search&_cdi=12926&_sort=d&_st=4&_docanchor=&_acct= C000059546&_version=1&_urlVersion=0&_userid=109269&md5=cb223b200e1f72121679b248a5a3b660>

Fenwick, J., Platteau, P., Murdoch, A.P., and Herbert, M. (2002). Time from insemination to first cleavage predicts developmental competence of human preimplantation embryos in vitro. *Human Reproduction, 17*, 407–412. Retrieved May 30, 2011, from: <http://humrep.oxfordjournals.org/cgi/content/full/17/2/407>

Freiberg, K.L. (1983). *Human Development: A Life-Span Approach, 2nd Edition.* Belmont, CA: Wadsworth Health Sciences.

Gillespie, C.C. (1974). *Dictionary of Scientific Biography (Vol. IX, p. 252).* New York: Charles Scribner's Sons.

Hatch, R.A. (1998). *Darwin's Friends & Contemporaries.* Retrieved May 30, 2011, from: <http://web.clas.ufl.edu/users/rhatch/pages/02-TeachingResources/readingwriting/darwin/05-Darwin-friends.htm>

Fisch, J.D., Rodriguez, H., Ross, R., Overby, G., and Sher, G. (2001). The graduated embryo score (GES) predicts blastocyst formation and pregnancy rate from cleavage-stage embryos. *Human Reproduction, 16*, 1970–1975. Retrieved May 30, 2011 from: <http://humrep.oxfordjournals.org/cgi/content/full/16/9/1970>

Jung, J. (1989). *Genetic Syndromes in Communication Disorders.* Austin, TX: Pro-ed, Inc.

Larsen, W.J. (2001). *Human Embryology (3rd Edition).* New York: Churchill Livingstone.

Lewis, B. (1990). Familial Phonological Disorders: Four Pedigrees. *Journal of Speech and Hearing Disorders, 55*, 160–170.

Mendel, G. (1865). *Experiments in Plant Hybridization.* Read at the meetings of February 8th, and March 8th, 1865. Retrieved May 30, 2011, from: <http://www.mendelweb.org/Mendel.html>

Moller, K. and Starr, C. (1993). *Cleft Palate: Interdisciplinary Issues and Treatment.* Austin, TX: Pro-ed, Inc.

O'Rahilly, R. and Muller, F. (1996). *Human Embryology and Teratology (2nd Ed.).* New York: Wiley-Liss.

Pearl, W.R. (1991). Single arterial trunk arising from the aortic arch. Evidence that the fifth branchial arch can persist as the definitive aortic arch. *Pediatric Radiology, 21*, 518–520.

Peterson, G.E. and Barney, H.L. (1952). Control methods used in a study of the vowels. *Journal of the Acoustical Society of America, 24*, 175–184.

Sebastian, A. (1999). *A Dictionary of the History of Medicine (p. 500).* New York: Parthenon Publishing Group.

Sinclair, D. (1976). Growth. In Hamilton, W. J. (1976). *Textbook of Human Anatomy (2nd Ed.).* Saint Louis; C.V. Mosby.

Timiras, P.S. (1972). *Developmental Physiology and Aging.* New York: Macmillan.

Tortora, G.J. and Grabowski, S.R. (1996). *Principles of Anatomy and Physiology.* New York: Harper-Collins.

Wu, B.L., Sanders, I., Mu, L., and Biller, H.F. (1994). The human communicating nerve. An extension of the external superior laryngeal nerve that innervates the vocal cord. *Archives of Otolaryngolology and Head Neck Surgery, 120*, 1321–1328.

IMPORTANT TERMS

Accretionary Growth: Increase in tissue size resulting from an increase in material between cells.

Alisphenoid Bone: A primitive osseous structure deriving from the first branchial arch. It will become the greater wing of the sphenoid bone.

Allele: One of two contrasting genetic traits attributed to a particular gene.

Apoptosis: A genetically encoded trait that limits growth in response to certain environmental conditions.

Autosome: Any of the twenty-two pairs of chromosomes in somatic cells that are not sex chromosomes. Autosomes are responsible for transmitting any trait that is not sex-linked.

Auxetic Growth: Increase in tissue size resulting from cell growth.

Blastocyst: A zygote once it has implanted itself in the uterine wall.

Branchial (Pharyngeal; Gill) Arch: One of six swellings of the ventral surface of the embryo destined to develop into structures of the distal respiratory and alimentary systems.

Branchial (Pharyngeal; Gill) Cleft or Groove: Spaces between the branchial arches on the outside of the embryo, destined to be foundations for laryngeal, oral, and nasal development.

Branchial (Pharyngeal; Gill) Pouch: A space corresponding to the branchial cleft, but located inside the embryo.

Catch-Up Growth: A period following growth inhibition by nutrition, illness, socioeconomic status, or psychological environment during which the child will grow at an accelerated rate until the normal developmental trend is recapitulated.

Crown-Rump Length: Overall fetal length measured from the crown of the skull to the end of the sacrum.

Daughter Cell: In mitosis, the first cell split into two identical cells, each having an identical chromosomal complement.

Developmental Anatomy: The study of individual growth from the single fertilized cell to birth.

Diploid: The total number of chromosomes inherited from both parents.

Dominant Trait: A genetic characteristic that expresses itself when in pairs or when combined with a recessive trait.

Ectoderm: The outer layer of embryonic germ cells.

Embryo: A developmental stage characterized by increase in cell size, number, and during which cells differentiate in to the stem cells for the various tissue types. This stage lasts from approximately the end of the third gestational week to the end of the eighth week.

Endoderm: The inner layer of embryonic germ cells.

Fetus: The developmental stage marked by very rapid overall growth as well as by continued differentiation and development of organs and systems. The fetal stage starts of the ninth gestational week and continues until birth, at thirty-eight to forty weeks.

Gamete: A germ cell from either parent, containing the haploid number of chromosomes and capable of combining to form a zygote.

Genetic Trait: Genetic possibilities, not all readily observable, contained in the genome of an organism.

Genetics: The science of heredity and the mechanics by which parents pass traits to their offspring.

Genotype: The genetic potential of an individual including the genetic potential of each parent and his or her ancestors.

Germ Cell: A gamete, either sperm from the father or ovum from the mother.

Gonadarche: Developmental stage following childhood during which the individual enters adolescence, a period marked acceleration in overall growth and by the onset of puberty during which hormones are produced by interaction of pituitary and gonadal functions.

Haploid: The number of chromosomes inherited from one parent; half the total number of chromosomes inherited by an individual.

Hemizygous: A term describing the genetic makeup of a cell or of an individual with only one gene representing a trait where the usual genetic compliment is two.

Heterozygous: A term describing the genetic makeup of a cell or of an individual with two contrasting alleles of a specific trait.

Homozygous: A term describing the genetic makeup of a cell or of an individual with two of the same alleles of a specific trait on a specific location of the same chromosome.

Meckel's Cartilage: The caudal cartilage of the first branchial arch (after Johan Friedrich Meckel, "The Younger," 1781–1833).

Mesencephalon: The middle part of the developing brain, destined to become the midbrain.

Mesenchyme: A type of embryonic tissue stemming from mesoderm; migratory mesenchyme cells move from their points of origin during the embryonic stage.

Mesoderm: The middle layer of embryonic germ tissue.

Morula: The embryonic stage between the zygote and the blastocyst, during which daughter cells have divided multiple, identical cells having the same characteristics and potentials. The name comes from Latin, *morulus*, meaning blackberry, which resembles the embryo at that stage.

Multigenic/Polyfactorial: An individual's developmental expression having more than one inherited source combined with exogenous factors.

Multiplicative Growth: An increase in the number of cells in a given tissue or organ.

Organic Brain Syndrome: A general loss of cognitive functions accompanying cerebral cortical neuron loss.

Ovum: The female gamete.

Palatopterygoquadrate Plate: The more rostral cartilage of the first branchial arch.

Phenotype: Manifested, observable traits in the individual.

Presbycusis: Hearing loss associated with advancing age.

Primitive Foregut: The distal part of the embryonic alimentary and respiratory system, formed about the fifth or sixth week of gestation and destined to develop into the pharyngeal cavities, the lower respiratory system and the esophagus.

Recessive Trait: A genetic trait that must paired with the same recessive trait in order to express itself in the phenotype.

Reichert's Cartilage: The cartilage of the second branchial arch (after Karl Bogislaus Reichert, 1811–1883).

Rules of Inheritance: Rules governing whether a genetic trait is expressed in a phenotype based on laws of probability. First published in the mid-nineteenth century by Gregor Mendel (1865) and based on his plant research.

Secondary Sexual Characteristics: Traits designating an individual as male or female, including development of external genitalia, distinctions in male and female body types (sexual dimorphism), and a decrease in vocal fundamental frequency (pitch).

Sex-Linked Chromosome: Chromosomes labeled "X" and "Y" that determine the gender of the individual. Females carry two X chromosomes and males carry an X and a Y.

Sigmoidal Growth Curve: A graphic projection of various aspects of growth as functions of time. the curve is so-names because it appears similar to the letter "S," with early slow acceleration followed by greatly increased acceleration, then a tapering off in the late stage.

Somatic Cell: Cells that form the body tissues, organs, and systems.

Somite: Roughly cubical paired swellings that begin to appear at the rostral end of the embryo during the third and fourth gestational week. Somites form the skin, bones, and musculature of the body wall.

Spermatozoan: The male gamete.

Teratology: The study of congenital anomalies.

Zygote: The combination of gametes; a fertilized ovum.

CHAPTER 4
Respiration

CHAPTER PREVIEW

The fourth chapter focuses on the power source for speech: the respiratory system. The respiratory system, as we all know, functions much more than as a bellows to power the vocal tract. Its most important function is to provide the exchange of life-sustaining gasses with the environment. This chapter will provide readers with a good deal of insight on this biological function, as well. Chapter 4 covers respiration and the respiratory system in terms of its biological functions first and then moves on to distinguish speech respiration from biological pulmonary ventilation. Gross and microscopic anatomy of the respiratory system and associated structures, neurological foundations essential to breathing, and developmental anatomy and growth of the respiratory system round out the material. This chapter is a valuable companion chapter to Chapter 7, "Deglutition," since the concerns and consequences of aspiration pneumonia are so pervasive in clinical concerns. Chapter sidebars describe respiratory disorders and their effects on speech.

CHAPTER OUTLINE

Definition of Respiration
  Cellular Respiration
  Internal Respiration
  External Respiration
  Pulmonary Ventilation
Physiology of Respiration
  Mechanics of Gas Exchange
  Boyle's Law and the Second Law of
    Thermodynamics
  Gas Constituents
  Alveolar Mechanisms
  Interactions of Thoracic Walls and Pleural Tissues
  Tidal Breathing
    Inspiration
    Expiration
    The Respiratory Cycle
    Respiratory Rates
    Respiratory Patterns
Respiratory Function in Speech
  Changes in Respiratory Cycles
  Active and Passive Expiration
  The Respiratory Pump
  Expiratory Pressures and Lung Volumes
  Speech Contexts and Respiratory Demands

Gross Anatomy of the Respiratory System
  Lungs
    Right and Left Lung Differences
    Alveolar Anatomy
    Pleural Membranes
  Respiratory Passages
    Bronchial Tree
    Pharynx
    Oral Cavity
    Nasal Cavity
Epithelial Tissues of the Respiratory System
Skeletal Muscles of the Respiratory System
Skeletal Framework of the Respiratory
  System
  Skeletal Framework
  Skeletal Support
Neurology of Respiration
Developmental Anatomy of the Respiratory
  System
Ontogeny of the Respiratory System
Chapter Summary
References and Suggested Reading

Definition of Respiration

The respiratory system permits the exchanges of life-sustaining gasses between the individual and the environment. It also provides the power for the production of speech. Gas exchange is the primary or biological respiratory system functions, whereas speech is a secondary or overlaid function.

Respiration may be considered on several levels, ranging from the molecular to the gross anatomical. At the smallest level, respiration is the metabolic process whereby nutrients and oxygen are combined to create energy for cellular activity. At the largest level, respiration is the contraction and relaxation of thoracic, cervical, and cranial musculature to move larger volumes of gasses in and out of the lungs and respiratory passages.

Cellular Respiration

Cellular respiration is one way a cell can obtain energy from blood oxygen and nutrients. Using this method, the cell converts of oxygen and carbohydrates, diffused through the cellular plasma membranes, into the energy required to maintain cellular physiology. The other way cells convert blood substances in to energy is through fermentation. Since fermentation is not a respiratory system function, and in fact, it requires no oxygen, this discussion focuses solely on cellular respiration.

Cellular respiration begins with the breaking up of *carbohydrates* contained in food. Its ultimate goal is to "burn" or oxidize glucose and produce carbon dioxide and water, accompanied by a release of energy. The digestive system breaks down the larger, complex molecules in food into more useable compounds and transports them to the bloodstream, which, in turn, delivers them to the cells.

As the name implies, a carbohydrate (literally, "Watered Carbon") is a molecule made up of hydrogen atoms bonded to carbon and oxygen. Carbohydrates include so-called "sugars," such as glucose, fructose or lactose, starches, and cellulose. These compounds can be relatively simple or very complex in composition. One of the simplest carbohydrates is glucose, composed of six carbon atoms, bonded to six oxygen atoms, and twelve hydrogen atoms ($C_6H_{12}O_6$). On most occasions, glucose is broken down from the far more complex chains of carbon, oxygen, and hydrogen found at the dining table. Table sugar (sucrose), for example, obtained from sugar cane or sugar beets, is a chain of twelve carbon, eleven oxygen, and twenty-two hydrogen atoms. One of the most complex carbohydrates is cellulose, composed of thousands of glucose molecules.

In order for the cell to use them, though, long chains of carbohydrates must be broken down into simpler compounds. The process begins with *glycolysis*. Glycolysis, or "the splitting of glucose," requires no oxygen in itself. However, the two subsequent steps in the cellular respiration energy-making process do require oxygen. These second and third steps are the *Krebs cycle* and the *electron transport chain*. Both processes take place in cellular mitochondria, and the net result is the conversion of carbohydrates and oxygen to carbon dioxide and water. Molecules of glucose and oxygen enter the cells, and carbon dioxide and water leave the cells by passing through the plasma membrane via a passive, non-energy consuming, transport mechanism. The chemical reaction formula for cellular respiration is:

$$C_6H_{12}O_6 + 6O_2 \rightarrow 6CO_2 + 6H_2O \uparrow$$

This formula indicates that six portions of glucose are combined six portions of oxygen, producing six portions of carbon dioxide and a like amount of water. The arrow pointing upward at the end of the formula indicates that energy is released by breaking the bonds that held the carbohydrate atoms together. The energy release is the purpose of cellular respiration.

Internal Respiration

Internal respiration, also called *tissue respiration,* is the exchange of blood gasses between the bloodstream and the cells. These compounds enter and exit the cells by diffusion across the plasma membrane, since the oxygen pressure in the bloodstream is greater than that of the cells. It will be noted that some definitions do not distinguish cellular respiration from internal respiration, considering both to be parts of a single metabolic process. Others consider cellular respiration to take place only within the plasma membrane, with internal respiration being a process of molecular diffusion across the plasma membrane. We take the latter approach.

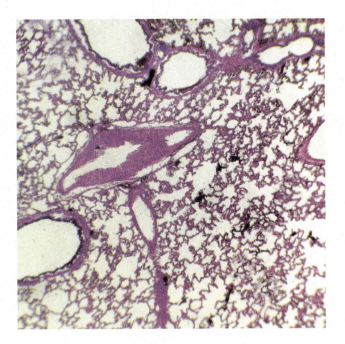

External Respiration

External respiration is the exchange of gasses between the lung's air spaces and the bloodstream. It is the process whereby oxygen and other gasses from the ambient air enter the circulatory system and by which waste gasses exit. The air we inhale contains about 20% oxygen. By the time it is exhaled, it contains about 16% oxygen (American Academy of Orthopedic Surgeons, 1987).

As is the case with internal respiration, the process of molecular transport in external respiration is one of diffusion across the thin alveolar-capillary membranes. These membranes are estimated to be less than one micron (e.g., Tortora and Grabowski, 1996). Oxygen, whether in its free state (O_2) or attached to carbon (CO_2) is carried by red blood cells, *erythrocytes*, to and from the tissues.

Pulmonary Ventilation

Pulmonary ventilation is the muscular process whereby gasses from the environment are moved in and out of the respiratory system. Of all the phases of respiration discussed in this chapter, pulmonary ventilation is the most readily observable and familiar.

Colloquially, pulmonary ventilation may be described simply as "breathing." It is accomplished by automatic or purposeful variation in lung volumes. In normal respiration to support metabolism, the rhythmic flow of gas in and out of the system is unconscious or involuntary. Breathing for speech, on the other hand, is usually conscious and voluntary.

Lung volume variations create pressure differentials between the ambient atmosphere and the interior spaces of the lungs and respiratory passages, causing air to flow in and out of the system in the classic, "In comes the good air and out goes the bad air," sequence. Mankind has learned to use the flow of air to create that wonderful human talent, spoken language.

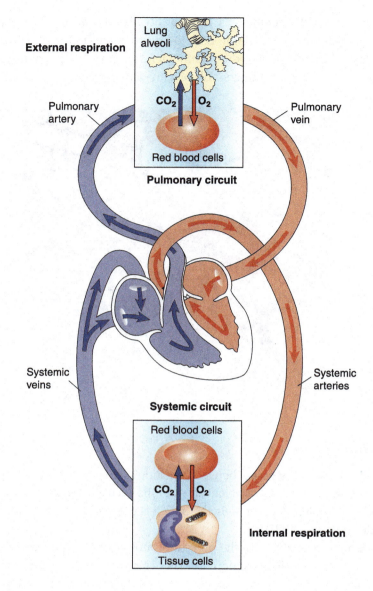

Physiology of Respiration

Mechanics of Gas Exchange

Boyle's Law and the Second Law of Thermodynamics

Respiratory function, whether for metabolic or for speech purposes, depends upon the ebb and flow of gasses in and out of the respiratory passages. The physical principles by which gas flows into and out of the respiratory system are those of the *Second Law of Thermodynamics* and *Boyle's Law*. The Second Law of Thermodynamics posits that heat cannot pass from a colder body to a hotter body without external help (Clausius, R., 1850). In thermodynamics, heat or temperature is energy and is roughly synonymous with pressure by virtue of combined gas laws. All this means a region of high pressure would be "hotter," just as a region of low pressure would be "cooler," and that high pressure air tends to flow to lower pressure regions until pressure between the two regions is equal.

Boyle's Law, named for Charles Boyle (1662), was the first principle of gas variable relationships to be discovered. It stipulates that the pressure of an ideal gas is inversely related to the volume of the container holding it. If the volume of a closed container increases, the pressure therein correspondingly decreases. Of course, if there is an opening in the container, and the volume expands, then, by the second law of thermodynamics, there will be an instantaneous flow of gas into the container. Gasses under relatively high pressure will tend to flow into volumes having lower pressures.

We will call the gas outside the respiratory system *atmospheric* gas to distinguish it from the gas inside the respiratory system. Under normal circumstances, the "container" for atmospheric air is the earth's atmosphere. The pressure of earth's atmosphere is the same as atmospheric pressure or barometric pressure, and varies with elevation, altitude, and weather conditions.

The respiratory system requires a constant ebb and flow of gas, in and out of the lungs and respiratory passages. Whatever the atmospheric air pressure, the respiratory system manages air flow through variation of the volumes of its gas containers, primarily the lungs, creating differences in gas pressures between the ambient air and the air inside the system. The difference in air pressures between the inside and outside of the respiratory system is called a *pressure differential*.

Gas Constituents

Under normal circumstances, atmospheric air is made up of oxygen and several other gasses. Air is about 78.08% nitrogen, 20.95% oxygen, 0.93% argon, and 0.03% carbon dioxide. Air pressure is a force spread out over the earth and almost everything on it, created by the masses of the combined gas molecules, and accelerated by gravity. Since there are more air molecules bearing down at lower elevations, atmospheric pressure is greater at lower elevations. Meteorologists have established a standard reference for atmospheric pressure. This is so-called "Standard Atmospheric Pressure," equal to 29.92 in. Hg. (101,325 Pa.). The standard reference pressure magnitude was established as the mean sea-level pressure at the latitude of Paris, France. Holding temperature and weather conditions constant, atmospheric pressure decreases with altitude, to the extent that atmospheric pressure at 10,000 feet is 20.57 in. Hg. (69,629 Pa.). In other words, the atmospheric pressure at 10,000 feet is about 69% of the atmospheric pressure at sea level.

Lower atmospheric pressure affects respiration in several ways. First, while the concentration of oxygen at high altitudes (over 8,000 feet) is the same as for sea level, the air pressure is substantially lower, meaning molecules of oxygen and air constituents are farther apart. This means that are fewer of them per unit volume, and the person breathing at high altitude needs to increase the respiratory rate, that is, breathe faster and deeper to maintain equivalent gas exchange. In addition, lower partial oxygen pressure in the rarefied atmosphere makes transfer of oxygen molecules across the alveolar walls more difficult, leading to lower concentrations of oxygen in the blood or *hypoxia*. The individual at altitude fatigues easier. After several days at high altitude, however, the body adapts. Lungs and alveoli increase in volume, and the body produces more red blood cells to make better use of the available oxygen.

Since air pressure is the driving force of the speech mechanism, speaking at altitude becomes more difficult than at sea level. Lowered air density means a speaker is able to utter fewer syllables per breath than at sea level (Rudmose, Clark, Carlson, Eisenstein, and Walker, 1946). Additionally, transient hypoxic periods brought about at high elevations are well-known to be accompanied by neurological disruptions, including dysarthria and dysphasia (e.g., Firth and Bolay, 2004).

Alveolar Mechanisms

At the proximal ends of the bronchial tree are the alveoli. An alveolus is a microscopic enlargement of the respiratory duct. It is here that gas exchange between the bloodstream and the respiratory passages occurs by means of an extremely thin barrier between the alveoli and the capillaries of the circulatory system. Oxygen and carbon dioxide move into and out of the bloodstream by diffusing across the alveolar capillary membrane, only a half-micrometer in thickness. A gas pressure differential is required for this step, as well as the diffusion of gasses from the bloodstream through the cell walls. The combined alveolar gas exchange surface is about 140 m^2.

Interactions of Thoracic Walls and Pleural Tissues

For air to flow into the respiratory system the air pressure inside the system must be less than that of the atmosphere outside. To create this pressure differential, that is, to decrease the gas pressure inside the lungs, an individual increases the lung volumes. This is accomplished by increasing the dimensions of the pleural cavities by using muscle contractions to increase the dimensions of the thorax. Since the ambient air pressure is momentarily greater than that inside the expanded lungs, at least for a moment, air flows from the higher pressure region to the lower pressure region until the pressures are equal. In comes the good air. If lung volumes are held constant, and neither expansion nor contraction occurs, there will be no pressure differential, no air will flow, and the individual will have "held" his or her breath.

Muscular efforts do not act directly on the lungs. They act first on the thoracic walls, creating expansion or contraction of the pleural cavities, the walls of which are invested with two pairs of endothelial pleural membranes or *pleura*. The space between the pleural membranes is completely filled with a *pleural fluid*, the surface tension of which binds the outer and inner pleural membranes together.

Inspiration begins with muscular contraction and expansion of the pleural cavities. This sets into motion a chain of pressure differentials, pleural pressures, the ultimate result of which is expansion of the alveolar spaces. Expansion of the thoracic wall draws the outer, or parietal, pleural layer outward, creating increased volume in the interpleural space. The principles of Boyle's law apply to the pleural fluid, so there is an accompanying decrease in pleural fluid pressure. The fluid pressure becomes more negative as continued thoracic expansion draws the outer, or parietal, pleural membranes away from the inner or visceral pleural membranes, increasing the volume of the intrapleural space. Negative pressure in the pleural fluid, naturally, draws the inner, or visceral, pleural membrane toward the outer, or parietal, membrane. Since the visceral pleura are attached to the outer surfaces of the lungs, the lungs must follow.

As the lung tissue follows its visceral pleural membrane, the substance of the lung expands, and as it does, so do its inner alveolar spaces. Expansion of the alveolar spaces results in a decrease in alveolar gas pressure, and air flows into the alveoli.

Expiration begins with relaxation of inspiratory muscles. These muscles have expanded the thorax and its contents, and now relax. Without the mechanical effort of inspiratory muscles to maintain thoracic walls in an expanded state, the natural tendency of the lungs to pull inward on the thoracic wall prevails, and the thorax contracts until tension between the inward pulling lungs and the outward pulling rib cage is in balance. During tidal breathing, whether air is flowing in or out of the system, the inspiratory muscles are the most active. Expiratory musculature is rarely activated during expiration. If forceful maximal expiration is desired, such as during great exertion, or if the mechanical effects of forced air are needed, muscles of expiration can come into play. Contraction of expiratory musculature is also required to compress the thorax beyond its resting state. When this occurs, the expiratory musculature forces act against the expanding forces of the rib cage.

Changes in thoracic dimensions are created by action of the respiratory muscles. Contraction of the muscles of inspiration increases thoracic volume, while contraction of the muscles of expiration, when they are activated, decreases the volume. As we shall see, it is sometimes difficult to distinguish muscles of inspiration from those of expiration.

Graphic representations of pressure differentials as functions of time follow a sigmoid, or "S-shaped," curve during breathing movements, and correlate very closely with muscular effort and movement speed. A sigmoid curve is one shaped like a stretched out letter "S," beginning with a slow acceleration rate, having a great increase in acceleration in the middle, and ending with a slow rate. Such a curve can describe the relationship between alveolar pressure and percent of vital capacity.

When inspiration begins, pressure in the alveoli is negative, lower than atmospheric pressure. This pressure differential creates an air stream, flowing into the respiratory passages and, ultimately the alveoli. As the respiratory cycle continues toward the peak of inspiration, the pressure differential between the atmosphere and the alveoli decreases. The difference between the alveolar pressure and atmospheric pressure becomes smaller as thoracic expansion slows, until there is no further expansion and equilibrium is achieved. At that instant, the instant that inspiration gives way to expiration, the pressure differential is zero: pressure in the alveoli and atmosphere are the same. The respiratory cycle continues, and during the expiratory phase of the cycle, the pressure differential between the alveoli and the atmosphere is positive. There is greater pressure the alveoli than in the atmosphere. Under these circumstances, air flows out of the system. The difference between alveolar and atmospheric pressures increases steadily until equilibrium is achieved again, and the next cycle begins.

During speaking of normal utterances, the inspiratory musculature is slowly relaxed against the elastic forces of the lungs. This allows maximum use of the volumes of air within the system as a power source for the vocal tract. Thus, during normal utterances, muscles of inspiration are active, even though air is flowing out of the system. When normal relaxation volumes are used up, the speaker inhales again. There are speech actions, such as shouting, adding emphasis, or expressing especially long utterances, during which the expiratory muscles play the role of squeezing the last possible volumes of air out of the system.

Tidal Breathing

The rhythmic, nearly periodic flow of air in and out of the respiratory system has long been likened to the rising and falling of the ocean's tides. For this reason, regular breathing is often called "tidal breathing." There is a regular, predictable pattern to tidal breathing, and it may be described as a series of respiratory cycles.

A respiratory cycle consists of one inspiration and one expiration. They cycle then repeats at a rate determined by the body's respiratory needs. If tissues demand more oxygen, the rate will increase, as the demands decrease, so does the rate of respiratory cycle repetition.

We begin examination of breathing with a frozen moment in the regular breathing sequence. This sequence is very complex, and involves interaction of multiple forces created by muscles and by the physical properties of the very tissues of the thorax and abdomen. A state of equilibrium between the expansive forces of the rib cage and the contracting forces of lung tissue and alveolar fluids is called the *resting state*. During the resting state, muscles of inspiration and expiration are inactive.

To set the sequence in motion, we start the action at the beginning of inspiration. During normal, tidal breathing, the individual contracts the muscles of inspiration and relaxes them. The expanded lung has a tendency to resist being expanded, due to the natural elasticity of its tissue and the surface tension of intra-alveolar fluids. This resistance to expansion causes the lung to pull the pleural linings and in turn, the thoracic walls back to their previous dimensions. Decreasing thoracic, interpleural and alveolar volumes momentarily create greater gas pressures inside the system than are outside. Greater internal pressures force the gas inside the system to flow out until the pressures inside and outside the system are equal.

Expiration may be passive or active, depending upon the needs of the breather. Passive expiration is most common and occurs as the inspiratory muscles relax. Expiration during tidal breathing and production of short, casual utterances is usually passive. Active expiration is brought about through contraction of the muscles of expiration.

Passive expiration is brought about by two non-muscular forces that cause the lungs to pull inward on the pleural cavities at all times, making active use of expiratory muscles unnecessary for tidal breathing. First is lung *elastance*, the natural elasticity of the lung tissue itself. Elastance is the tendency of lung tissue to return to its original shape after having been stretched. In a manner similar to a stretched spring, lung elastance increases as lung volumes increase during pleural cavity expansion. Once inspiratory muscular effort relaxes, the elasticity of lung tissue causes the lungs to return to their pre-stretched shapes. The other property that causes the lungs to contract is the surface tension of the fluids inside the alveoli. Surface tension is created at the interface of liquid and gaseous molecules, a condition that occurs in the interior volumes of each alveolus. In the alveoli, surface tension causes the distended alveoli to collapse, reducing their volumes and increasing their internal gas pressures.

Instances such as those occasioned by forceful expulsion of matter from the airway, blowing up an air mattress, cheering on the home team, or speaking long utterances without pause require assistance of the muscles of expiration. These muscles compress the pleural cavities, extending the duration and, depending upon the circumstances, the magnitudes of positive pressure differentials between the alveoli and the atmosphere, forcing air out of the system. Contraction of expiratory muscles is always required for compression of the pleural cavities beyond resting state.

The timing of respiratory cycles is of interest. During *eupnea* or normal tidal breathing, the inspiration phase of the respiratory cycle typically takes up the shortest amount of time, being slightly less than one-third of the cycle period. The reason for this is that during expiration, aerodynamic events and autonomic control cause the vocal folds to adduct slightly, increasing resistance to and slowing airflow out of the system. The longer expiratory phase of the respiratory cycle is fortuitous, as it allows more time for gas exchange. Further, the end of the expiratory phase includes is a period during which air flow pauses. This pause is approximately equal in duration to the phases during which air flows into or flows out of the system.

Respiratory Rates

Respiratory rates are the number of respiratory cycles produced by an individual in one minute. Rates vary with age and constitutional condition, being slower in athletic adults and faster in infants. The average respiratory rate for adults is 12 to 20 cycles per minute, while infants may have rates as high as 70 cycles per minute.

The degree of exertion has a well known effect on respiratory rates, as well, with the autonomic nervous system signaling an increase rate to meet increased oxygen demands. Autonomic nervous system control of respiratory cycles supersedes voluntary control, such as that engaged during speech, to the extent that an exhausted speaker, or a speaker starved for oxygen, will tend to speak in very short utterances while breathing cycles are rapid (tachypnea).

NORMAL RESPIRATORY RATES
(American Academy of Orthopedic Surgeons, 1987):

CYCLES PER MINUTE:

10	20	30	40	50	60	70	80
30 to 70		-------infants------					
25 to 30**		~~~1 to 5~~~					
20 to 24***		~~5 to 12~~					
12 to 20****	adults						
10	20	30	40	50	60	70	80

**Children aged 1 to 5 years have rates as low as 20 and as high as 40.

***Children aged 5 to 12 have rates as low as 16 and as high as 36.

****Adults have rates as low as 10 and as high as 30

Respiratory Patterns

Patterns of respiratory activity differ among various individuals and with the onset of various pathological conditions. In normal respiration, patterns may be described in terms if those parts of the body that evince the most observable movement. In pathological conditions, respiratory patterns are described in terms if their timing, volumes, and some other features, in addition to movements.

The most common normal respiratory pattern is the *diaphragmatic/abdominal* pattern, so called because the most obvious movement is distention of the abdomen during the inspiratory phase of the respiratory cycle. Abdominal distention is produced by the contraction of the diaphragm, the dome-shaped muscle separating the thorax from the abdomen. As diaphragmatic muscle fibers contract, the diaphragm becomes flatter, increasing the superior-inferior dimensions of the thorax and, in turn, the pleural cavities. The flattening diaphragm displaces the abdominal viscera lying just below, and since they can only move ventrally, the belly expands.

In the thoracic respiratory pattern, the most obvious movement is seen in the thorax. A predominantly thoracic pattern is quite common, and even if the predominant respiratory movement is abdominal, some movement of the thorax can be expected in normal respiration. In cases where the attention is placed on the act of breathing, such as during a clinical evaluation, it is not uncommon to observe the breather changing subtly from a diaphragmatic/abdominal pattern to a thoracic one. For whatever reason, some individuals believe that the thoracic pattern in more "correct," and will adopt it, even though the abdominal pattern is more efficient.

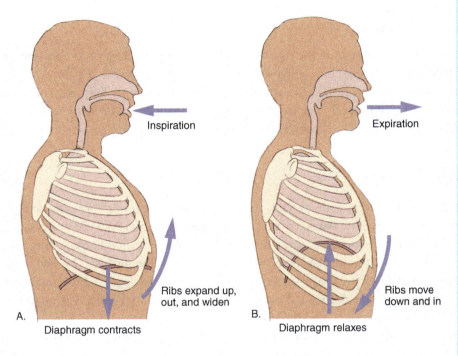

A. Diaphragm contracts — Inspiration — Ribs expand up, out, and widen

B. Diaphragm relaxes — Expiration — Ribs move down and in

Thoracic movement during respiration takes place through contraction of eleven paired external intercostal muscles, and is rather complex. Those intercostal muscles between the first six ribs move the rib cage superiorly and anteriorly in what has been called a "pump-handle" movement. The "pump-handle" moniker derives from the resemblance of sterno-costal movement to that of an old fashioned water pump, moving up and forward through its stroke. Contraction of the intercostal muscles between ribs seven through ten creates a lateral and superior thoracic expansion, called the "bucket-handle" movement. The "bucket-handle" analogy presumably follows the "pump-handle" analogy, wherein, after the pump handle has been used to draw two buckets of water, they are placed on either side of the water bearer, who will lift their handles upward and laterally for the walk back home from the well. In thoracic breathing, both "pump-handle" and "bucket-handle" movements occur.

A clavicular respiratory pattern is seen when the breather needs to move maximal volumes of gas in and out of the respiratory system. Such conditions exist when the breather is physically exerted, needs to blow maximum air volumes for mechanical purposes including speech or singing, or under disease conditions, when normal respiratory patterns provide insufficient gas transfer. During clavicular breathing, the secondary or accessory respiratory muscles are brought into play, moving the upper, apical regions of the pleural cavities to the maximum extent by lifting the clavicles and scapulae. A clavicular pattern alone or when combined with a "pump-handle" pattern is a relatively inefficient means of moving gas for tidal respiration or for speech purposes.

PAHTOLOGICAL RESPIRATORY PATTERNS

Apnea is complete cessation of breathing. Apnea can be transient, followed by spontaneous or induced resumption of ventilation, or it can precede death.

Biot's Respiration (after Camille Biot, ninteenth-century French physician) is a pattern of rapid, deep, regular inspirations followed by 10 to 30 second periods of apnea. It is associated with central nervous system dysfunction and offers a poor prognosis.

Cheyne-Stokes (after John Cheyne and John Stokes, ninteenth-century Scottish and Irish physicians) respiration results from central nervous system disturbance. The pattern is one of rapid, irregular breathing starts with shallow volumes, and sometimes progressing to deeper, then shallow volumes again. This pattern lasts for 30 seconds to 2 minutes, and alternates with periods of *apnea* (no breathing) for 5 to 30 seconds. It is distinguished from Biot's respiration by its irregular patterns of hyperpnea. Another pathological respiratory pattern is *neurogenic hyperventilation*. The pattern is one of deep, rapid respiration (hyperpnea and tachypnea).

Dyspnea: Difficulty breathing.

Hyperpnea: Deep, abnormally large volume breathing.

Kussmaul (after Adolf Kussmaul, nineteenth-century French pathologist) respiration is seen in diabetic attacks (ketoacidosis) or "air hunger." It was named. This pattern is one of deep sighing breathing.

Neurogenic hyperventilation can be caused by head trauma or diabetic coma and may be a grave sign of impending death, especially if the cycles become very shallow, "agonal" respirations.

Tachypnea: abnormally high frequency of respiratory cycles.

Pathological respiratory patterns may be the result of central nervous system dysfunction or metabolic emergencies. The essential and inseparable relationship of respiration to life implies that an individual under great stress will show changes in respiratory timing.

Respiratory Function in Speech

The power to drive the vocal tract comes, in part, from the conversion of the potential energy of compressed air in the respiratory system into kinetic energy by relaxation of inspiratory forces or by enlistment of expiratory forces. Expenditure of muscular energy is required to position the articulators, but respiratory air flow assists even this effort. In this sense, the driving forces of the speech mechanism operate in a manner similar to a mechanical air compressor, drawing atmospheric air into a storage vessel and storing it until the appropriate pneumatic tool is attached and operated by a user.

Respiration During Speech

During speech, the timing of the respiratory cycle changes. The inhalatory phase of the respiratory cycle occupies a much shorter portion of the respiratory period. The speaker makes a very quick inspiration followed by a prolonged expiration, depending upon the length of the utterance required. The volume of air drawn into the system during speech inspiration depends upon how long the speaker perceives the upcoming utterance is to be and is of no fixed magnitude. In fact, intra-individual differences between speakers may be counted upon to thwart attempts to define volumetric speaking parameters. A general rule, though, is that the longer the intended utterance is perceived to be, the more air will probably be drawn into the system.

In examining respiratory function during speech, an important consideration is the force used to drive the vocal folds during voicing efforts. Of course, since the voice is not used in tidal breathing the force required for driving the vocal folds is not a consideration. This force, when distributed over the surface of the tissue underlying the glottis, is known as *subglottal pressure*. It is difficult to measure precisely the amount of subglottal pressure necessary for speech at various amplitudes.

For normal conversational speech, approximately 6 cm. H_2O (588.4 Pa.) of subglottal pressure is required. For louder speech, greater pressure is required, and for softer speech, less is required. To cease phonation when the vocal folds are vibrating, pressure must drop below 4 cm. H_2O (392.3 Pa.). During normal vowel production, trans-glottal pressures are in the range of 7 to 10 cm. H_2O (686.5–980.7 Pa.) (Netsell and Hixon, 1978).

Most speakers are able to produce and maintain a relatively constant subglottal pressure magnitude using a combination of relaxation and muscular forces. After a relatively quick inspiration, relaxation pressure provides more than enough force to create the required subglottal pressure. When volumes of air expired reach about 55% of vital capacity, the balance of forces shifts from relaxation pressure alone to a gradually increasing demand for muscular effort to maintain the required subglottal pressure. Ultimately, if the utterance is prolonged past resting state, all of the force required to drive the vocal tract will come from muscle contraction.

Lung Volumes and Expiratory Pressures

Several volume measures are useful when describing respiratory functions. Respiratory volumes are measures with a device called a *spirometer*, and the act of measuring respiratory volumes is called *spirometry*. In spirometry, an individual places a mouthpiece in the oral cavity, holding it tight enough with the lips so that no air can escape outside the tube to which the mouthpiece is attached. A special clamp is placed on the nostrils, pinching them closed so that no air can escape through the nasal cavity. After the proper attachments, the individual is instructed to perform various respiratory tasks, the purposes of which are to make volumetric measurements.

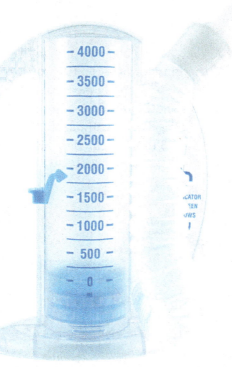

It is, of course, impossible to accurately measure respiratory volume exchanges during speech with the mechanical apparatus just described. The presence of a mouthpiece prohibits it. However, several important "non-speech" respiratory volumes can be measured. These volumes vary among individuals. As a general rule, respiratory volumes will be greatest for athletic adult males. Since respiratory air pressures drive the speech system, it is apparent that diminished volumes, particularly when evident in the same individual over time, and will provide a basis for understanding and accommodating speech disorders.

The most important of these non-speech respiratory volumes is *vital capacity*. Vital capacity is the total volume of air that can be expelled from the lungs after a maximum inspiration. This volume represents the effective gas moving ability of the respiratory system, and in principle, the maximum volume that can be used for speech. Both inspiratory and expiratory muscles are brought into play at their maximum range during measurement of vital capacity. Since accurate measurement of vital capacity depends upon the subject drawing the greatest volume of air possible in and then forcing out the greatest volume of air.

Normal vital capacity varies with body weight, height, race, and gender (Mosby, 2011; Miao1, Zhang, He, Yan, Wang, Cao and Fu, 2009), with normal values and is greatest in adults between twenty and thirty years of age. Vital capacity is not the same as total lung capacity. Total lung capacity includes residual volume, or the volume of air that remains in the lungs after a maximum exhalation. Normally, vital capacity is about 75 to 80% of total lung capacity, and varies from about 3 to 5 liters in adults.

Tidal volume is the volume of air exchanged during quiet breathing. This volume is much less than vital capacity, approximately .5 liters, and muscular effort involves only the inspiratory muscles. The greatest difficulty involved in accurate tidal volume measurement involves the presence of the measuring mechanism and its effect of changing what is usually an automatic movement. For speech purposes, tidal volumes are most important when considered in relationship to volumes produced with more effort.

Inspiratory reserve volume, also called "supplemental air" is the difference between the peak inspiratory tidal volume and the peak of forced inspiration. In other words, it is the amount of air one can draw into the system beyond that of tidal inspiration. If a speaker wishes to prolong an utterance beyond a few syllables, the starting place is to draw volume greater than tidal inspiration. Normal inspiratory reserve volumes are 30 to 50% of vital capacity, or around 1.9 to 3.1 liters, varying with the age, race, gender, height, and body weight of the individual.

Expiratory reserve volume, also called "complemental air," is the volume of air that can be forcefully expelled beyond a tidal expiration. The muscles of expiration must be contracted to their fullest extent, and the lungs compressed to the maximum to properly measure expiratory reserve. Referring again to the prolonged utterance, a speaker who runs short of air during normal conversation may tend to use expiratory reserve volumes to squeeze in the most syllables without stopping to breathe again. Normal expiratory reserve volume is 20% of vital capacity, or .8 to 1.2 liters, depending, again, on the constitutional makeup of the individual.

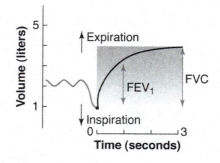

Residual air volume is the amount of volume left in the lungs after a maximal expiration. Residual volume is created by the adhesion of the pleural membranes to one another, causing the lungs to maintain the general interior dimensions of the pleural cavities. This volume is constant, and is only released when a lung collapses into the pleural cavity. Typical residual volumes are 20 to 25% of total lung capacity.

Gross Anatomy of the Respiratory System

The respiratory system consists of the lungs and the respiratory passages. The lungs are the organs in which gas exchange takes place to support metabolism, and form what is usually termed the *respiratory part* of the system. The lungs also serve as a storage space for pulmonic air to be used in the production of speech. The respiratory passages are conduits through which gasses flow into and out of the lungs, and are called the *conducting part* of the system. In that sense, they're just like the ducts that let air flow through your home or car heating system . . . except these ducts are alive.

Respiratory structures are located in the thorax, head, and neck. Those located in the head and neck share some structures with the digestive system. Shared structures include those in the oral cavity and oropharynx. This fact is of utmost importance when there may be danger that substances introduced into the digestive system may enter the respiratory system. Since the respiratory system is contained in the cranial, cervical, and thoracic regions, its skeletal components are those of the skull, neck, and chest.

The respiratory system may be clinically divided into lower and upper sections. Forming a muscular and cartilaginous gateway between the lower and upper sections is the larynx. The larynx serves a twofold purpose of protecting the lower airway from the intrusion of foreign matter and producing several sound sources for speech. Lower respiratory structures, including the trachea, bronchial tree, and lungs, are inferior to the larynx. The upper respiratory section consists of the larynx and structures superior to it. The larynx is the subject of its own chapter in this volume.

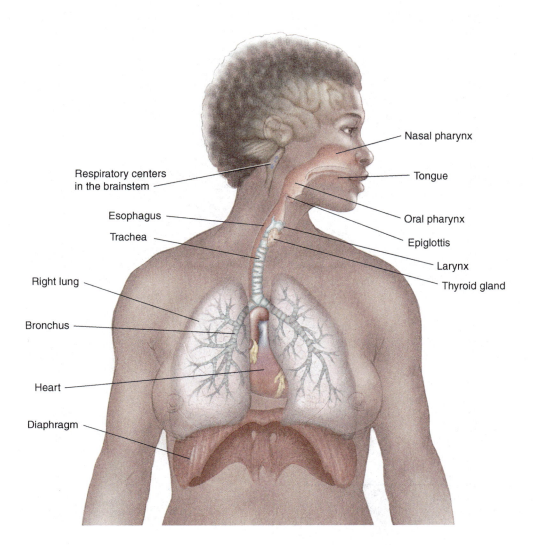

Nasal pharynx

Respiratory centers
in the brainstem

Tongue

Esophagus

Oral pharynx

Trachea

Epiglottis

Larynx

Right lung

Thyroid gland

Bronchus

Heart

Diaphragm

Lungs

The two lungs are contained in two pleural cavities, one on the right and left sides of the thorax. Each lung has a roughly pyramidal shape, with a base inferiorly and an apex superiorly. The *costal* surface of each lung lies just beneath the ribs, and the medial surface lies between the two lungs. The lungs expand and contract with the movements of the thorax, creating cycles of position changes inside and outside the chest.

Each lung has three rather sharply defined edges or borders. An anterior border defines the anteromedial extent of a lung, a curved posterior border defines the posteromedial extent, and an inferior border defines the base of each lung.

The tissue of both lungs contains about 700 million alveolar air sacs, resulting in a spongy texture. Early in life, the lungs are pink in color. Exposure to airborne pollutants, including tobacco smoke and exhaust debris from industrialized areas, causes the lungs of the adult to gradually lose their pink color and become blackened.

Between each lung, next to the medial surface is the mediastinum, a complex thoracic epithelial tissue septum containing the heart, bronchial tree, great cardiac vessels, the trachea and esophagus, thoracic lymph duct, the vagus and phrenic nerves, and the thymus glands. The base of each lung is concave, with the margins lower than the center. This shape allows for the dome of the diaphragm to closely fit in the underside of the thoracic surface. During inspiration, when the dome of the diaphragm flattens, the inferior border of the lung moves inferiorly 5 to 7 cm. moving back upward on expiration.

The lungs are free-floating in the pleural cavities, and are only attached at the mediastinal points where the bronchial tree branches enter. This part of each lung is called its *hilus* or root. Each root contains vital structures

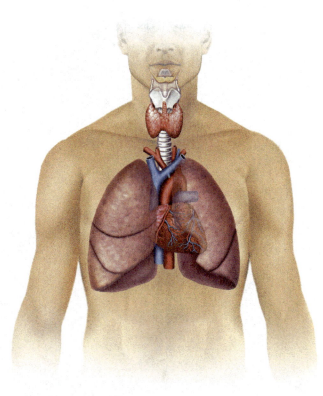

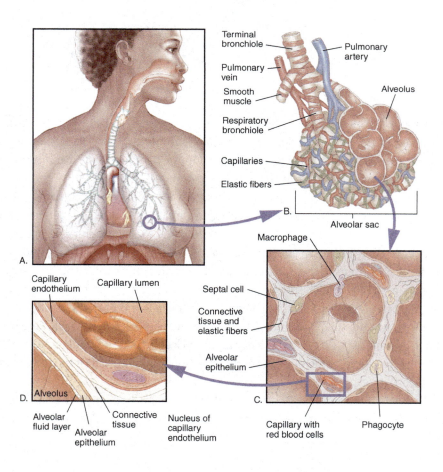

A.

B.

Terminal bronchiole

Pulmonary artery

Pulmonary vein

Smooth muscle

Respiratory bronchiole

Alveolus

Capillaries

Elastic fibers

Alveolar sac

Macrophage

Septal cell

Connective tissue and elastic fibers

Alveolar epithelium

C.

Capillary with red blood cells

Phagocyte

Capillary endothelium

Capillary lumen

Alveolus

D.

Alveolar fluid layer

Alveolar epithelium

Connective tissue

Nucleus of capillary endothelium

for support of the organ, including blood and lymph vessels, nerves, and the bronchial tree. The lungs are also separated from one another by the mediastinum and by the pleural membranes, so that the failure on a single lung need not affect the function of the other lung.

Right and Left Lung Differences

There are distinct differences in the right and left lungs, resulting mainly from the location of the heart to the left of the center of the thorax. The left lung is, therefore, slightly smaller than the right lung and has only two lobes as compared to the right lung's three lobes. The right lung is, however, a bit shorter in its superior-inferior dimensions because the mass of the liver, situated in the upper right abdominal quadrant, causes the dome of the diaphragm to protrude higher into the thorax under the lung on that side.

The left lung, as mentioned, has two lobes—a superior lobe and an inferior lobe. These are separated by an oblique fissure. It also has a space for the heart, the cardiac impression, and a relieved anterior area, the cardiac notch, at the inferior and medial border of the inferior lobe. The right lung has three lobes, a superior lobe, a middle lobe, and an inferior lobe. The middle lobe appears smaller than the superior and inferior lobes, and it does not extend as far dorsally as the other two. An oblique fissure separates the superior and inferior lobes on the upper, dorsal costal border, where the middle lobe is not present, and the middle and inferior lobes on the anterior and medial surfaces. A horizontal fissure separates the middle and superior lobes. Like the left lung, the right lung has a space for the heart, the cardiac impression, but there is no cardiac notch in its anterior surface. Although the fissures of the lungs lend them to separation by lobes, a more functional division of lungs, at least for medical diagnostic and treatment purposes, is described in terms of *bronchopulmonary segments*, ten of which serve distinct regions of each lung.

Pleural Membranes

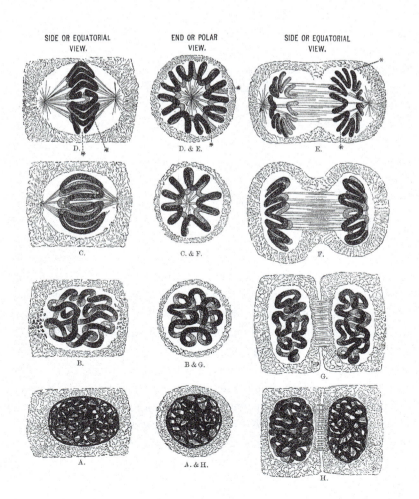

The pleural cavities are lined with two *pleural membranes,* or *pleurae,* continuous with each other at the lungs' roots. The two membranes might be considered a single membrane were it not for some structural and functional differences. Anatomically, the *visceral pleurae* form the outer surfaces of the lungs, including the surfaces inside the fissures, and are mechanically inseparable from the lung's surfaces. The *parietal pleurae* are thicker than the visceral pleurae and form the linings of the thoracic cavity, including the top of the diaphragm. Functionally, the visceral pleurae are relatively insensitive to pain, whereas the parietal pleurae are highly sensitive. The parietal pleural membrane has spaces (stromata) in its tissue for the absorption of pleural fluid. Fluid passes through these stromata to be transported by the lymphatic system (Miserocchi, 1997).

The space between the pleural membranes is called the *pleural sack.* It contains a small amount of serous fluid, called *pleural fluid,* which lubricates the membranes and causes them to adhere to one another through surface tension, but at the same time, to move in relationship to one another. Pleural fluid is secreted and absorbed by the visceral pleural membranes in much the same manner as lymph circulates (Hamm and Light, 1997). Fluid enters the pleural sac in the apical region and flows down to the mediastinal and diaphragmatic areas, where it is absorbed. In healthy individuals, there is a balance between the secretion and absorption of pleural fluid, maintaining a volume of about 1 ml. in the pleural sack.

Respiratory Passages

The respiratory passages conduct gasses between the lungs and the atmosphere, and open to the outside environment at the nose or mouth. Speech produced with air flowing out of the respiratory system is referred to as *egressive.*

The most commonly encountered way of considering the bronchial tree is as a progression from larger to smaller diameter branches or generations. This scheme may be useful for examining chest x-rays, but for the purposes of considering respiratory function in speech, it is more relevant to consider the branches as beginning at the smallest diameter branches and ending at the largest diameter the trachea. This approach is called a *convergent* approach (Horsfield, 1997) and is more consistent with the fact that egressive air, such as that expired during speech, begins under pressure in the alveolar storage spaces and moves into progressively larger tubes on its way to the vocal tract and atmosphere.

The respiratory passages include the converging structures of the bronchial tree, most of which is embedded in the substance of the lung, the trachea ("wind pipe"), the pharynx, oral, and nasal cavities. The respiratory system is divided into a lower section and an upper one for clinical purposes. The boundary between the two divisions is at the upper extent of the trachea or lower border of the larynx.

Bronchial Tree

The bronchial tree is commonly so named for the fact that it appears to be an inverted tree, branching from a single tracheal "trunk" into progressively smaller ventilation tube "branches." The first branching of this tree is a division into two parts, the *main stem bronchi,* each of which serves a lung. The pattern of splitting into two branches continues developmentally, and may be traced to the most proximal divisions of the respiratory passages.

Proximal to the main stem bronchi, most of the rest of the bronchial tree branches are embedded in the lung substance, but there is a short part outside the substance of the lungs. The parts of the bronchial tree embedded in the substance of the lungs are called intrapulmonary, and those parts outside the lungs are called extrapulmonary.

In a convergent model, air under pressure higher than that of the atmosphere moves from the alveolar spaces through the smallest, or first-order, bronchial passages, flowing through progressively larger diameter tubes until it reaches the main conduit, the trachea. Bronchial branches become larger to accommodate the higher volumes of air required to evacuate and supply respectively larger lung volumes.

The smallest branches of the bronchial tree are called the respiratory bronchioles. These are about 0.5 mm in diameter and lack cartilage. They are completely encircled by a ring of smooth muscle, which control the diameter of the tube's opening. Depending upon respiratory demands, these passages dilate or constrict, allowing greater or lesser impedance to air flow. Respiratory bronchioles communicate directly, without branching, to larger bronchioles, about 1 mm in diameter. These larger bronchioles have a cartilaginous skeleton consisting of loosely scattered plates proximally, becoming more ring-like distally.

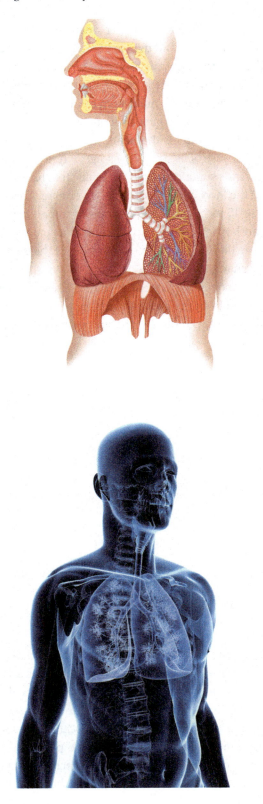

Bronchioles converge at larger diameter tubes, called segmental bronchi, each of which serves a single bronchopulmonary segment. There are at least ten bronchopulmonary segments in each lung, each functioning as a nearly separate anatomic unit (Tompsett, 1965). Since each bronchopulmonary segment is served by a single segmental bronchial tube a blockage of that tube affects the entire segment, and pathological conditions affecting a single segment may be isolated from the rest of the lung.

Branches of the bronchial tree become larger in diameter. Segmental bronchi converge at the lobar bronchi, each of which serves a single lobe, three on the right and two on the left. Lobar bronchi converge at main stem bronchi at approximately the level of the fifth thoracic vertebra in adults. These main stem bronchi are bifurcations of the trachea. The cartilaginous support at the bifurcation, called the carina, has two inferior openings and one superior opening.

There is one bronchus for each lung. The two main stem bronchi are not identical. On the right side, the main stem bronchus takes a nearly straight superior course as it emerges from the right lung, while the one on the left takes a definite lateral course to allow space for the heart. Because of this asymmetry, aspirated substances are more likely to enter the right lung because of its straight passage, when the individual is standing or seated upright.

The trachea or "windpipe" extends superiorly from the main stem bronchi to the most inferior of the laryngeal cartilages, the cricoid cartilage, at the level of the sixth cervical vertebra. This anatomic landmark is also called the *cricoid line*. The trachea consists of 15 to 20 cartilaginous rings, incomplete posteriorly, and held together by connective tissue and muscle. The cartilaginous rings help it maintain its patency or openness so air can freely flow to and from the lungs, and at the same time, allow for free movement of the trachea within the thorax as other organs expand and contract or otherwise change location.

The trachea lies anterior to the esophagus for its entire course. The esophagus is located between the spine and the trachea. This main passage to and from the lower respiratory tract needs to be flexible because it moves about freely during breathing, swallowing, and according to changes in body posture.

TRADITIONAL VIEW OF BRONCHIAL TREE ORGANIZATION

The usual view of the bronchial tree begins the "trunk," and branches by doubling at every division through progressive smaller "branches." This orientation may derive from the archetypical "tree" shape, albeit inverted, or it may be because the respiratory cycle is usually described as beginning with inspiration. In any case, the usual orientation of the bronchial tree is presented here for clarification:

"Trunk," the largest diameter tube: Trachea

Branches: First division: Two Primary or Main Stem Bronchi
Second Division: Secondary or Lobar Bronchi
Third Division: Tertiary or Segmental Bronchi
Fourth Division: Terminal Bronchi

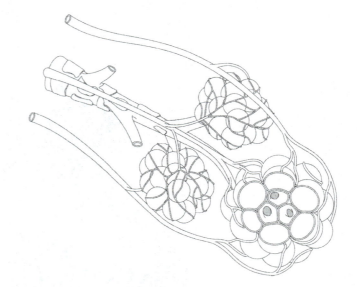

The upper respiratory passages include structures that comprise the *vocal tract*, and are of special interest to students of speech. The vocal tract is that part of the peripheral speech mechanism by which pulmonic air, that is, air from the lungs, is manipulated to create the sounds of speech. It is generally held to extend superiorly from the glottis, or vocal folds, to the lips, if the sound emanates from the mouth, or to the nares, if the phoneme is nasal. Thus, it includes the pharynx, oral, and nasal cavities. Speech sounds emanating from the oral cavity are referred to as *oral* sounds, while those emanating from the nasal apertures are called *nasal* sounds. Upper respiratory passages proper are generally held to begin slightly inferior to the glottis at the upper border of the trachea.

In the simplest of mechanical models, the vocal tract is a tube, approximately 17 cm. in adult length, closed at the glottal (proximal) end and open at the lips or nostrils, with movable parts that allow it to change shape along its extent. Since it is part of the respiratory system, tidal air flows in and out of the vocal tract constantly and automatically, with nearly periodic cycles. However, when employed for speech, automatic function of the respiratory system gives way to voluntary activation, and the cyclic nature of the air flow is temporarily replaced with short, deep inspirations and prolonged expirations. Moveable structures within the vocal tract manipulate, reroute or obstruct the flow of air to produce the sounds of speech.

Movements of vocal tract organs create constrictions or blockages so that energy generated by pulmonic air pressure creates three general types of sound: a *quasiperiodic* phonatory sound source, created through successive opening and closing cycles of the vocal folds; a *continuous aperiodic* sound source produced through force of pressurized air flowing through tight constrictions; and an *transient aperiodic* sound source created by temporary blockage and quick release of the air in the tract. Changes in vocal tract shape produce distinctive changes in the way its interior tube resonates to sound sources produced by the larynx and other moveable structures. Great flexibility enables a speaker to change locations of these sound sources as well as to change overall vocal tract shape and thereby produce the many sounds human beings discriminate as speech.

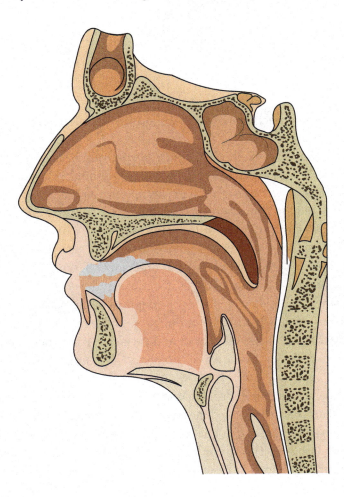

Pharynx

The pharynx is a muscular tube extending from the trachea to the oral and nasal cavities. It serves a dual role as a gas conduit between those respiratory structures and as an alimentary passage for food and drink from

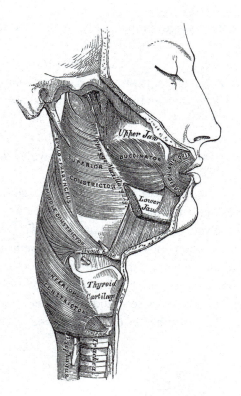

the mouth to the esophagus. The pharyngeal walls are formed of muscle and fiber and are about 12.5 cm. in length in the adult.

The principal muscles of the pharynx are the *superior pharyngeal constrictor, middle pharyngeal constrictor,* and *inferior pharyngeal constrictor.* These muscles help the pharynx perform is digestive and respiratory functions. Other pharyngeal muscles include *stylopharyngeus, palatopharyngeus,* and *salpingopharyngeus.*

The pharynx may be divided into three sections to simplify study. The first section, starting at the proximal end, contains the muscles and cartilages of the larynx or "voice box." This part of the pharynx is also called the *laryngopharynx.* Distal to the laryngopharynx, the pharyngeal tube makes a sharp posterior turn and continues superiorly past the posterior part of the oral cavity. From edge of the laryngeal vestibule to the horizontal plane of the oral roof, the pharyngeal tube is called the *oropharynx.* The pharyngeal tube continues a short distance further, ending at the posterior openings of the nasal cavities, the choanae, where it is referred to as the *nasopharynx.* Unobstructed continuation of the oral and nasal cavities and the pharynx make it possible to breathe through either the mouth or nose and to use either structure as part of the speech mechanism. The connection of the oral cavity and the pharynx also makes it possible for food or liquid to enter the respiratory tract.

The laryngopharynx widens at its inferior connection with the trachea and is continuous with the oropharynx superiorly. It lies behind and partially around the larynx. The laryngopharynx is composed almost entirely of the larynx, an organ that consists of three anatomic valves situated in a vertical series so as to protect the airway from intrusion by unintended objects. The larynx has special significance in the study of human communication and is covered specifically in Chapter 5. However, its relevance to respiratory function is significant and is covered here.

By virtue of its three valve design, the larynx can completely separate the upper and lower airways. This separation has the obvious effect of isolating and sealing the lower airway against unwanted intrusion, or *aspiration,* of material. In addition to keeping unwanted material out, the laryngeal valves can keep air trapped within the lower part of the system. Sealing impounded air in the lower airway increases thoracic rigidity to support the momentary requirements of lifting, defecation, child birth, and, through the sudden release of pressurized air, to forcefully eject unwanted materials from the airway during coughing.

The most inferior of the three laryngeal valves, the *glottis*, predominates laryngeal speech functions. The glottis is formed by the *vocal folds*, paired horizontal structures at about the level of the fifth cervical vertebra. They are the anatomical landmarks delineating the inferior border of the laryngeal *ventricle*, a lateral bulge in the laryngeal cavity named for the seventeenth-century anatomist, Giovanni Morgagni.

The vocal folds are composed of several layers of epithelium and connective tissues, supported by cartilages and moved by a complex set of skeletal muscles. Free medially, they are highly mobile posteriorly but fixed anteriorly, forming a V-shaped chink in the pharyngeal passage. The vocal folds are under voluntary control, so they may be approximated in a variety of postures, including a loose adduction for voice production or phonation, a tight adduction and sudden abduction for production of a glottal plosive /?/, and a Y-shaped configuration for production of glottal fricatives /ɦ/ and /h/. In addition to its vocal fold function, another contribution of the laryngopharynx to speech is found in its distinct resonating characteristics.

The glottis also plays a dynamic role in pulmonary ventilation. During inspiration, the vocal folds are widely abducted, allowing the least impedance to inspired air. During expiration, the folds adduct slightly, having the effect of delaying the egress of air and maintaining alveolar pressure slightly longer for maximum gas transfer.

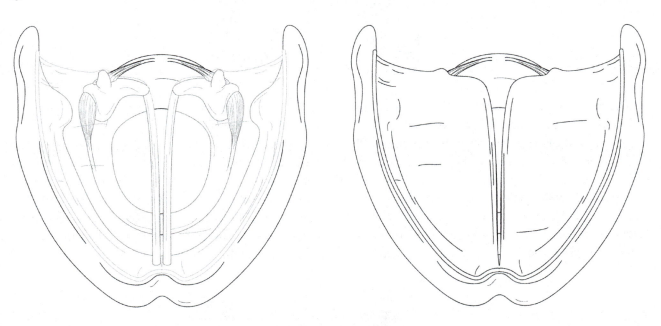

At the top of the laryngeal ventricle of the vestibule are two loose, horizontal, laterally opposed *ventricular folds*. Like the vocal folds, these are free medially and form the second protective valve. They are formed of two *vestibular ligaments*, two thin *ventricularis* muscles, and are covered with the same epithelium that lines most of the pharyngeal tube.

The opening of the larynx into the rest of the pharynx is called its *additus*, or *vestibule*. The vestibule takes a sharp posterior turn, so that its opening is oblique to the coronal plane through the remainder of the pharynx. It is composed of an accumulation of loose epithelial tissue, the *aryepiglottic fold*, supported by thin striated musculature and several cartilages, the *epiglottis*, anteriorly, and the *arytenoid, corniculate,* and *cuneiform* cartilages posteriorly. This sphincteric valve at the distal end, where the laryngopharynx is continuous with the oropharynx, closes as a part of a complex orchestration of muscles during the pharyngeal stage of swallowing. Its closure forces material to be swallowed to continue its inferior progression into the esophagus and keeps it from detouring anteriorly into

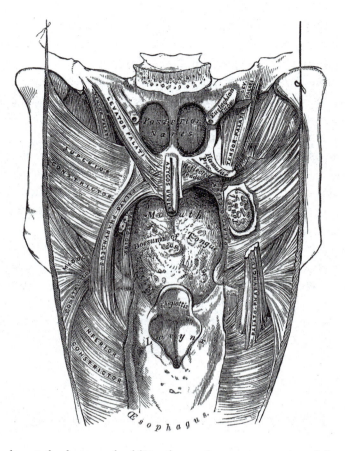

the laryngopharynx. The valve at the laryngeal additus forms the most superior of the three protective laryngeal structures. The lower border of the vestibule is marked by the paired *ventricular folds*.

Distal to the laryngopharynx is the oropharynx, the middle division of the pharynx, occupying the space posterior to the oral cavity. The oropharynx and laryngopharynx together are popularly called the "throat." The oropharynx extends from the inlet of the larynx to the inferior extent of the nasopharynx on a plane parallel to the palatine processes of the maxilla. It lies posterior to the buccal cavity (between the insides of the cheeks and the teeth), inferior to the soft palate, and superior to the laryngeal inlet. The posterior faucial arch, or oropharyngeal isthmus, is the theoretical boundary between the oral cavity and the oropharynx. It is formed by the imminences of the palatoglossus muscles, originating within the substance of the velum and inserting into the fibers of the middle pharyngeal constrictor muscle.

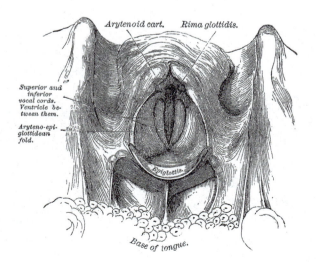

The speech function of the oropharynx is to act as a resonating tube in the creation of the spectral characteristics that distinguish the voice and other glottal sounds. Since the posterior part of the tongue is anchored in the oropharynx, the pharyngeal tube is anything but static in its internal dimensions during speech. It moves with the motions of the tongue, altering its resonating characteristics accordingly. Thus, the full spectra of the speech acoustics are, at least in part, functions of the oropharynx.

The nasopharynx is the most distal or superior part of the pharynx and is very small, extending from a horizontal plane at the palatal processes of the maxillae, approximately the "roof of the mouth," to the transverse plane at the choanae. The nasopharynx contains the eustachian tubes and the pharyngeal tonsil, also called the "adenoid." The Eustachian tubes, named for the sixteenth-century Italian anatomist, Bartolomio Eustachius (Walsh, 1909), connect the nasopharynx to the middle ear or tympanic cavity. They are also called "auditory tubes," or most descriptively, "pharyngotympanic tubes." Connection between the tympanic cavity and the nasopharynx allows pressure equalization between the middle ear and the atmosphere.

Oral Cavity

The oral cavity, known colloquially as the "mouth," has several important and well-known biological functions in its dual roles as an external opening for both the respiratory system and the digestive (alimentary) system. These roles make the oral cavity of great significance to students or practitioners of speech-language pathology. The oral cavity is examined in particular detail in Chapter 6, which covers speech articulation.

The oral cavity's respiratory function is that of providing an alternate vent to the environment. This alternate vent is brought into play in cases when the nasal cavity fails to pass a sufficient volume or air for respiration, such as when the individual is exerted and has greater gas exchange requirements, when it is plugged with mucous secretions secondary to irritation or disease, or when injury compromises its patency.

Another biological function of the oral cavity is to provide an entryway for food and drink. Digestion of food begins in the oral cavity, where solid material is broken down by the act of mastication, or chewing. Chemicals in salvia, secreted by salivary glands around the oral cavity, also break down food. The oral cavity is also the location of the sensory end-organs for taste. These organs are located on the tongue and soft palate in fairly distinct regions for particular sensations of taste, including sweet, sour, salty, bitter, and umami (savory taste). Taste and smell play important roles in nutrition.

The dual roles of the oral cavity in the intake of both food and air can become complicated in cases where anatomical structure or where muscular function are compromised. Speech-language pathologists and others are trained to treat disorders of chewing and the initial stages of swallowing (mastication and deglutition) and to help the patient protect the airway from intrusion by food or liquid. Chapter 7 addresses the swallowing processes.

The well known communicative function of the oral cavity is to modify the breath stream to create speech sound types called *phonemes* for speech communication. Most phonemes are produced by articulations of oral cavity structures, and phonemes are at least partially described in terms of which vocal tract organs approximate. The oral cavity is able to create the sounds of speech through juxtaposition of its moveable structures (lips, tongue, and velum) with its relatively fixed structures (teeth, hard palate, superior alveolar ridge) in a process called speech articulation. Typical names for oral speech articulations are "labio-dental" or "velar."

The term *articulation* has three definitions in the anatomy and physiology of speech. Articulation may refer to the connections of bones, to the relative position of the speech organs required to produce a given phone, or to a clear enunciation of speech. Chapter 6 is specifically devoted to speech articulation.

Structures of the Oral Cavity

The distal opening of the oral cavity is at the lips. These form a muscular sphincter to either admit or prevent admission to and escape from the oral cavity. They also have communicative functions in the production of *labial* speech sounds and facial expression. The muscular and flexible tongue is both an organ of digestion and of speech, and

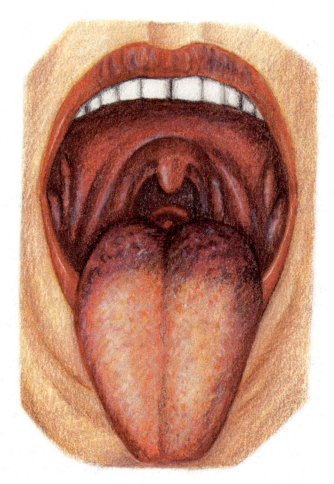

occupies the floor of the oral cavity. In the posterior end of the oral cavity, a flexible velum can direct air through the nasal or oral cavities as needed for both respiratory and speaking purposes. At its tip is the uvula. The uvula and the velum contain touch receptors that can trigger the post-oral stages of deglutition.

There are two dental arches: one superior and one inferior. The superior dental arch arises from the alveolar process of the maxilla, and the inferior arch arises from a similar process of the mandible. Around the dental arches, the roof of the oral cavity is soft epithelial tissue covering parts of the maxillae and palatine bones, while the oral floor is covered by similar, but more varied soft epithelium, founded by the moveable, horseshoe-shaped mandible, containing the muscular tongue, several salivary glands, and supporting structures.

At the posterior end of the oral cavity are the posterior and anterior *faucial arches*, or *fauces*, forming, with the root of the tongue, the oropharyngeal isthmus. The fauces are elevations in the oral and pharyngeal epithelium created by muscles underlying the tissue. The muscles which underlie the anterior arch, also called the *glossopalatine arch*, are the twin *palatoglossus* muscles. These have fibers that are continuous with one another and are heavily invested with elastic collagen fibers (Kuehn and Azzam, 1978). Their name is descriptive in that they attach at the posterior sides of the tongue and at the palatal aponeurosis. When they contract, they shorten the distance between the back of the tongue and the velum, either raising the posterior tongue or pulling the velum downward. The paired muscles that form the posterior or *pharyngopalatine faucial arch* are called *palatopharyngeus*. They have the same origin as palatoglossus, but insert into the muscles of the pharynx. Their contraction shortens the distance between the velum and the lateral pharyngeal walls and usually takes place during deglutition, indirectly causing superior displacement of the laryngopharynx.

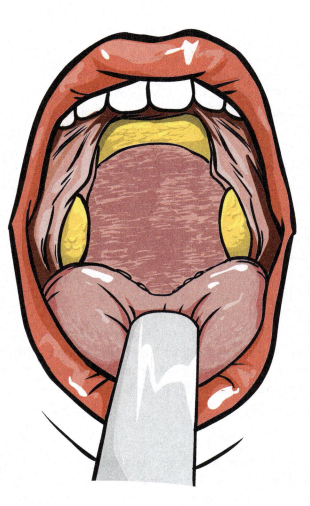

As noted, the palatoglossus and palatopharyngeus muscles converge superiorly at the velum. In so doing, they create two roughly triangular recessed areas, the *triangular fossae*, on either side of the posterior oral cavity. Each triangular fossa contains a readily visible mass of lymphoid tissue called a *palatine tonsil*. The two palatine tonsils are most visible, and are the ones most often referred to as simply, "the tonsils." They are enclosed in capsules, and their surface is encrypted with deep invaginations into their substance.

Besides the palatine tonsils, two other lymphoidal masses form a *tonsillar ring*, known as *"Waldeyer's ring,"* after German anatomist Heinrich von Waldeyer-Hartz (1836–1921). The ring is formed by the palatine tonsils and the less visible *lingual* and *pharyngeal tonsils*. All of these masses protect the airway from bacterial invasions, apparently by collecting and analyzing bacteria and other substances entering the airway and triggering an immune system response.

In the posterior nasopharynx, superior to the velum, is the pharyngeal tonsil, also called the *adenoid* or *adenoid pad*. It is formed of ciliated columnar epithelium and is not encapsulated.

At the root of the tongue, just posterior to the terminal sulcus, is the lingual tonsil. It is of the same non-keratinized pseudostratified epithelium as the palatine tonsils, and like them, has crypts in its surface, making it more susceptible to infection. The lingual tonsil is not encapsulated.

Tonsils may be quite large in youth, occupying and obstructing a large area of the respiratory tract or alimentary tract. They are also susceptible to infection. For these and other reasons, palatine tonsils may be surgically removed in an operation known as a *tonsillectomy*. The procedure sometimes includes an *adenoidectomy*, or "T and A." Removal of the pharyngeal tonsil (adenoid) sometimes results in a temporary velopharyngeal incompetence that causes hypernasality of speech.

Nasal Cavity

The nose and nasal cavity form the primary biological entrance to the respiratory system. The nose is specially adapted for the purpose of admitting life-sustaining air into the body and expelling respiratory exhaust from the body. The nose is the ideal exit and entrance for air for several reasons. First, it provides a rigid opening for air intake. This rigidity ensures it will be open when air is needed. Rigidity is provided by the nasal bones and the bones which form the sides of the cavity and by the wall separating the twin nasal passages, called the nasal septum. Another reason the nose is a good starting place for respiratory air is that the nasal cavity warms air and adds moisture as needed. Finally, there are hairs and mucous tissue in the nasal cavity which filter dust and clean the air.

The nasal cavity also provides a receptive site for smell, or *olfactory* sense. End organs in the most superior part of the nasal cavity communicate directly with the olfactory bulbs of the brain to begin a chemical analysis of the entering air.

The speech function of the nasal cavity is to provide an alternate resonating cavity for the sounds emanating from the larynx. By action of the muscular valve of the velopharyngeal sphincter, the nasal cavity can be coupled or uncoupled to the rest of the vocal tract, changing vocal tract resonating characteristics by creating a different shape to the resonating tube system. There is an acceptable degree of nasal resonance or *nasality*, in speech, dictated by dialectical, coarticulatory, and social constraints.

Speech that has too much nasal resonance is termed *hypernasal*, whereas speech with insufficient resonance is *hyponasal*. Hypernasality can result from velopharyngeal *incompetence*, wherein the muscular action of the velopharyngeal valve is incomplete or slow, or it can be the product of velopharyngeal *insufficiency*, a condition in which there is not enough tissue present to separate the nasal cavity from the vocal tract. Hyponasality is usually the result of obstruction of the nasal cavity.

Parts of the Nose and Nasal Cavity

The nasal cavity is formed of two cavities, or passages, separated by a *nasal septum*. The superior region of the cavity is the end organ for the sense of smell, and is sometimes referred to as the *olfactory* region, while the rest of the nasal cavity may be called the *respiratory* region. At the distal end are the *nares* or *nostrils*. The *choanae* are the posterior opening of the nasal passages, and open into the nasopharynx. Between the two nares, the choanae and the passages to which they give entrance is the nasal septum. It is formed of cartilage and bone. The cartilages of the nasal septum are the *septal cartilage* and the greater *alar* cartilage. The bones of the nasal septum include the *perpendicular plate of the ethmoid bone* and the *vomer*. Each nasal passage has three *nasal turbinates*, or *conchae*, formed of thin bone and covered with mucosal epithelium. These protrude medially into the nasal cavities. The *inferior turbinate (concha)* is a separate bone, whereas the middle and superior turbinates are parts of the ethmoid bone. Two lateral walls, formed of the nasal bones and the alar cartilages, project from the face and meet to form the *dorsum* of the nose. The root of the nose is the point at which the nasal bones articulate with the face. Here, the bones include small parts of the maxillae and the frontal bone. At its distal end, the nose has an *apex*.

Tissues of the Respiratory System

Tissues of the respiratory system are of two types, depending upon whether they are parts of the lungs, or the *respiratory* portion of the system, or parts of the respiratory passages, or the *conducting* portion of the system. Respiratory tissues of the lungs form the separation or gateway between the air in the alveolar sacks and the hemoglobin in the blood. To serve the function of gas exchange, they must be very thin.

Four thin layers of respiratory epithelial tissue separate the air in the alveoli from the blood in the pulmonary capillaries. First, lining the alveoli, is a layer of epithelium consisting of two types of cells, *Type I* cells and *Type II* cells. Simple squamous Type I pulmonary epithelial cells provide the thin surface through which gas molecules may pass, and Type II septal cells are ciliated cuboidal cells that lubricate the alveolar lining. The next layer of alveolar tissue is a basal lamina to support the Type I cells. This basal layer is situated next to the third layer, the basal

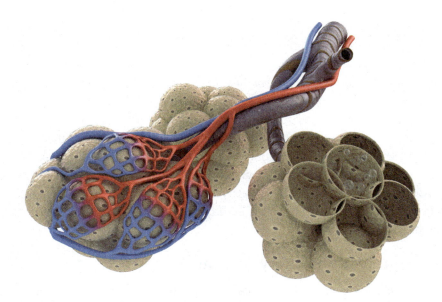

membrane of the pulmonary capillary endothelium. The pulmonary capillary endothelium is the fourth layer, lining the insides of the tiny venules and arterioles that form the capillary network surrounding the alveoli.

The lungs are said to have a double blood supply. Bronchial arteries, branching from the aorta, supply the tissues of the lungs to support their metabolism. The other blood supply to the lungs is deoxygenated blood from the heart, sent to the lungs to be replenished with oxygen. This is supplied through the much larger pulmonary arteries and returned to the heart through pulmonary veins.

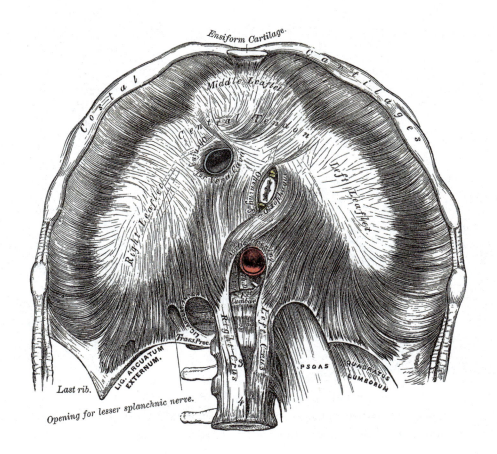

The respiratory passages have a much more complex tissue organization. Lining the bronchioles, bronchi, trachea, pharynx, and nasal cavities are linings of pseudostratified ciliated columnar epithelium, interspersed with mucus-secreting goblet cells to clean and humidify the air. Larger respiratory passages are surrounded by layers of smooth muscle that can contract or relax to control the caliber of their interiors and held patent along their extents by cartilaginous rings. The smaller conducting passages are lined with squamous epithelium, and their cartilaginous support is from disconnected places. Bronchioles less than 1 mm in diameter have no cartilage. In upper respiratory passage areas that are exposed to wear, such as the shared passages of the respiratory and digestive tracts or the contact surfaces of the vocal folds, interior tissues are formed of layers of stratified squamous epithelium. A protective tonsillar ring of lymphoidal tissue, mentioned above, is found in the pharynx, forming part of the immune system. Vascular complexes in the nasal cavity warm or cool the air in various places and hair follicles line the distal end of the nasal cavity to filter the air. In the superior part of the nasal cavity, sensory epithelial cells are triggered by air chemistry.

Skeletal Muscles of the Respiratory System

The muscles of the respiratory system act on the skeleton or soft tissues of the respiratory system to allow it to move gasses in and out of the lungs and respiratory passages. They do this by increasing or decreasing the internal dimensions of the thorax.

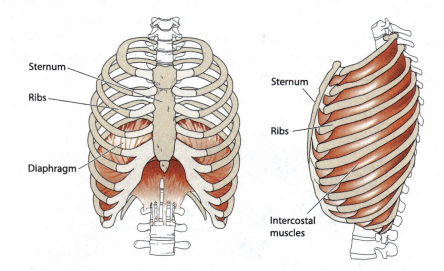

Muscles of respiration may be grouped into *"Primary"* and *"Secondary"* or *"Accessory"* groups. Primary muscles of respiration function during the normal pulmonary ventilation. Secondary muscles of respiration come into play when large amounts of gas need to be moved. In such cases, ventilation is said to be forced. Such large amounts are necessary when the individual is exerted or when the individual is ill with some disease that compromises respiratory function. Prolonged coughing also requires the use of secondary muscles.

Respiratory muscles can also be grouped according to which way they direct the gasses. Some bring gas into the system, while others force gas out of the system. Muscles that bring gas into the system are called *"Muscles of Inspiration."* Those that force gas out of the system are *"Muscles of Expiration."*

Inspiration: Primary Muscles

Muscles of inspiration function by expanding the interior of the thoracic cavity. The primary muscles of inspiration function in tidal breathing or in conversational speech. Expansion in tidal breathing occurs in several dimensions: inferiorly, anteriorly, and laterally, through *diaphragmatic* and *costal* movements.

Diaphragmatic or Abdominal Movement

Contraction of the *diaphragm* increases the volume in the inferior part of the thoracic cavity. This dome-shaped muscle is probably the most important primary muscle of inhalation, because the most prominent expansion of the thorax in breathing is vertical. Located immediately inferior to the lungs, contacting the inferior part of the parietal pleural membrane, the diaphragm is generally recognized as the anatomical border between the thoracic and abdominal regions of the body.

In general form, the diaphragm is composed of muscle tissue around its periphery connected to a large central tendon. Peripherally, muscle fibers originate from the xiphoid process of the sternum, the deep aspects of the lower six costal (rib) cartilages, the bodies of the first two or three lumbar vertebrae and three ligaments on either side of the spine, called the lateral and medial arcuate ligaments. The central tendon is connected to the cardiac membrane, the *pericardium*. Openings in the diaphragm allow passage of the aorta, thoracic lymphatic duct, vena cava, esophagus, and vagus (cranial) nerve.

When the fibers of the diaphragm contract, they cause the entire muscle to flatten. Since the muscle forms the floor of the thoracic cavity, this flattening increases the top-to-bottom volume of the thorax, and, by the fact that they are attracted to the thoracic wall, the lungs become larger, too.

Of course, flattening the floor of the thorax means flattening the ceiling of the abdomen. This causes the organs in that region to become compressed, and their only space for movement is in the anterior and lateral directions, since they are trapped on the top, bottom, and back. Breathing action of the diaphragm is inferred from observing the movements of the abdomen, and the respiratory movement is called *diaphragmatic-abdominal*. Most people breathe with a predominantly diaphragmatic-abdominal movement pattern.

Costal Movement

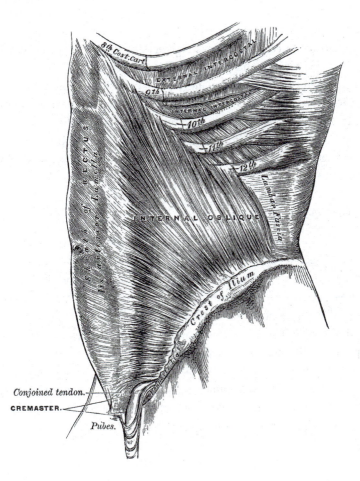

Costal, or rib, movement expands thoracic volume by moving chest walls outward laterally and anteriorly. Such movement is minimal or even absent in quiet breathing, but becomes more prominent as breathing becomes deeper or more effortful. Costal movement is accomplished mainly through contractions of *intercostal* muscles. As their name implies, these muscles are located in the areas between pairs of ribs, the intercostal space.

There are eleven pairs of *external, internal* and *innermost* intercostal muscles, located on each side of the thorax. Deep to the internal intercostal muscles are eleven pairs of innermost intercostal muscles, also called *intrercostales intimi* or *infracostal* muscles. Superficial to the innermost intercostal muscles and separated from them by a neuro-vascular bundle, containing an intercostal nerve, artery, and vein are the internal intercostal muscles. Fibers of the innermost and internal intercostal muscles course posteromedially. The most superficial of intercostal muscles are the external intercostal muscles. Their fibers angle anteromedially.

Each pair of intercostal muscles is attached at one end to lateral side of the costal groove of an upper rib, beginning at the lateral border of first rib, to attach to the upper border of the rib immediately below. Close to the sternum, intercostal muscle fibers give way to membranous tissue, the intercostal membranes.

Rib cage muscle functions in breathing are still the subject of some controversy. Convention has long held that the external intercostal muscles, acting in coordination with the diaphragm, are muscles of inspiration and that their internal counterparts are muscles of expiration (Hamberger, 1749). More recent evidence (De Troyer, Kirkwood, and Wilson, 2005) has suggested that the role of intercostal muscles is far more complex, and that both sets of intercostal muscles play inspiratory and expiratory roles. These roles depend upon mechanical advantages created by the courses of their fibers at their specific locations in the intercostal spaces. Rostrally, the external inter-costal muscle have a mechanical advantage for inspiration, especially in the dorsal aspects, but that advantage gives way to an advantage for expiration in the caudal intercostal spaces and especially for the ventral fibers. Conversely, the internal intercostal muscles have mechanical advantages for expiration in the dorsal caudal intercostal spaces, but that advantage diminishes rostrally and ventrally, in reverse pattern to that of the external intercostal muscles, and becomes an inspiratory advantage in the rostral spaces, especially ventrally, near the sternum. In addition to creating costal movements, the intercostal muscles help keep the internal thoracic viscera from protruding through the intercostal spaces when air pressure and pleural fluid changes occur.

Costal movements between ribs one and six are anterior and superior and are known colloquially as "Pump-Handle Movement." This name is meant to remind one of an old manual water pump handle, with the sternum moving upward and outward as the pump draws well water into its chamber. "Bucket-Handle Movement" is the lateral thoracic expansion brought about by movement of ribs seven through ten. This descriptive term is intended to remind one of the lateral and superior movement of the handles of two buckets, filled with water from the well, as a water bearer lifts them to bring them back to the house.

Inspiration: Secondary Muscles

When an individual is in a state of exertion or under disease conditions, the secondary or accessory muscles of inspiration come into play. Secondary muscles of inspiration pull the thoracic walls a little more upward and laterally to move the maximum amount of air under forced or stressful conditions. They are usually classified as muscles of other structures, such as the upper extremity, neck, or shoulder girdle, but can be "recruited" to assist in breathing when necessary. The list of accessory respiratory muscles is arguable, particularly if posture and physical condition are considered. For the purposes of this examination, we will list any thoracic or abdominal muscle that may be recruited to expand the thoracic wall a secondary muscle.

Secondary muscles of inspiration can be divided into ventral and dorsal groups to facilitate study. They are all paired, and include *pectoralis major, pectoralis minor, serratus anterior,* and *sternocleidomastoid* in the ventral group. The dorsal group includes the *costal levators, scalenus muscles, serratus posterior superior, latissimus dorsi, rhomboids,* and *trapezius muscles.* The status of serratus posterior superior and latissimus dorsi as respiratory muscles is subject to debate.

Ventral secondary muscles of inspiration originate at the ribs or sternum and insert into the humerus, the bone of the upper arm, and the scapula, or shoulder bone. These muscles assist breathing by raising the ribs upward and laterally with the shoulder and upper extremity fixed.

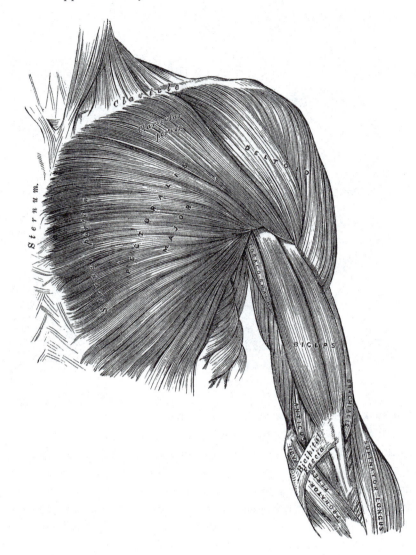

Pectoralis major is a superficial muscle located on the anterior chest. More properly considered a muscle of the upper extremity than of the thorax, it is listed here because its contraction can either adduct the upper extremities or raise the upper ribs. The muscle originates in the medial portion of the clavicle, the anterior sternum and first through sixth costal cartilages, and its fibers converge to insert in a very narrow area in the intertubercular groove of the upper humerus. Pectoralis major has been shown to function during defensive respiratory movements such as coughing (Bolser and Reier, 1998).

Pectoralis minor is a similar shaped muscle, located deep to pectoralis major, though it is much smaller. Its origins are the third, fourth, and fifth ribs and their intervening fascia. Fibers course superiorly and laterally to insert in the coracoid process of the scapula. With the scapula and humerus fixed, contraction of the pectoralis muscles pulls upward on the third, fourth, and fifth ribs, providing a little extra lateral and superior thoracic expansion.

Serratus anterior is arranged in a series of finger-like digitations. It originates on the first eight ribs and inserts into the scapula. Contraction of serratus anterior draws the shoulders forward, or with the shoulders fixed, draws the ribs toward them, creating lateral thoracic expansion.

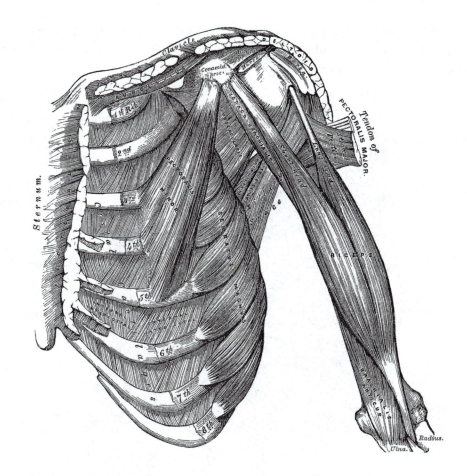

The *sternocleidomastoid* muscles are the large muscles on either side of the neck. They originate, as their name suggests, at two heads: the sternal head is at the manubrium of the sternum, and the clavicular head is at the medial one third of the clavicle. With the skull fixed or neck extended, contraction of both sternocleidomastoid muscles pulls upward on the clavicles and sternum, providing a little extra anterior and superior thoracic expansion.

Dorsal secondary muscles of inspiration originate at the vertebrae and insert into the ribs. The *costal levators*, also known as *levatores costarum,* are pairs of small muscles, originating at the transverse processes of each thoracic vertebra and insert into the lower rib, between its angle and tubercle. These muscles cover the posterior surfaces of the external intercostal muscles and provide some lifting power over the ribs in the dorsal thorax.

The anterior, posterior, and medial *scalenus* muscles are of three types, all originating at the transverse processes of cervical vertebrae and insert in the upper ribs. *Scalenus anterior* originates at the anterior tubercles of the third through sixth cervical vertebrae and attaches to the inner border of the first rib. *Scalenus medius* originates at the posterior tubercles of the second to sixth cervical vertebrae and inserts at the first rib as well. *Scalenus posterior* originates at the posterior tubercles of the transverse processes of cervical vertebrae four, five, and six and inserts in the second rib. Contraction of the scalenus muscles can rotate or flex the neck, or with the neck fixed, raise the second rib.

There are two pairs of flat, angular *serratus posterior* muscles, one superior and one inferior. *Serratus posterior superior* originates at the nuchal ligament of the cervical spine and at the fascia of the spinous processes of the upper two or three thoracic vertebrae and passes inferiorly and laterally to insert in ribs two through five, just ventral to their angles. Its attachments suggest that contraction of serratus posterior superior pulls ribs two through five upward and dorsally, creating a little dorsal thoracic expansion. However, some authorities have suggested that it and its inferior counterpart have no function in breathing, and are more important as organs of proprioception (Vilensky, Baltes, Weikel, Fortin, and Fourie, 2001).

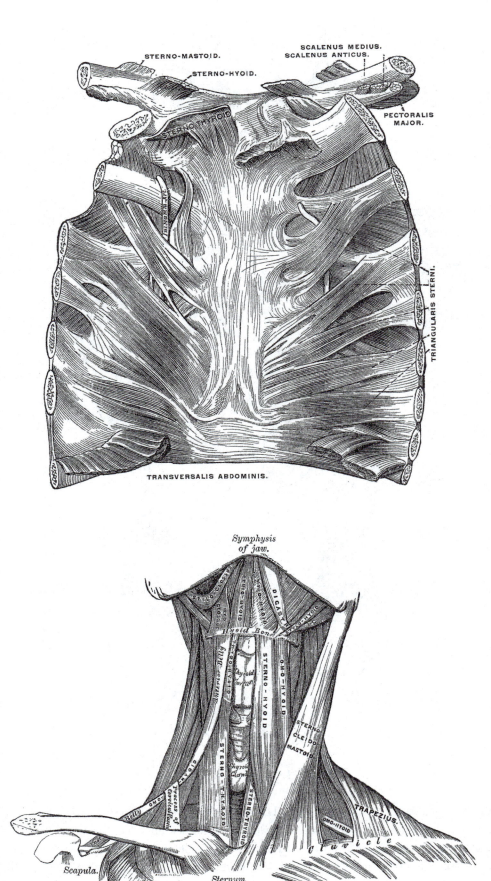

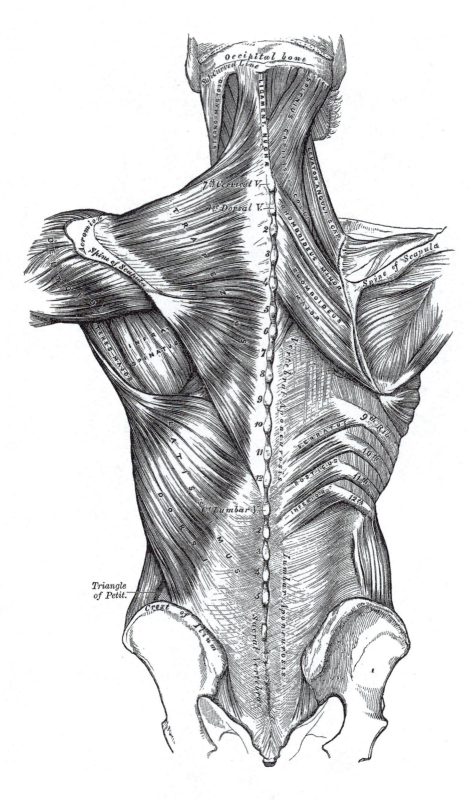

Latissimus dorsi is a large muscle with a broad origin in the lower and lateral back and a narrow insertion on the upper humerus. Its origins are in the fascia of the lower six thoracic and all of the lumbar vertebrae, a small area on the iliac crest, the lower three or four ribs, and the inferior angle of the scapula. From these points, fibers take a rather desultory path to the humerus to insert in its intertubercular groove. As is the case with the serratus posterior muscles, some controversy exists as to the role of latissimus dorsi in respiration. With the shoulders fixed, or in extreme cases, elevated, it can clearly lift the lower three or four ribs, thus leading to its inclusion in this section. In its role of drawing the trunk and shoulder girdle together, it also appears to function as a muscle of expiration.

The *rhomboid* muscles, *rhomboid major* and *rhomboid minor*, originate at the lower end of the nuchal ligament of the cervical spine and at the spinous processes of the upper thoracic vertebrae and insert at the vertebral border of each scapula. In some cases, they are fused to form a single muscle on each side. Contraction of their fibers adducts the shoulder blades and draws them superiorly toward the cervical and upper thoracic spine. This action, in conjunction with spinal extension, "opens" the ventral chest and allows more expansion.

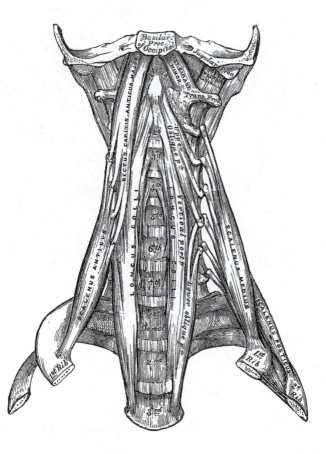

The *trapezius* muscles originate medially along a wide dorsal swath, from the external occipital protuberance at the back of the skull to the last thoracic vertebra. They insert into the medial aspects of the scapulae, from the acromion and down the scapular spine to their inferior borders. Contraction of these broad shoulder muscles adducts and elevates the shoulder girdle to open the ventral thorax and assist breathing.

Expiration: Primary Muscles

Expiration is usually a passive function. During normal expiration, the thorax is allowed to contract through its natural elasticity. In other words, action or contraction of the muscles of expiration is minimal or nonexistent during normal breathing. Instead, the muscles of inspiration relax, allowing the thorax to return to its natural resting shape passively. Muscles of expiration are only used then it is necessary to contract the thorax beyond its neutral or relaxed state or during effortful expiration.

Muscles of expiration function by opposing the actions of inspiratory muscles, that is, by contracting the thorax. They do this by forcing the abdominal viscera posteriorly and superiorly, returning the convex shape to

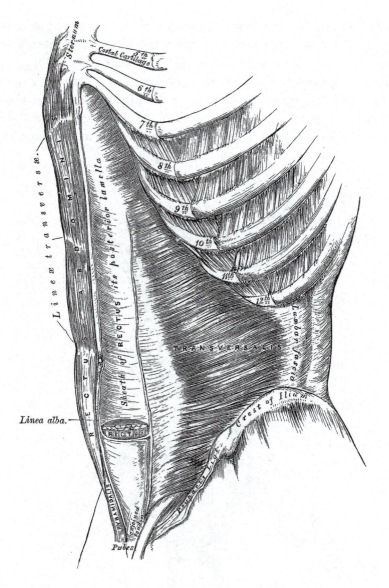

the diaphragmatic dome, and by squeezing the rib cage inward medially and inferiorly. In speech, the muscles of expiration may be used when emphasis or stress is required, when loudness requirements exceed the magnitude of subglottal pressure provided by normal recoil force, or when the speaker is trying to utter a particularly long passage without stopping. Normally, and perhaps ironically, the muscles of *inspiration* are most often active during speech, gradually relaxing their tonus (contractile state) and allowing the thorax to slowly return to its resting state and prolong the flow of air out of the system.

Primary expiratory muscles may be active for mild sustained voluntary expiration, when recoil force is not sufficient. Primary muscles of expiration are all paired and include *intercostals, rectus abdominus, transversus abdominus, external obliques,* and *internal obliques.*

The *internal* and *external intercostal muscles* can function as muscles of expiration and as muscles of inspiration. These dual roles were discussed above, in the section on inspiratory muscles, as were the origins and insertions of the intercostal muscles. Briefly, the expiratory roles of the intercostal muscles depend upon their segmental locations. The external intercostal muscles have an expiratory mechanical advantage in the caudal and ventral intercostal spaces, and internal intercostal muscles have mechanical advantages for expiration in the caudal and dorsal intercostal spaces. The paired innermost intercostals are deep to, and parallel in course with, the internal intercostal muscles and are only well developed in the caudal spaces. Their function as respiratory muscles has yet to be distinguished, but their similarity to the internal intercostal muscles would seem telling. In addition to creating costal movements, the intercostal muscles help maintain the structural integrity of the thoracic wall in the costal spaces.

The *rectus abdominus* muscles are two flat sheets, forming the most obvious musculature of the anterior abdomen. They attach at their inferior ends to the top of the pubic bone of the pelvis and at the superior ends to the fifth, sixth, and seventh costal cartilages. Three tendons interrupt the course of rectus abdominus fibers, creating six distinct segments, the lower one at the umbilicus, the upper one at the xiphoid process and the third in between. The tendons allow the segments to function independently if needed. The entire muscle group is contained in the *rectus sheath*, formed by the aponeuroses of the internal oblique and transversus abdominus muscles. Contraction of rectus abdominus forces the contents of the abdominal cavity posteriorly and superiorly, pushing against the diaphragm. This action decreases the vertical dimension of the thorax and forces air out of the respiratory system.

Transversus abdominus is a paired, flat muscle, forming the muscular sides of the abdomen. It wraps around the lateral abdomen, and has three points of origin: the spinal column, through attachment to the thoracolumbar fascia, the iliac crest, and the lower six costal cartilages. Its fibers interdigitate with the diaphragm in the thoracic region. It inserts and joins its contralateral counterpart at the *linea alba*, a tendonous meeting of the two halves of the rectus sheath. Transversus abdominus assists expiration by squeezing the abdominal contents medially, much as does the rectus abdominus, but from the sides, instead of the front.

The *external oblique* muscles are large, superficial ventral muscles, extending from the lower thoracic region to the mid abdomen. Their origins are at ribs five through twelve and are digitations that merge with those of the serratus anterior and latissimus dorsi muscles. Most external oblique fibers give way to a large aponeurosis that becomes one with the rectus sheath medially. A few fibers insert into the iliac crest, and these have a free edge posteriorly. Contraction of the external oblique muscles assists expiration by bringing the lower ribs and the rectus sheath together, contracting both the abdomen and the lower thorax.

The *internal oblique* muscles lie deep to the external obliques for most of their mass and extend from the spinal column and iliac crest, superiorly and medially around the sides of the abdomen to meet at its front. Fibers originate in a broad area, ranging from the thoracolumbar fascia to the lower three or four ribs and parts of the pelvis, including the iliac crest and the inguinal ligament. They meet anteriorly as an aponeurosis, continuous with the rectus sheath in its upper portion and free in its lower quarter. The internal obliques assist expiration by opposing the diaphragm, much as do the external obliques.

Expiration: Secondary Muscles

Secondary (accessory) muscles of expiration push the thoracic walls inward actively to force the maximum amount of air out of the lungs or when disease or disability require assistance in the breathing process. The most common occasion in which the secondary expiratory muscles are recruited is coughing. Secondary muscles of expiration include: the *transverse thoracis* muscles, *latissimus dorsi*, and *quadratus lumborum*.

The *transverse thoracis* muscles are arranged in several slips, all lying in the interior of the rib cage. They take origin at the sternum and insert in the ribs. The uppermost slips of muscles take a superior course from the sternum to insert in the upper ribs, while the most inferior slip fibers take a lateral course and are continuous with fibers of the transverse abdominus muscle. Together with the innermost intercostal muscles and the *subcostal* muscles, the transverse thoracis forms the innermost layer of thoracic musculature. The subcostal and innermost intercostal muscles are as variable from person to person and even from opposite sides of the same thorax as they are from anatomist to anatomist in their identification. Contraction of the upper transverse thoracic muscle fibers draws the ribs inferiorly, compressing the thorax slightly, and contraction of the subcostal muscles probably produces the same results as contractions of the internal intercostal muscles.

Latissimus dorsi can be either a muscle of inspiration or expiration. Technically a muscle of the shoulder, in its role of depressing the shoulder girdle against the rib cage, it can push slightly more air out of the system. Attachments of latissimus dorsi are described in the section on inspiratory muscles.

Quadratus lumborum attaches to the iliac crest at one end and the twelfth rib at the other end. It also has attachments to the lumbar vertebrae, at the anterior faces of their transverse processes. The lateral arcuate ligament mentioned above as an attachment of the diaphragm, arches over the anterior aspect of the muscle as it connects the diaphragm to the first lumbar vertebra. Contraction of the quadratus lumborum fibers assists expiration by drawing the last rib toward the pelvis, thereby compressing the lower thorax slightly. It may also assist inspiration by fixing the twelfth rib against diaphragmatic contraction.

Other Muscles of Respiration

So far, the discussion of respiratory muscles has focused on the thorax and abdomen. However, the discussion would be incomplete without mention of the roles of certain cervical, oral, and facial muscles play during pulmonary ventilation. These muscles are familiar as parts of the peripheral speech mechanism, but here we are examining the roles of these muscles apart from their communicative functions. Rather than acting on the thorax change the force or direction of airflow, these muscles act on the airways in ways similar to those of the smooth bronchial musculature to increase or decrease airway resistance. In contrast to smooth bronchial musculature, these striated muscles can be contracted voluntarily.

Cervical muscles in this "other" group are those muscles that adjust the glottal opening. The *posterior cricoarytenoid* muscles abduct the vocal folds, opening the airway, and the *lateral cricoarytenoid, oblique arytenoid, and transverse arytenoid* muscles perform the opposite act, adducting the vocal folds and increasing upper airway resistance. Combined, these muscles vary upper airway resistance, allowing increased airflow during inspiration and increased resistance during expiration. Details of laryngeal muscles are presented in Chapter 5.

Oral muscles perform the same acts, and allow greater or lesser oral airway resistance, depending upon the circumstances. The *intrinsic tongue muscles,* detailed in Chapter 6, can elevate the tongue apex and blade against the palate to increase resistance to oral ingress or egress, assisted by the *palatoglossus* muscle, which can draw the tongue upward at its posterior end. These same paired palatoglossus muscles can also pull the velum downward, opening the velopharyngeal sphincter and assisting airflow through the nasal cavity. The *genioglossus* muscle, forming the tongue's muscular bulk, originates at mental spine in the interior of the mandible. Its contractions protrude the tongue, creating an occlusive presence in the oral cavity, while the combined contractions of the *styloglossus* and palatoglossus muscles retract the tongue.

In the facial region, *orbicularis oris*, which forms a sphincter at the distal end of the oral cavity, can provide grater resistance to air flow. Its multiple antagonists in the facial region, also detailed in Chapter 6, lessen the resistance by pulling the oral opening to varying extents laterally, superiorly, and inferiorly. At the distal end of the nose are muscles that vary airflow resistance by dilating or constricting the nares. The muscle that decreases airflow resistance is *dilator naris*, which originates at the anterior maxilla and inserts into the alar cartilages. Those that increase airflow resistance are *compressor naris*, which originates on the maxilla, lateral to the nares, and *depressor septi*, originating on the anterior maxilla, near the upper incisor teeth and inserting into the cartilaginous part of the nasal septum.

Skeleton of the Respiratory System

The skeleton of the respiratory system has both internal and external components, and might be separated into skeletal *framework* and skeletal *support* groups for ease of study. These groups would be partially overlapping. Skeletal framework components would then include those bones and cartilages that provide patency to the airways, including the skull's facial skeleton, hyoid bone, laryngeal cartilages, tracheal and bronchial tree cartilages. Skeletal support components would be those that form the *axial skeleton* of the human body. They suspend and protect the soft tissues of the system, including the skull, spinal column (backbone), rib cage, clavicles (collar bones), scapulae, (shoulder bone), and pelvis (hip bones).

Skeletal Framework

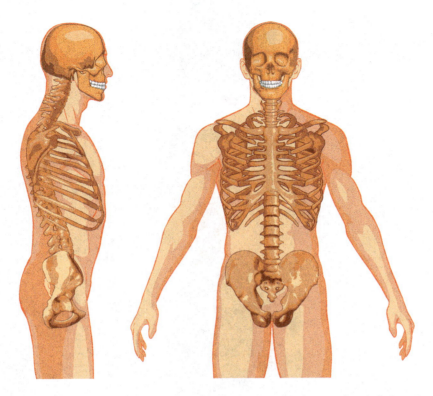

Structural patency and rigidity of the respiratory passages is essential to keep a functioning airway through which gasses can pass unobstructed. The skull provides for this at the rostral end of the respiratory systems. In the laryngopharynx, hyaline and elastic cartilages ensure patency when the larynx is open and rigidity when it is closed. These cartilages will be the subjects of a lecture in the next section. The hyaline cartilages of the tracheal rings and of the smaller branches of the bronchial tree may also be considered parts of the internal respiratory skeletal structure.

In the skull, the nasal cavity provides the primary opening for air egress or ingress. Bones and cartilages of the nasal cavity are less flexible than the tissues of the oral cavity, and maintain better patency. More caudally, pharyngeal bones include the *sphenoid bone* and *vomer*, forming the roof of the nasopharynx, the anterior portion of the occipital bone and cervical vertebrae, forming the dorsal support. Ventrally, the skeletal framework contains some gaps, because the oral cavity's soft tissues intervene. Oral soft tissues are, however, supported by the maxillae and palatine bones, forming both the floor of the nasal cavity and roof or the oral cavity, and the mandible, supporting the floor of the oral cavity. Between the mandible and the laryngopharynx is the horseshoe-shaped *hyoid bone*, open at its posterior end, providing a skeletal anchor for the tongue, above and suspension for the thyroid cartilage, below, as well as airway patency for the rostral tract.

The mandible, hyoid bone, and thyroid cartilages begin a pattern of incomplete bony or cartilaginous rings, open posteriorly, interrupted by the cricoid cartilage, but continuing with the sixteen to twenty tracheal cartilages that form the main trunk of the bronchial tree. The more rostral of these structures, extending caudally from the skull to the larynx, are branchial arch derivatives. At its connection with the trachea, the most caudal laryngeal cartilage, the cricoid cartilage, forms a complete ring, and then the open pattern returns and maintains in the bronchial tubes.

The trachea articulates with the cricoid cartilage via the cricotracheal ligament and continues caudally with a bifurcation at about the level of the fifth thoracic vertebra in the adult, and the third thoracic vertebra in the infant. Postural changes, as well as respiratory and digestive activities, cause the flexible tracheal tube to move about in the thorax, so that the location of its bifurcation is variable.

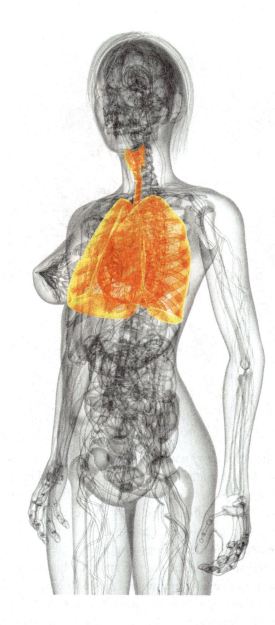

Tracheal rings are flat and thin, with a flat outer surfaces and concave inner surfaces. Each is about 4 mm wide and 1 mm deep, and is attached to the ones above and below by fibrous connective tissue. In some instances, rings may be continuous with the ones above or below. The rings are incomplete posteriorly, having a flattened posterior hiatus filled by fibrous tissue and smooth muscle.

The incomplete pattern of cartilaginous skeletal support continues with the bronchial tubes, even as they become smaller in diameter. Two main stem bronchi are bifurcations of the trachea. Cartilaginous skeletal support for the bronchial tree continues as described except for the smallest, terminal respiratory bronchioles. These tubes, about 0.5 mm in diameter, have no skeletal support.

Skeletal Support of the Respiratory System

The respiratory system is supported in the body and protected from injury by the bony plates of the skull, spinal column, shoulder girdle, rib cage, and pelvis. These bones also serve as attachments for respiratory musculature.

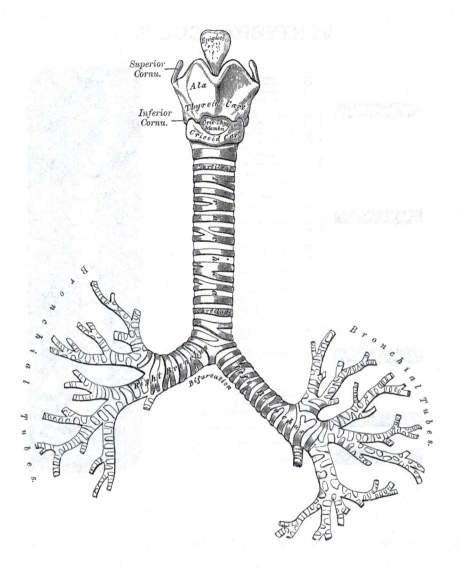

The Skull

The bones of the skull provide skeletal framework for upper respiratory structures in addition to their contributions to airway patency. Soft tissues of the nasal and oral cavities have all or most of their attachments to bones of the facial skeleton. Skull bones are described in detail in Chapter 6, which covers speech articulation.

Spinal Column

The spinal column, or vertebral column, is a stack of thirty-three irregular bones called *vertebrae*, extending from the base of the skull to the pelvis. It forms the core of the long axis of the body, and is a central structure that supports other skeletal elements, including the skull, shoulder girdle, and pelvis.

Together with the skull and thoracic bones, the spinal column forms the *axial* skeleton. The shoulder girdle and pelvis, in turn, support the extremities. Combined, bones of the extremities form the *appendicular skeleton.*

Although there are distinctive differences among the types of vertebrae, there are also some basic similarities. Each vertebra, except the first cervical, has a thick body which articulates with the vertebrae above and below it, and bears the weight of the anatomy superior to it. Vertebrae are relatively small at the rostral or superior end of the column and become progressive larger toward the pelvis. The larger size of the inferior vertebrae enables them to bear the collective greater weight imposed by the superior body parts.

VERTEBRAL COLUMN

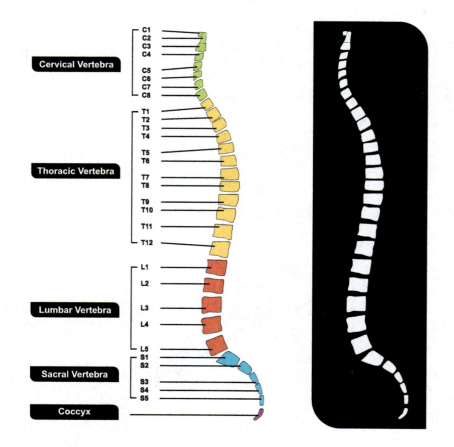

Dorsal to the body of each vertebra is a *neural arch,* formed by two dorsally extending *pedicles* and two *laminae* (missing in the first cervical). The laminae meet and fuse in the midline. The arches and bodies of each vertebra form a vertebral foramen, and when these are stacked, they form the *vertebral canal.* The vertebral canal contains the spinal cord, membranes, and the roots of the spinal nerves.

Vertebrae have two *transverse processes* extending laterally from the points where pedicles and laminae fuse. Transverse processes serve as attachments for muscles of the spine. In all but the first cervical vertebra, a *spinous process* extends dorsally from the point where the laminae fuse.

Most of the vertebrae articulate at three points. One of these is between vertebral bodies, and the other two are at the articular processes on the superior and inferior surfaces of the neural arch pedicles.

Articulation between vertebral bodies is cartilaginous or amphiarthrodial, with an *intervertebral disc* and system of ligaments designed to allow limited movement and to absorb vertical shock. The cartilaginous intervertebral disc has a tough ring around its edge and a gelatinous center.

The *articular processes* are associated with the neural arch of each vertebra, dorsal to the vertebral body. These joints are synovial and contain an articular capsule, covered by a ligament. The dorsal aspects of vertebrae interarticulate, with superior facets of one vertebra articulating with inferior facets of the one above.

Other spinal ligaments include *anterior* and *posterior longitudinal* ligaments, extending the full length of the column, a *nuchal ligament,* connecting the superficial aspects of spinous processes from the occipital bone to C-7, a *supraspinous ligament* between neighboring spinous processes of vertebra from C-7 to the sacrum, *ligamenta flava,* connecting the vertebral laminae.

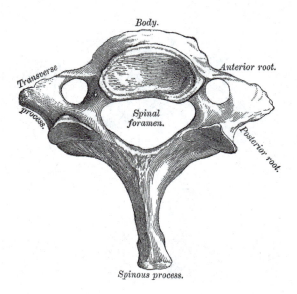

Collectively, the vertebral column has a series of curves that are of structural and developmental significance. The *primary curves* are present at birth. These create dorsal concavity or *kyphosis* to the spine and reflect the flexed posture of the fetus. Six weeks to three months of postnatal development sees the typical newborn lift the head to look around, creating the first *secondary* curve in the cervical area. This cervical spinal extension pattern, creating a concave curve in the dorsal region, is called *lordosis*. Another lordosis appears when, at about six months of age, the child begins to sit up.

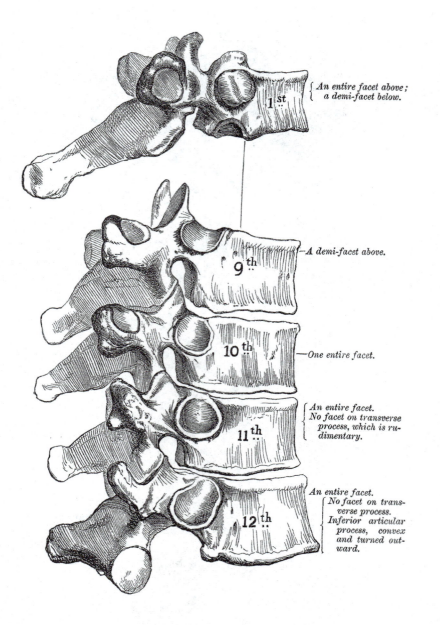

*An entire facet above;
a demi-facet below.*

1st

A demi-facet above.

9th

One entire facet.

10th

*An entire facet.
No facet on transverse
process, which is ru-
dimentary.*

11th

*An entire facet.
No facet on trans-
verse process.
Inferior articular
process, convex
and turned out-
ward.*

12th

Vertebrae are of five distinct types, depending upon in which body region the vertebrae are located. The actual number varies from person to person, but there are usually seven *cervical*, twelve *thoracic*, five *lumbar*, five *sacral*, and four *coccygeal* vertebrae. The sacral vertebrae are usually fused to form the *sacrum* and coccygeal vertebrae are fused to form one or two sections of the *coccyx*.

Most of the vertebrae are identified according the first letter of their regional name and their numbers in the stack, beginning from the top. For example, the first cervical vertebra may be referred to as "C-1" while the second is "C-2." The ninth thoracic vertebra is "T-9."

There are seven cervical vertebrae, somewhat dissimilar in form. Five of these, C-2 through C-6, are distinguished by bifid (split) spinous processes. All have *transverse foramina* in their transverse processes. The vertebral arteries pass through these stacked foramina, from C-6 to C-1, on their way to the brain. In some instances, the vertebral artery passes through the left transverse foramen of C-7. Three cervical vertebrae have special names to go with their special shapes in addition to their "C-" names. These are the first cervical vertebra, called "atlas," the second cervical vertebra, called "axis," and the third cervical vertebra, called "vertebra prominens." Each is distinguished by morphological peculiarities.

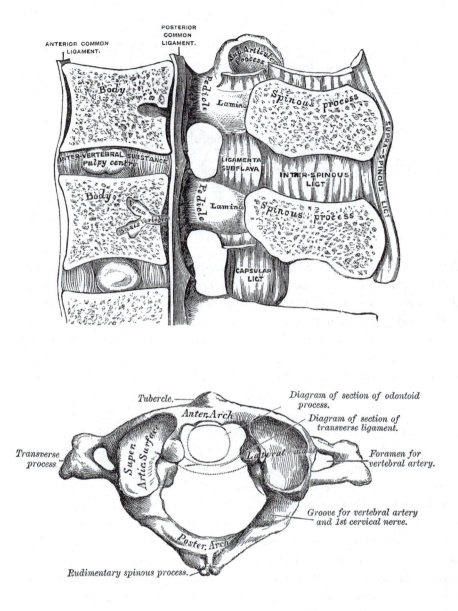

The first cervical vertebra is called "*the Atlas*," after the mythical Greek Titan, condemned by Zeus to spend eternity holding up the celestial sphere. Since C-1 is literally a platform for the skull, it is not hard to appreciate the reason for its name. Unique among the other vertebrae, the Atlas has neither a spinous process nor a vertebral body. In place of the vertebral body, the Atlas has an *anterior arch*, extending ventrally from the transverse processes, its two halves fusing in an anterior tubercle. A *posterior arch* gives the vertebra its roughly oval appearance of bone surrounding the oval *spinal* (also called *vertebral*) *foramen* through which the spinal cord passes. The posterior arch also has a tubercle at which its halves fused during development. Medial to its transverse processes, C-1 has two kidney shaped articular facets at which the vertebra articulates with similar facets on the inferior surface of the skull's occipital bone. The Atlas articulates with the occipital bone at the *atlanto-occipital* joint. There is very little movement at this joint. What little there is consists of "nodding" anterior/posterior flexion/extension movement and lateral flexion. Similar facets on the inferior surface of the Atlas provide smooth articular surfaces with those of C-2, the axis at the *atlanto-axial* joint. As mentioned above, the transverse processes have foramina through which the vertebral arteries pass.

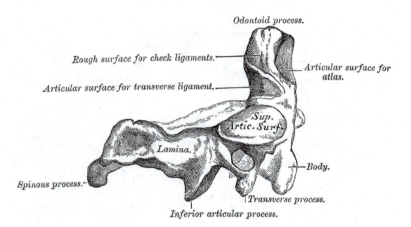

Odontoid process.

Rough surface for check ligaments.

Articular surface for atlas.

Articular surface for transverse ligament.

Sup. Artic. Surf.

Lamina.

Body.

Spinous process.

Transverse process.

Inferior articular process.

The second cervical vertebra is called the *"axis"* because most of the skull's rotation in relationship to the neck occurs at the joint between axis and Atlas, the atlantoaxial joint. It has an abbreviated vertebral body, with an *odontoid* ("shaped like a tooth") process projecting superiorly into the spinal foramen of C-1. Articulation between C-1 and C-2 is limited by a complex system of ligaments within the anterior portion of the spinal foramen. Rotation of C-1 about the odontoid process is limited in extent by an *alar* ligament, whereas a *transverse* ligament checks posterior movement into the spinal foramen. Other movements are restricted by ligaments between the axis and the skull. The axis has superior and inferior articular facets on either side of the vertebral body for articulation with superior and inferior vertebrae, transverse processes, and a bifid spinous process.

C-7 is most often called "vertebra prominens," because its long, non-bifid spinous process is usually the one most easily felt (palpated) beneath the skin in the lower cervical or upper thoracic region. The term "Vertebra Prominens" refers to the vertebra with the most palpable spinous process in the region, and, while this is most often C-7, in a minority of cases the term refers to C-6 or to T-1 (Stonelake, Burwell and Webb, 1987).

Twelve thoracic vertebrae are distinguished by their downward projecting spinous processes and their articular facets for the twelve ribs. The spinous processes project inferiorly, each upper vertebra overlapping the one below, forming a bony shield for the medial thorax. Ribs articulate on both sides of thoracic vertebrae, the first rib articulating with T-1, the second with T-2, and so on through T-12. The points of articulation are *costovertebral joints* between the vertebral bodies and rib heads, and *costotransverse joints* between the rib's tubercle and the transverse process of the corresponding vertebra. Costovertebral joint facets have a complex interarticulatory pattern that is shared between adjacent vertebrae. Thus, a given rib's head articulates with one facet in the upper body of its corresponding vertebra and with a facet in the lower body of the vertebra immediately superior.

The five lumbar vertebrae are distinguished by their lack of costal articular facets and by their great size. While a neural arch is still present, the adult spinal cord ends at the space between L-1 and L-2. From there on, the neural arch contains the spinal nerves of the *cauda equina*. The great size of the lumbar vertebrae enables them to bear the weight of the superior anatomy. While it seems far removed from the speech mechanism, the lumbar spine provides attachments for deep respiratory muscles, not the least of which is the diaphragm.

The five sacral vertebrae are fused and form the dorsal part of the pelvic girdle. Together, they form a triangular bone with four pairs of holes arranged longitudinally on either side of a sacral crest. The sacral crest is the remnant of the sacral spinous processes, and the sacral transverse processes have become indistinguishable through their fusion. A sacral canal provides a continuing passage for the lumbar and sacral spinal nerves of the *cauda equina*.

Three or four coccygeal vertebrae form the coccyx. These are fused to the sacrum. In some cases, the first coccygeal vertebra is separate from the other three.

Pelvis

The pelvic girdle provides attachments for certain muscles of the abdominal wall that function in breathing. It also forms a floor for the abdominal viscera, against which the diaphragm, the primary muscle of inhalation, pushes during breathing. The adult pelvic girdle is a ring of bone at the caudal end of the axial skeleton. It is formed by the sacrum in the dorsal middle, articulating with paired hip bones. During the developmental stage, three bones formed the hip bones: two iliac bones on the upper part of each side, two ischial bones on the lower part of each side, and a pubic bone connecting them anteriorly. These bones later fused to form the hip bones. The *iliac crest* remains a prominent anatomical landmark on either side of the waist.

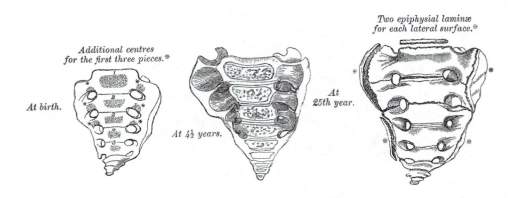

Rib Cage

Twelve paired ribs and the sternum form a protective cage around the heart and lungs. The ribs are flexible and articulate with the vertebrae posteriorly and sternum anteriorly. Ribs become larger from the first through the seventh ribs and then become smaller. Ribs serve as a means of protection for the soft tissues of the thorax and as

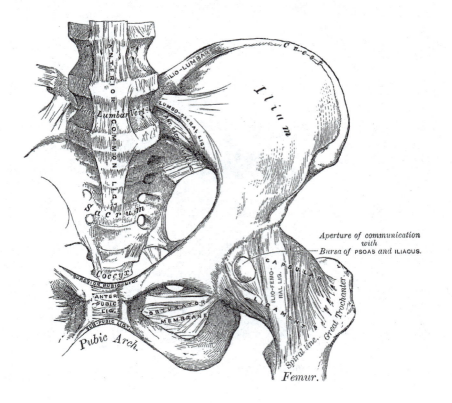

a skeletal framework to support muscles and soft tissue movements during chest expansion and contraction. In anatomy, the term "costal" applies to the ribs.

Each rib has a *head, neck,* and one or two *tubercles* at its vertebral end, continuous with a flat, flexible body having a sharp *angle* as it leaves the spine and straightening out into a more gentle curve as it extends ventrally. At the ventral ends of the rib bodies are costal cartilages. These articulate directly or indirectly with the sternum with *chondro-sternal* articulations. At their dorsal ends, ribs articulate with corresponding thoracic vertebrae at two points, mentioned above. Ribs 1 and 10–12 articulate at their heads with only their corresponding vertebrae, T-1,

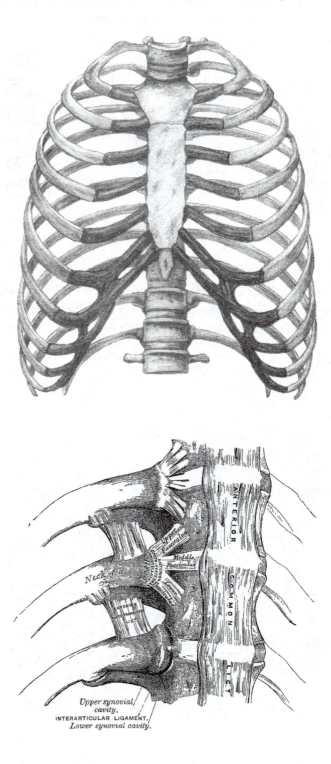

T-10, T-11, and T-12, whereas the other ribs articulate at their heads with the upper body of their corresponding vertebra and the lower body of the superior continuous one.

Only the first ten rib cartilages articulate with the sternum ventrally. Ribs one through seven articulate directly with the sternum and are called "true" ribs, whereas ribs eight through ten articulate with the costal cartilage of rib seven by means of interchondral membranes. Ribs eight through ten are thus called "false" ribs because they articulate indirectly with the sternum. The eleventh and twelfth ribs have no anterior articulation, and have thus earned the nickname "floating ribs." These ribs are not directly involved in thoracic expansion for breathing.

Sternum

The sternum is also called the "Breast Bone." It forms a bony shield in the center of the thorax. The upper part of the sternum is called its *manubrium*. Its superior margin in concave, vertically, creating the *jugular* or *sternal notch*, and serves as the anatomical marker for the beginning or *inlet* of the thorax. The first rib articulates at this point by means of a cartilaginous joint, secured by a tough *sternocostal ligament*. This joint allows very little movement of the first rib. Ribs 2–10 articulate anteriorly with the body of the sternum at small synovial joints, again secured by sternocostal ligaments. These joints allow gliding movements superiorly and laterally. Below the sternal body is a projection of bone called the *xiphoid process*. It has no articulation with the rib cage, but is an important anatomical landmark and a location to be avoided during chest the compressions of cardiopulmonary resuscitation efforts.

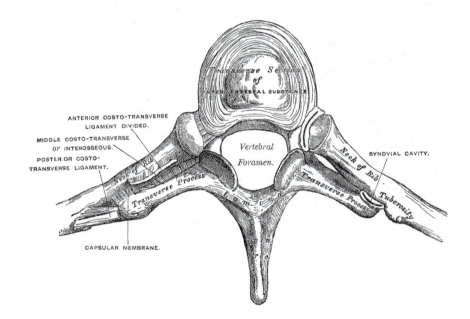

Clavicles

The clavicles are also called the "collar bones," and can be easily seen beneath the skin on either side of upper thorax. They are the only horizontal long bones in the human body. From the sternal end, the clavicles curve forward in a horizontal course, then curve inward as they reach their articulations with the scapulae. These bones serve as attachments for some accessory muscles of respiration and are active in most upper extremity movements. Elevation and retraction of the clavicles and shoulder girdle expands the chest during deep inspirations, and a respiratory pattern in which such movement predominates is called *clavicular*. They articulate with the sternum, at the *sternoclavicular* joint, at their proximal ends, and with the acromion, a process of the scapula at their distal ends with synovial joints. The distal articulation is called the *acromioclavicular* joint.

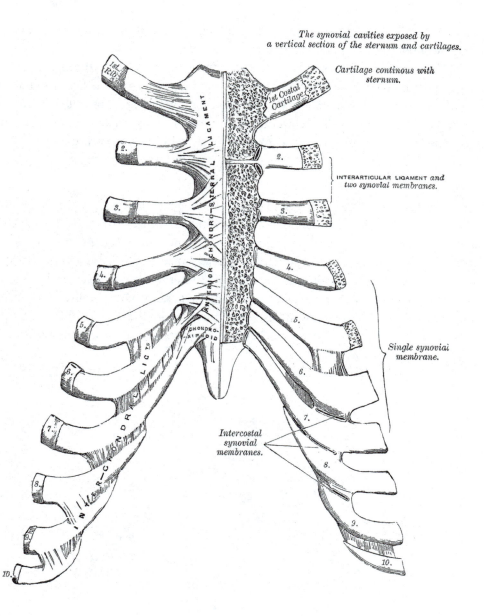

The synovial cavities exposed by a vertical section of the sternum and cartilages.

Cartilage continous with sternum.

INTERARTICULAR LIGAMENT and two synovial membranes.

Single synovial membrane.

Intercostal synovial membranes.

Scapulae

The scapulae are often called the "shoulder bones." These triangular bones form the dorsal pat of the shoulder girdle, and bridge the gap between the *humerus*, or arm, and clavicle. The medial border, also called the vertebral border, is nearest and parallels the spinal column. A superior border, continuing the horizontal line of the clavicles, contains a pronounced scapular notch, and a lateral, also called axillary, border extends at a sharp angle from the medial border to the point at which the humerus and clavicle articulate. The humerus articulates at the *glenoid* cavity or fossa, whereas the clavicle articulates with the acromion. A large scapular spine divides the dorsal surface about one-fourth of the way from the superior border to the inferior angle. The broad wide bodies of the scapulae serve as attachments for numerous muscles, many of which can be recruited as accessory muscles of respiration, and function in concert with muscles that move the clavicles in that action. Though they may assist in respiration, all of these muscles are properly considered to be muscles of the shoulder girdle. Elevation, extension, and abduction of the shoulder girdle expand the thorax, while depression, flexion, and adduction compress it.

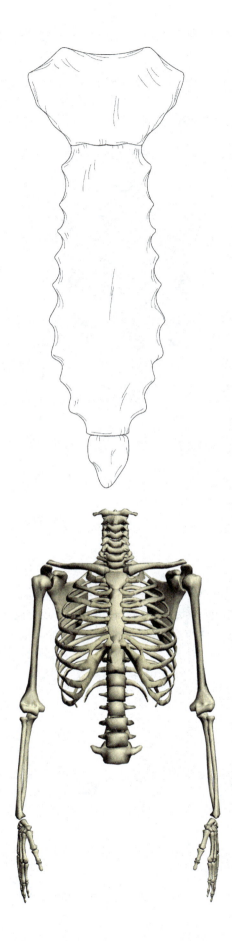

Neurology of Respiration

Innervation of the respiratory system may be examined in terms of peripheral and central divisions. Peripheral nervous support consists of spinal and cranial nerves and the autonomic nervous system. Central mediation is accomplished through brain and spinal cord centers.

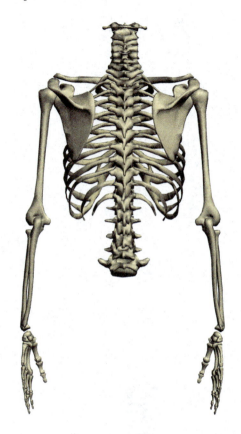

Peripheral Innervation

The respiratory system is innervated by autonomic and somatic branches of the peripheral nervous system, mediated by all levels of the central nervous system. Such diverse nervous system support ensures that the system can be functional unconsciously or consciously and can keep up with changing metabolic demands.

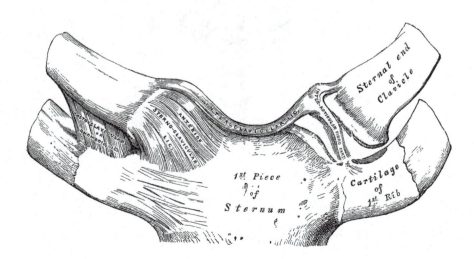

Special sensors, called *glomus bodies*, located in carotid sinuses, carotid bodies, aortic bodies, and jugular veins, send signals to the central nervous system via the glossopharyngeal and vagus nerves. Changes in the tension between oxygen and carbon dioxide or changes in the blood acid/alkali (pH) balance trigger these signals, which are then transmitted back to brainstem centers to modify respiratory rates and depths as needed.

The lungs receive parasympathetic innervation from the vagus nerve, and sympathetic innervation from the second through eighth thoracic ganglia of the sympathetic trunk. Efferent innervation is to the smooth muscles of the bronchial tree and arteries, dilating and constricting them as needed to achieve optimal air flow. Afferent innervation is from the bronchial epithelium and, it is suspected, the alveolar walls. Accumulated substances in the bronchial tubes trigger afferent signals, which, in turn, stimulate efferent responses such as coughing. Evidence in animals suggests that afferent fibers from the alveoli monitor their stretch status and stimulate control of the ventilation response.

Autonomic innervation to and from the trachea is supplied by the vagus nerve, for the smooth *trachealis* muscle, and by the middle cervical ganglia, which have collateral connections to the vagus nerve.

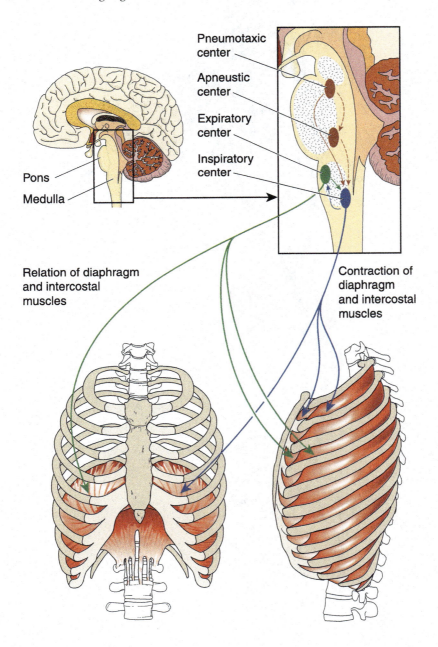

Pneumotaxic center

Apneustic center

Expiratory center

Inspiratory center

Pons

Medulla

Relation of diaphragm and intercostal muscles

Contraction of diaphragm and intercostal muscles

The larynx receives its autonomic innervation from the superior laryngeal and recurrent laryngeal branches of the vagus nerve. Fibers of the accessory nerve join those of the vagus to provide somatic innervation of voluntary laryngeal musculature. The superior laryngeal nerve divides into internal and external branches at the level of the hyoid bone. The external branch descends lateral to the larynx and supplies the cricothyroid muscle. The internal branch divides again, with a superior division supplying the lower pharynx and structures of the laryngeal vestibule, and an inferior division supplying the aryepiglottic folds and posterior glottal mucosa.

The rest of the pharynx receives its sensory and autonomic smooth muscle innervation from the pharyngeal plexus, a group of nerves derived from the combined fibers of the vagus and glossopharyngeal nerves and the cervical sympathetic ganglion. Nasal cavity epithelial tissue sends nondiscriminatory afferent somesthetic (touch) information to the central nervous system via the ophthalmic and maxillary branches of the trigeminal nerve.

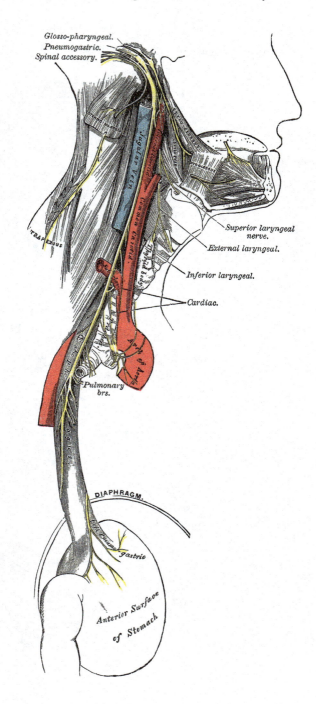

Striated respiratory musculature may be under voluntary or involuntary control. During normal breathing cycles, for example, the individual is only rarely aware of their contractions and relaxations. For most speech purposes, however, respiratory muscle contractions and relaxations result from conscious central programming, with complex and diffuse origins, in concert with oral and facial muscle actions.

The diaphragm receives its innervation through the *phrenic* nerve, a combination of fibers from cervical spinal nerve segments three through five. Intercostal muscles and other muscles of respiration receive their innervation through thoracic and lumbar spinal nerves. Diaphragmatic and intercostal contractions can be purposeful and directed consciously, but multiple connections with the autonomic nervous system enable unconscious respiratory responses to metabolic needs when necessary.

Automatic respiratory contractions to accommodate metabolic demands are the result of the *ventilation response*. Exercise and high blood carbon dioxide levels are two of the most common triggers of the ventilation response. Its resultant muscular contractions take precedence over other functions, such as speech, as anyone who has tried to speak when short of breath can testify. This fact is of clinical significance when considering the speech of persons having respiratory difficulties.

Central Innervation

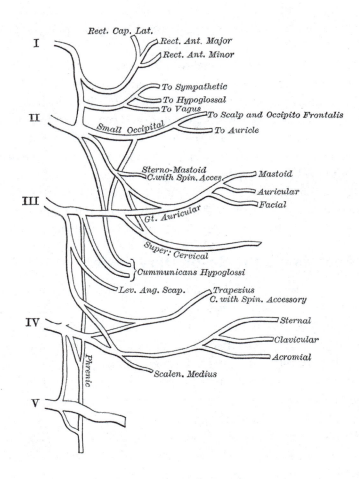

Central nervous system mediation for breathing is complex, as would be expected with such a vital and variable system. The respiratory system must strive to meet the full array of metabolic demands under conditions ranging from sleep to extreme exertion, and in all postures, from prone or supine to erect, and from flexed to extended. In addition, the system must function involuntarily for vegetative purposes, as well as voluntarily for protective, speech or other directed purposes. Breathing must coordinate with swallowing, sucking, coughing, vomiting, and conditions requiring thoracic rigidity, such as heavy lifting, defecation, or childbirth. To fulfill these conditions, central nervous innervation for respiratory needs originates in a diffuse network of brain locations.

For tidal or vegetative breathing, *dorsal* and *ventral respiratory centers* in the medulla oblongata appear to maintain a pattern of respiratory muscular contraction and relaxation, much like those for other vital functions, such as walking or chewing. These patterns are regulated, as we have said, by changes in blood gasses, chemistry, or mechanical reflexes. Brainstem centers associated with contraction of laryngeal adductors appear coordinated with inhibition of breathing cycles, accounting for changes in respiratory behaviors during sucking, swallowing, straining, and coughing, and the like.

Non-metabolic or voluntary requirements of the respiratory system trigger activation in other, more diffuse central nervous system centers (von Euler, 1997). One must imagine regulation of breathing under such diverse circumstances, as fear or rage, during sleep, or simply blowing out birthday cake candles. Under these circumstances, various centers in the cerebral cortex play their roles in regulation of breathing. Upper motor neurons activate concerted voluntary muscle contractions, following the planning of those movements by cells in premotor centers. Central control of respiration brings more into play than simple voluntary muscle contraction. The vitality or literally, spirit, brought forth by respiratory movements is also a factor. In addition to cortical centers, other non-cortical centers project their influence. Rostral brainstem centers in the pons and midbrain, deep prosencepahalic centers such as the amygdaloid body or corpora striatum, diencephalic structures, such as the hypothalamus, become active as the situation warrants. Cerebellar centers respond to cerebral and peripheral input for coordination of muscle group contractions (Dick, Orem, and Shea, 1997).

Respiration for speech brings into play a laterality of function not evident in other voluntary activities (Evans, Shea, and Saykin, 1999). Thus, there are differences in left and right hemisphere activities during speaking that are not observed in other voluntary respiratory activities. Such laterality is consistent with the role of the respiratory system as a servo for spoken language, with cerebral cortical centers playing roles in conception, formulation, expression, and monitoring of the speech signal, itself created by respiratory action. Inspiratory and expiratory patterns in speech differ widely from those of other respiratory system activities. An essential difference is the lack of periodicity of the respiratory cycle during speaking. Another change is in the relative durations of inspiration and expiration. More complex are differences are the resulting from conscious actions and integration of respiratory efforts in segmental aspects of speech, such as the formulation, planning, and execution of syllables, and non-segmental linguistic acts, such as syllable stress modification, pitch modulation, prolongation of nuclei, or pausing.

Developmental Anatomy of the Respiratory System

The distal portions of the respiratory system, including the oral and nasal cavities, are lined with tissues originating in the ectodermal germ layer. These tissues fold inward from the nasal placodes, opening at the rostral outer surface of the developing embryo and invaginating down to meet the primitive foregut. The proximal interior of the respiratory system, beginning at the laryngeal entrance and continuing downward, is endodermal in origin. Muscles and connective tissues originate in the intervening mesoderm.

The first signs of respiratory development are the appearances of two thickened areas in the ventral part of what will differentiate into the embryonic digestive and respiratory systems. Sometime between the 20th and 22nd day after fertilization, a ventral bulging of the primitive foregut endoderm appears in the area of the fourth to sixth branchial pouches in the form of two longitudinal tracheo-esophageal folds. This bulging forms a groove of tissue, called the *laryngotracheal groove*. As its name implies, the laryngotracheal groove contains the cellular components of the future respiratory system. The folds fuse in midline, forming a distinct tube, although still in communication with the foregut at its rostral end. Over the next few days, the groove deepens and extends caudally. Its issues differentiate into three important structures: the *laryngeal additus*, the *lung bud,* and the *laryngotracheal diverticulum*. The laryngeal additus will form the structures of the larynx, to be discussed more completely in the next chapter. The lung bud will become the two lungs, and the laryngeotracheal diverticulum will form a wall of tissue separating the respiratory system, on the ventral side of the wall, from the alimentary canal, on the dorsal side.

The lung bud appears at the caudal and of the laryngotracheal groove, and by day 28 after fertilization, it splits into two pouches, these being the precursors of the two lungs. The two pouches elongate and differentiate, forming three branches on the right and two on the left, in a pattern that will persist in the formation of the lung lobes.

Bronchial passages begin to branch, most often in a two-way split, or *dichotomous* pattern. Branching begins with the first tracheal bifurcation, and persists through up to twenty-four orders of branching, to end with the future respiratory bronchioles. After branching, as the system develops, the tubes increase in diameter, with consequent thinning of their walls. The smallest tubes will enlarge at their closed proximal ends, making the thinnest walls of all. These enlarged ends will form three to six terminal alveolar sacks, the tissues of which will differentiate in to type I and type II alveolar cells by about twenty-six weeks. At the same time, vascular development or branching, *angiogenesis*, accompanies the branching of the respiratory passages, ultimately forming the thin-walled alveolar capillaries, through which will pass blood gasses. By the 30th week, type II cells are secreting surfactant, the important fluid that lubricates alveolar tissues and assists in ventilation by resisting the stretch of inspiration.

Mature alveoli do not appear until just before birth, during the 36th week. This fact presents a major respiratory challenge for premature infants, but even in full-term infants, alveolar maturation and proliferation continues at a high pace through the third year. At this point, the basic pattern of organogenesis is complete, even though respiratory development continues into adulthood.

Before birth, the lungs of the fetus are filled with fluid originating in the alveolar cells. This fluid is removed from the lungs mechanically, during delivery, and by absorption through the circulatory and lymphatic systems after birth.

Ontogeny of the Respiratory System

A child does not begin to breathe until after birth. At that time, chemical receptors in the peripheral and central nervous systems trigger the muscular contractions of the first breath. The placenta, heretofore mediating blood gas exchange, suddenly ceases that function and turns it over to the respiratory system, which will carry on until death. Chemical receptors, inactive before birth, appear to become responsive after birth, triggering variations in ventilation. Once engaged, they become sensitive to increased levels of carbon dioxide, decreased levels of oxygen and increasing blood pH levels, in a manner similar to that in the adult. The first breath's inspiration is a strong one, necessitated by the need to overcome the resistance of remaining lung fluids and by the need to inflate the lungs and establish residual volume. Expiratory volume in that first breath is about half of its inspiratory volume, with some inspired air remaining in the lungs to provide the needed residual.

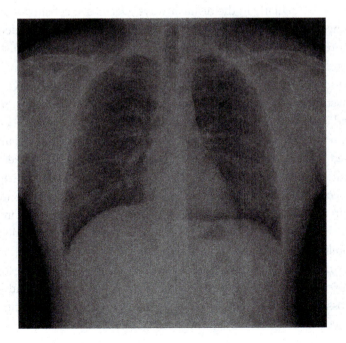

After birth, alveoli continue to multiply and mature. At birth, about twenty-four million immature alveoli will mature and multiply tenfold, reaching a peak at about age twelve. The typical healthy youth has enough respiratory capacity to suffice for many times more than resting demand. Total lung capacity normally increases steadily to puberty, although lung disease and exposure to irritants can compromise this capacity.

Respiratory capacities and functions diminish with advancing age. The rate and extent of this process varies individually, owing to its multiple mechanisms. Loss of nervous and muscular tissues, combined with decreases in alveolar surface area and respiratory system tissue elasticity, reduce the responsiveness and mechanical functions of the system in general. A history of disease and exposure to irritants exacerbates the process.

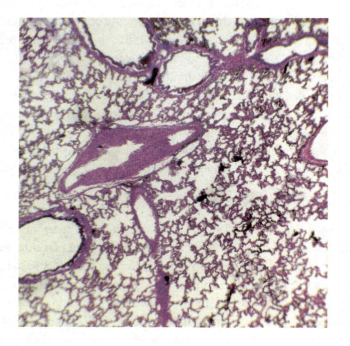

General decrease in nervous system function is at the root of certain aspects of decreased respiratory function in the long-lived individual. One consequence of decreased innervation is an accompanying diminution of protective laryngeal functions, particularly in the distal airways. Decreased sensitivity to irritation, leads to decreased likelihood and effectiveness of coughing as a protective mechanism. Sequellae of decreased protection are chronic inflammation of the entire system, owing to aspiration of irritants and septic organisms from the outside as well as internally, from the mouth and stomach.

Loss of muscle mass and function, in some ways a result of decreased innervation, is another mechanism of decreased respiratory system effectiveness in the long-lived person. The mechanism of age-related muscular dysfunction is complex, and includes metabolic as well as lifestyle factors. After about age forty, the ability to synthesize muscle-building amino acids decreases. This leads to reductions in masses of muscle groups, including those of respiration. Decreased muscle mass means decreased muscle strength. Less exercise complicates the scenario. Combined with a sedentary lifestyle, the result is progressively weaker muscular support for ventilation. Speech-related signs of weakened respiratory support for speech include weak phonation and short utterances.

Also coincident with aging is a general loss of tissue flexibility and structural integrity. Stiffening of the body tissues adds to the effects of decreased neural and muscular support as the individual ages. Bones, cartilages, and alveoli lose their ability to spring back from the deformations of breathing. Loss of thoracic mobility means lower volumes of gasses flowing in and out of the system. In the alveoli, tissues become less able to transfer blood gasses. Alveolar tissues lose their flexibility and become distended, creating "dead space" wherein no gasses are transferred, adding to loss of respiratory effectiveness. Respiratory bronchioles having no cartilaginous support are subject to narrowing or collapse.

CHAPTER SUMMARY

Chapter 4 has examined respiration on several levels, ranging from the molecular to the gross anatomical. At the smallest level, respiration is the metabolic process whereby nutrients and oxygen are combined to create energy for cellular activity. At the largest level, respiration is the contraction and relaxation of thoracic, cervical, and cranial musculature to move larger volumes of gasses in and out of the lungs and respiratory passages for breathing, for protective reflexes, and for voluntary purposes, such as speech.

Cellular respiration is the conversion of oxygen and carbohydrates, diffused through the cellular plasma membranes, into the energy required to maintain cellular physiology. Internal respiration, also called tissue respiration, is the exchange of blood gasses between the bloodstream and the cells. External respiration is the exchange of gasses between the lung's air spaces and the blood stream. Pulmonary ventilation is the visible muscular process whereby gasses from the environment are moved in and out of the respiratory system.

Respiratory function, whether for metabolic or for speech purposes, depends upon the ebb and flow of gasses in and out of the lungs and respiratory passages. Under normal circumstances, atmospheric air is made up of oxygen and several other gasses. Air is about 78.08% nitrogen, 20.95% oxygen, 0.93% argon, and 0.03% carbon dioxide. The physical principles by which these gasses flow into and out of the respiratory system are those of the *Second Law of Thermodynamics* and *Boyle's Law*.

For air to flow into the respiratory system the gas pressure inside the system must be less than that of the atmosphere outside. To create this pressure differential, that is, to decrease the gas pressure inside the lungs, an individual increases the lung volumes. The opposite pressure differential, higher pressure inside than out, is necessary to make gas flow out of the system.

Oxygen and carbon dioxide move into and out of the bloodstream by diffusing across the alveolar capillary membrane, only a half-micrometer in thickness. Gas pressure differentials are necessary at this level as well. The combined alveolar gas exchange surface in both lungs is about 140 m^2.

There is a regular, predictable pattern to tidal breathing, and it may be described as a series of respiratory cycles. Respiratory rates are the number of respiratory cycles produced by an individual in one minute. Rates vary with age and constitutional condition, being slower in athletic adults and faster in infants. In normal respiration, patterns may be described in terms if those parts of the body that evince the most observable movement. In pathological conditions, respiratory patterns are described in terms if their timing, volumes, and some other features in addition to movements. During speech, the timing of the respiratory cycle changes. The inspiratory phase of the respiratory cycle occupies a much shorter portion of the respiratory period. Respiratory volumes are measures with a device called a spirometer, and the act of measuring respiratory volumes is called spirometry.

Respiratory system anatomy consists of the lungs and respiratory passages, with skeletal, muscular, and nervous system support for their functions. The two lungs are contained in two pleural cavities, one on the right and left sides of the thorax. Each lung has a roughly pyramidal shape, with a base inferiorly and an apex superiorly. The respiratory passages conduct gases between the lungs and the atmosphere, and open to the outside environment at the nose or mouth openings. Tissues of the respiratory system are of two types, depending upon whether they are parts of the lungs, or the respiratory portion of the system, or parts of the respiratory passages, or the conducting portion of the system.

Muscles of respiration may be grouped into primary and secondary groups. They may also be grouped into inspiratory or expiratory groups, depending upon whether they bring air into or force air out of the system, respectively. Primary muscles of inspiration are most active during normal pulmonary ventilation, with recoil elasticity of the system providing enough energy to force air out of the system. Muscles of expiration assist as needed. Secondary muscles of respiration come into play when large amounts of gas need to be moved.

The skeleton of the respiratory system has both internal and external components, and might be separated into skeletal framework and skeletal support groups for ease of study. These groups would be partially overlapping.

Innervation of the respiratory system can be examined in terms of its peripheral and central divisions. Peripheral nervous support consists of spinal and cranial nerves and the autonomic nervous system, while central mediation is accomplished through brain and spinal cord centers.

The first signs of respiratory development are the appearances of two thickened areas in the ventral part of what will become the embryonic digestive and respiratory systems. The first breath inspiration is a strong one, necessitated by the need to overcome the resistance of remaining lung fluids and by the need to inflate the lungs and establish residual volume. Development of the entire system continues after birth and into puberty, although it is most accelerated during the first three years. Respiratory capacities and functions diminish with age. The rate and extent of this process varies individually, being brought about by multiple factors.

REFERENCES AND SUGGESTED READING

American Academy of Orthopedic Surgeons (1987). Emergency Care and Transportation of the Sick and Injured *(4th Edition)*. Park Ridge IL: American Academy of Orthopedic Surgeons.

Bolser, D.C., and Reier, P.J. (1998). Inspiratory and expiratory patterns of the pectoralis major muscle during pulmonary defensive reflexes. *Journal of Applied Physiology, 85,* 1786–1792.

Des Jardins, T. (2002). Cardiopulmonary Anatomy & Physiology *(4th Edition)*. Albany, NY. Delmar.

De Troyer, A., Kirkwood P.A. and Wilson, T. A. (2005). Respiratory action of the intercostal muscles. *Physiology Review, 85,* 717–756.

De Troyer, A., and Legrand, A. (1999). Spatial distribution of external and internal intercostal activity in dogs. *The Journal of Physiology, 518.1,* 291–300.

Dick, T.E., Orem, J.M., and Shea, S.A. (1997). Behavioral control of breathing. In R.G. Crystal, J.B. West, P. J. Barnes & E.W. Weibel (Eds.). *The Lung: Scientific Foundations* (2nd Edition) pp. 1821–1837. Philadelphia: Lippencott-Raven.

Evans, K.C., Shea, S.A., and Saykin, A.J. (1999). Functional MRI localisation of central nervous system regions associated with volitional inspiration in humans. *Journal of Physiology, 520.2,* 383–392.

Firth, P.G., and Bolay, H. (2004). Transient High Altitude Neurological Dysfunction: An Origin in the Temporoparietal Cortex *High Altitude Medicine & Biology, 5, 1,* 71–75.

Gayan-Ramirez, H.J., Dekhuijzen, R., and Decramer, M. (1993). Respiratory function of the rib cage muscles. *The European Respiratory Journal, 6,* 722–728.

Hamberger, G. E. (1749). *De Respirationis Mechanismo et usu Genuino.* Iena. Cited in Alexandre Legrand and André De Troyer (1999). Spatial distribution of external and internal intercostal activity in dogs. *The Journal of Physiology, 518.1,* 291–300.

H. Hamm, R.W. Light (1997). The pleura: the outer space of pulmonary medicine. *European Respiratory Journal, 10,* 2–3.

Harmsen, A.G., Muggenburg, B.A., Snipes, M.B., and Bice D.E. (1985). The role of macrophages in particle translocation from lungs to lymph nodes. *Science 230,* 1277–1280.

Hixon, T.J., and Collaborators (1987). *Respiratory Function in Speech and Song.* Boston: College Hill.

Horsfield, K. (1997). Pulmonary airways and blood vessels considered as confluent trees. In R.G. Crystal, J.B. West, P. J. Barnes, & E.W. Weibel (Eds.), *The Lung: Scientific Foundations* (2nd Edition) pp. 1073–1079. Philadelphia: Lippencott-Raven.

Johansson, A., and Camner, P. (1980). Are alveolar macrophages translocated to the lymph nodes? *Toxicology, 15 (3),* 157–162.

Kuehn, D.P., and Azzam, N.A. (1978). Anatomical characteristics of the palatoglossus and the anterior faucial pillar. *Cleft Palate Journal, 15,* 349–359.

Miao1, G.E., Zhang, Y., He, Jin-wei, Yan, Yan-chun, Wang, X., Cao L.1, Fu, H. (2009). Normal reference value of forced vital capacity of Chinese younger men and geographical factors. *Journal of Chinese Clinical Medicine; 4,* (4) n.p. Retrieved May 31,2011, from: http://old.cjmed.net/html/2009444_356.html?PHPSESSID=6ce3d2fb2b69 e86047d12f4f34a061fb

Miserocci, G. (1997). Physiology and pathophysiology of pleural fluid turnover. *European Respiratory Journal, 10,* 219–225.

Netsell, R., and Hixon, T.J. (1978). A noninvasive method for clinically estimating subglottal air pressure. *Journal of Speech and Hearing Disorders, 43,* 326–30.

Rudmose, H.W., Clark, K.C., Carlson, F.D., Eisenstein, J.C., and Walker, R.A., (1946). The effects of high altitude on speech and hearing. *The Journal of the Acoustical Society of America, 18,* 250–251.

Stonelake, P.S., Burwell, R.G., and Webb, J.K. (1988). Variation in the vertebral levels of vertebra prominens and sacral dimples in subjects with scoliosis. *Journal of Anatomy, 159,* 165–172.

Tompsett, D.H. (1965). The bronchopulmonary segments. *Medical History, 9,* 177–181.

Tortora, G. J., and Grabowski, S.R. (1996). *Principles of Anatomy and Physiology.* New York: Harper-Collins.

Vilensky, J.A., Baltes, M., Weikel, L., Fortin, J.D., and Fourie, L.J. (2001). Serratus posterior muscles: anatomy, clinical relevance, and function. *Clinical Anatomy, 14,* 237–241.

Vital capacity. (n.d.) *Mosby's Medical Dictionary, 8th edition.* (2009). Retrieved May 31, 2011 from http://medical-dictionary.thefreedictionary.com/vital + capacity

Von Euler, C. (1997). Neural organization and rhythm generation. In Crystal, R.G., West, J.B., Barnes, P. J. & Weibel, E.W. (Eds.), *The Lung: Scientific Foundations* (2nd Edition) pp. 1711–1724. Philadelphia: Lippencott-Raven.

Walsh, J.J. (1909). Bartolomeo Eustachius. In The Catholic Encyclopedia. New York: Robert Appleton Company. Retrieved June 1, 2011, from New Advent: http://www.newadvent.org/cathen/05626d.htm

IMPORTANT TERMS

Apnea: A state in which there are no respiratory cycles; cessation of breathing.

Aspiration: Intrusion of matter into the respiratory tract.

Asthma: A condition in which the respiratory passages become obstructed, either by blockage with mucus, by edematous swelling or by bronchospasm.

Bellows: A mechanical device to create compressed air.

Boyle's Law: A physical description relating the volume of a fixed quantity of gas to its pressure (Robert Boyle, 1662).

Bronchial Tree: An anatomical term describing the patulous configuration of the respiratory passages from the trachea (trunk) to the terminal bronchi (smallest branches).

Bucket Handle: Descriptive term referring to the lateral movements of ribs 7–10 during breathing.

Carbohydrate: An organic compound containing carbon, hydrogen, and oxygen atoms; Literally: "Watered Carbon."

Cellular Respiration: The process by which glucose is oxidized within the cell, producing energy, carbon dioxide, and water.

Cheyne-Stokes Respiration: A pathological pattern of respiration, suggesting central nervous system dysfunction. The pattern consists of periods of regular respiratory cycles, increasing in frequency and amplitude, followed by periods of apnea.

Chronic Bronchitis: Long-term inflammation of the bronchial tree.

Clavicular Pattern: A breathing pattern used by individuals who need to move maximum air volumes.

Diaphragm: The striated muscular boundary between the thorax and the abdomen.

Diaphragmatic/Abdominal Pattern: A breathing pattern produced when diaphragmatic contraction compresses the abdominal viscera from above, causing them to move the anterior abdominal wall anteriorly.

Dyspnea: Difficulty breathing.

Erythrocyte: A red blood cell.

Eupnea: Normal tidal breathing.

Expiratory Reserve: The maximum volume of air that can be exhaled beyond a tidal exhalation.

External Respiration: The process whereby gasses diffuse through the cell wall and exit or enter the bloodstream.

Glucose: The simplest, most abundant, and most important of the monosaccharide carbohydrates; structurally, a straight-chain, six-carbon, pentahydroxy aldehyde.

Glycolysis: Oxidation of glucose to produce energy, water, and carbon dioxide.

Hyperpnea: Deep respiration.

I-Fraction: The ratio of the duration of inhalation to the duration of the entire respiratory cycle.

Inspiratory Reserve: The maximum volume of air that can be inhaled beyond a tidal inhalation.

Internal Respiration: The progress whereby blood gasses cross the lung alveolar walls to enter or leave the bloodstream.

Kinetic Theory of Gasses: The physical principle that gasses will tend to equalize their pressures.

Kussmaul Respiration: A pathological respiratory pattern typified by deep sighing and seen in ketoacidosis or air hunger. *(Adolf Kussmaul: 1822 German pathologist and medical pioneer)*.

Lung Capacity: Potential amount of air contained by the lungs; also called respiratory capacity or forced inspiratory volume.

Lung Volume: Actual space occupied by the air in the lungs at a given time.

Pulmonary Ventilation: The muscular process whereby gasses from the environment are moved in and out of the respiratory system.

Respiratory Volume: Any of several measures of gas volumes in the respiratory system measured by spirometry.

Lungs: The thoracic organs wherein gasses are transferred between the respiratory system and the bloodstream.

Neurogenic Hyperventilation: A pathological respiratory pattern consisting of full inhalations (hyperpnea) and rapid cycles (tachypnea). It is a grave symptom of central nervous system dysfunction.

Olfactory: Referring to the sense of smell.

Pleura: The epithelium surrounding the superficial surfaces of the lungs and the internal lining of the pleural cavities.

Pleural Fluid: The clear fluid in the interpleural space.

Pneumothorax: Collapse of a lung in the pleural cavity.

Pulmonic: Referring to the lungs.

Pump Handle: A descriptive term referring to the upward and forward displacement of the thorax during inhalation; created by movements of the first six ribs.

Rales: Lung sounds created when collapsed alveolar and spaces and terminal airways open suddenly. They are observed through stethoscopic auscultation during inhalation.

Residual Volume: That volume of air that remains in the lungs following maximum exhalation. Residual volume is created by the attraction of the lung surface to the thoracic lining.

Respiratory Passages: The anatomical conduits whereby respiratory gasses enter and leave the respiratory system.

Respiratory Rate: The number of respiratory cycles completed during 1 minute.

Respiratory Pattern: The observed displacements and timing of thoracic and abdominal movements during pulmonary ventilation.

Ronchi: Lung sounds created by fluid accretion in the airways. They are observed through stethoscopic auscultation during inhalation or exhalation.

Second Law of Thermodynamics: Principle of energy transfer that stipulates that heat (or gas pressure) cannot pass unassisted from a lower energy region to one of higher energy.

Septum: An anatomical wall separating two cavities.

Speech Breathing: Respiratory patterns used to produce spoken language.

Subcutaneous Emphysema: A crackling or popping sounds created by palpating tissue that has air infused into the intercellular spaces by traumatic mechanism.

Tachypnea: Rapid respiratory rate.

Thoracic Pattern: The pattern of respiration produces by contraction of the external intercostal muscles.

Tidal Breathing: Life-sustaining breathing cycles.

Tidal Volume: The volume of gas moved during tidal breathing.

Tonsillectomy: Surgical removal of the palatine tonsils, The procedure sometimes includes an *adenoidectomy*, and is called "T and A."

Trachea: The largest of the lower respiratory airways, serving as the "trunk" of the bronchial tree. The first or most superior tracheal cartilage articulates with the most inferior laryngeal cartilage, the cricoid.

Vegetative Breathing: Life-sustaining breathing cycles.

Vital Capacity: The maximum volume of air that can be exhaled following a maximum inhalation.

CHAPTER 5
Phonation

CHAPTER PREVIEW

Chapter 5 focuses on the larynx, origin of the major acoustic driving source for speech: the voice. We begin with a description of the larynx's phonatory or voice function to provide a basis for study of its form as well as to provide a firm foundation for understanding of subsequent material. The larynx is a series of upper respiratory valves, supported by cartilage and operated by striated muscles. Its respiratory functions are covered in Chapter 4, and this chapter reviews those respiratory functions in brief. A full appreciation of laryngeal form and function in speech requires an understanding of its importance as a respiratory structure, since respiratory difficulty is very often manifested as an unusual or changed voice. Like many other animals, mankind has learned to modulate the flow of pulmonary air through the laryngeal valves in various ways to produce sounds for communication. These sounds include not only the voice, but fricative and plosive sounds as well. After the treatment of laryngeal functions, the chapter's following sections examine the gross anatomy of the larynx relative to other cervical structures. We take successively closer looks at laryngeal components, from cells and tissues to a skeletal framework with complex articulations and the muscles and nerves that make them work. The chapter ends with description of changes in the larynx across the life span, to form a foundation for clinical treatment. Chapter sidebars describe commonly encountered phonatory disorders, their causes, and effects on speech.

CHAPTER OUTLINE

General Description and Location

The larynx (pronounced "lare-inks" or /lɛrɪŋks/) is a multiply articulated, highly flexible structure, located at the rostral end of the respiratory system. Its inferior border represents the boundary separating the lower and upper respiratory tracts, and its epithelial tissues, musculature and cartilaginous framework form a series of valves designed to modulate ingress and egress of the respiratory system. Mankind and other animals have learned to use the larynx to produce sounds for communication and to supplement its various non-communicative functions.

Laryngeal Function in Speech

The vocal tract begins in the larynx, at the *glottis*. The glottis, about two centimeters superior to the lower border of the larynx, is the meeting point for the adduction of twin, horizontally opposing *vocal folds*, sometimes called the "vocal cords." By opening and closing in various postures, the vocal folds can produce all three of the acoustic driving sources for speech, the plosive, fricative and uniquely, the phonatory sources.

While the plosive and fricative sources can, and frequently are, produced at other vocal tract sites, only the larynx is capable of producing the quasiperiodic phonatory source, the voice, essential for production of vowels, approximants and nasals, and a distinctive phonetic feature for the voiced obstruent consonants. The larynx' position at the most proximal end of the vocal tract affords it the entire tract for a distinctive resonance.

The space between the abducted vocal folds is called the glottis, but the term is sometimes applied to the vocal folds and their intervening space as well. The vocal folds are mobile only at their posterior ends, so the glottal opening widest at its posterior end and narrows anteriorly as the folds meet in a V-configuration. The space immediately below the glottis is called *subglottal*, and the space immediately superior to the glottis is called *supraglottal*.

The Phonatory Source

Energy tapped from pulmonic air in the larynx creates the sound of the voice, and producing this sound is called *phonation*. The vocal sound is also called the *phonatory* or *glottal* source, and is produced through successive pulses of air, impounded and released by self-sustaining cyclic action as its flow is first obstructed by the adducted folds, then released, when subglottal pressure reaches sufficient magnitude to overcome the obstructive force and push the loose superficial layers of vocal fold epithelium aside. Tissue elastance and aerodynamic forces bring the tissues back together for the next cyclic pulse, as long as the vocal folds are adducted.

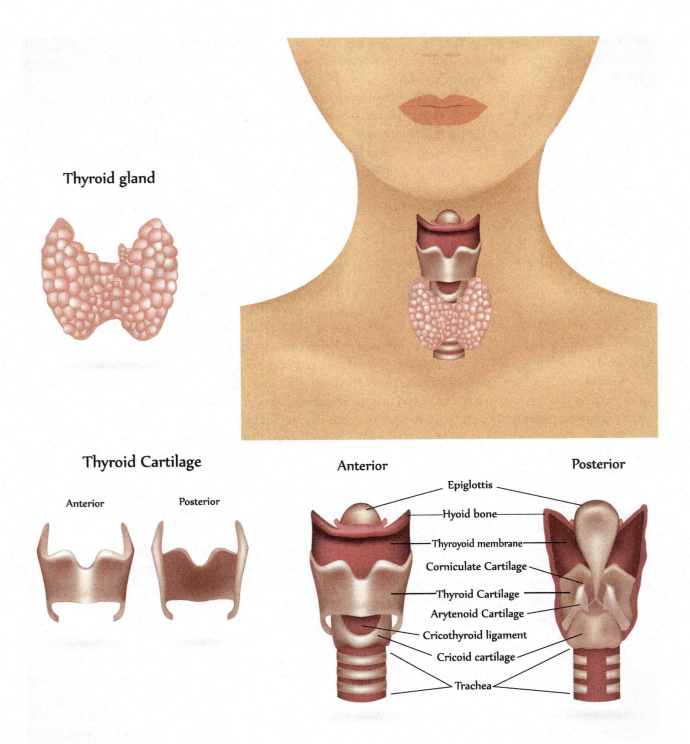

Thyroid gland

Thyroid Cartilage

Anterior

Posterior

Anterior

Posterior

Epiglottis

Hyoid bone

Thyroyoid membrane

Corniculate Cartilage

Thyroid Cartilage

Arytenoid Cartilage

Cricothyroid ligament

Cricoid cartilage

Trachea

Through interaction of thoracic and laryngeal musculature, a speaker can vary the frequency, amplitude, and spectrum of the phonatory source to suit a wide range of communicative needs. Changes in these voice parameters can result from speaker intent or from environmental demands. Glottal frequency changes are perceived as changes in vocal pitch. For example, they are required for certain suprasegmental aspects of speech, including changes in inflection to signal that a question has been asked. Amplitude changes are perceived as loudness changes. Increases in amplitude are, of course, necessary in noisy environments, while decreases, as mentioned above, may be required in quiet environments. Amplitude changes are also used to signal *syllabic stress*, often accompanied by changes in glottal frequency. Spectral changes, perceived as voice quality changes, include whispering, glottal pulsing ("fry"), or the strained voice of an angry speaker. Singing imposes additional and obvious demands for changes in frequency, amplitude, and quality.

As the speaker produces a voice, the phonatory source emanates upward, though the upper respiratory passages, at the speed of sound, resonating in various spectral contours depending upon the momentary configuration of the vocal tract. The cyclic pulsing of laryngeal tissues can be easily felt by gently placing fingers beside the "Adam's Apple," or *laryngeal prominence*, in the middle of the anterior neck, and saying, "Aaaah."

The phonatory source is essential to the production of most speech sounds. In English, it is the basic driving source for vowels, approximants, nasals, and the voiced obstruent consonants. In fact, it is easier to consider those phonemes that *do not* require the phonatory source than to consider those that do. In English, that group is small, consisting of only the voiceless obstruent consonants.

Acoustics of the Phonatory Source

The sound produced by air pulsing through the vocal folds is almost periodic, and human vocal frequencies can range from around 87 Hz, at the low end, to 1047 Hz, at the high end, depending upon the speaker's gender and vocal intentions. (This does not include falsetto voice in adult males [Nair, 2006]). Normal voluntary or involuntary variations in muscle tonus throughout the respiratory system cause variations in cyclic frequency and amplitude, and are the reason the sound is said to be an "almost" periodic, or *quasi* periodic, source.

Natural peak-to-peak variations in glottal frequency are called *jitter*, while variations in amplitude are referred to as *shimmer*. These variations are produced by inconsistencies in vocal fold tension and subglottal air pressure. Excesses in either jitter or shimmer may be signs of underlying pathology.

Since it is not currently possible to report the exact acoustics of the phonatory source in a living human being, independent of the vocal tract, its exact characteristics must be estimated indirectly. Indirect evidence includes *stroboscopic* and *kymographic* photography of the phonating glottis, recording of electrical impedance changes across the cervical region over time with an *electroglottograph*, and inferring acoustic parameters at the glottal opening from measurements of acoustic parameters at the oral opening, such as volume velocity changes over time (e.g., Sulter and Wit, 1996).

The glottal wave form thus measured is said to be "triangular." This characterization refers to a graphic plot of the changing glottal opening area over time with its slow opening and rapid closing phases followed by a brief closed phase. The resultant graphic form resembles a series of crudely drawn triangles having no bottom legs, and of course, no negative values. The changing *glottal area* graph is grossly symmetrical to a graphic plot of changing *volume velocity* (Flanagan, 1958; Stevens and House, 1961), with volume velocity changes following roughly the same contours as glottal area changes over time.

The term *volume velocity* bears some explanation. In general, volume velocity is air volume displacement per unit of time. As it applies to phonatory function, there are actually two volume velocity waves. One is pulmonic and the other is acoustic.

The pulmonic volume velocity wave is created by the natural egress of air from the respiratory system during laryngeal valving. It can be readily understood that as the glottal opening enlarges during the glottal cycle, the volume of pressurized pulmonic air passing through increases. Similarly, as the glottal opening closes during the next phase of the cycle, the volume decreases. Changing volumes passing through the glottal opening relate to changing in vocal tract air pressure, as well, with the greater volumes accompanied by greater air pressures. Normally, air flows from the glottal opening through the vocal tract at a rate of approximately 5 cm /sec. (Minifie, 1973).

The other manifestation of volume velocity observed during glottal cycling is created by an *acoustic glottal wave*. Rapid variation in air pressure produced by the opening and closing glottal cycles creates sound, or acoustic energy, which propagates in waves through the vocal tract. Like the pulmonic glottal wave, the acoustic glottal wave is one of volume velocity variations coincident with sound pressure level changes, as is the case with all air propagated sound waves.

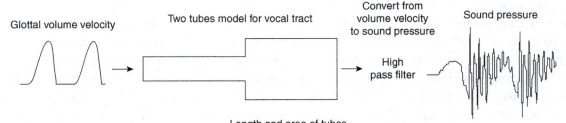

The glottal acoustic wave forms at the same instant as the pulmonic air pressure wave, but propagates through the vocal tract at the speed of sound, approximately 34,400 cm/sec at sea level. It, too, has a volume velocity, inferred to be *quasi*-triangular in its wave form, in this instance with both positive and negative sound pressure level values. The waves are said to be quasi-triangular because true triangular sound waves have only odd numbered harmonics, whereas the natural variability of the glottal source allows ample opportunity for other harmonics to form.

Phonation and the Glottal Cycle

During sustained glottal cycles, little puffs of air or, put more technically correct, successive pressurized volumes, are emitted up into the vocal tract by the rapid opening and closing of the glottis. These puffs excite the air in the vocal tract, and create a nearly periodic complex tone, called the glottal source, phonatory source or, simply, the voice.

The phonatory acoustic driving source is produced by alternating compressions and rarefactions of air molecules in the vocal tract. To begin voice production, a speaker simultaneously relaxes the thorax and loosely adducts the vocal folds. The adducted folds block egress of pulmonic air until their resistance is overcome by the rising air pressure. Typically, only the loose, superficial layers of the vocal folds vibrate during phonation, since they are the first to be affected by subglottal air pressure. As soon as these layers are forced apart, a volume of pressurized air passes through the aperture, and pressure below the glottis drops. Once the pressure has dropped appropriately and airflow conditions are met, the folds return to their closed state. As soon as the vocal folds return to the loosely adducted position, air pressure begins to rebuild. The whole glottal cycle begins anew. Repeated opening and closing of the glottis in *glottal cycles* creates the variations in air pressure. Glottal cycles repeat, over and over, many times per second until the speaker stops phonating.

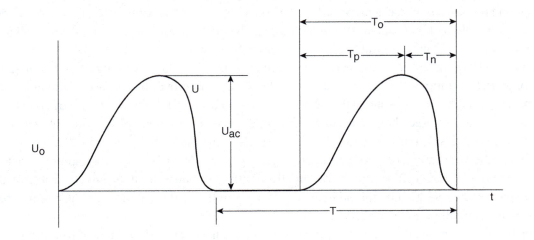

Mechanically, the glottal cycle operates by means of combined *myoelastic and aerodynamic* forces (Van den Berg, 1958). Myoelasticity is the natural tendency of muscles and tissues of the adducted vocal folds to resist deformation. Superficial epithelial and connective tissues are not subject to active contraction, and retain their natural elasticity as they are medially compressed. Their natural elasticity is, however, subject changes accompanying variations in longitudinal forces exerted during pitch changes (Haji, Mori, Omori, and Isshiki, 1992). Muscle tissue, on the other hand, contracts actively and vocal fold muscles tend to resist deformation more as muscle tissue contracts. The degree to which a muscle is contracted is referred to as its tonic status. As long as the tonic status of the muscles underlying the vocal folds is maintained, that is, the muscles are held contracted loosely, they will naturally resist displacement at lower subglottal pressures, allowing the superficial mucosa to cycle freely.

The Bernoulli's principle upon which the glottal cycle functions is the *Bernoulli Effect* (named for French mathematician Daniel Bernouilli, 1700–1782). This fluid dynamic principle dictates a drop in the internal pressure of a liquid medium, such as air, as its volume velocity increases as a result of a pressure imbalance. It is a manifestation of the principle of energy conservation, wherein the total energy in a closed system must remain constant. Potential energy, in this case represented by air pressure, is converted to kinetic energy as volume velocity increases because of the imbalance in air pressure on either side of the glottis. But, since the principle of energy conservation dictates that total energy must remain constant, potential energy, or pressure, must decrease as kinetic energy increases. Through the Bernoulli principle, vocal fold tissues are drawn back toward the midline by the suction of flowing low pressure air, aided by the natural tonus of the activated muscles of glottal adduction and natural elasticity of vocal fold tissues.

This principle of phonatory glottal closing is called the *Myoelastic-Aerodynamic Principle of Phonation* (Van den Berg, 1958). The Myoelastic-Aerodynamic principle of phonation was elaborated by Hiroto (1966) with a *"Mucoviscoelastic"* principle. Hiroto and others recognized that the loose epithelial and connective tissues covering the medial edges of the vocal folds move independently of the main vocal fold mass, and thus would have cyclic phase differences. Instead of a single cycling mass, current examination of glottal movement focuses on a *mucosal wave* and *mucosal upheaval* (Haji, Mori, Omori, and Isshiki, 1992; Yumoto and Kadota, 1998). The lower, or leading edges of the vocal folds, being affected by air flow first, are slightly out of phase with the upper or trailing edges. Because these differences exist in the coronal or vertical plane, they are often called *vertical* phase differences.

A single glottal cycle may be considered as having an *open phase*, during which the vocal folds are separated, creating an opening between them, and a *closed phase*, during which the folds are touching on another in the midline, and there is no glottal opening. The open phase takes up the most time, occupying about 90% of the entire glottal period, depending upon certain particulars of the speaker, including cyclic amplitude, frequency, and gender. The closed phase occupies about 10% of the period.

The open phase of the glottal cycle may be broken down further into an *opening* phase, during which the vocal folds are moving away from the midline, and a *closing* phase, during which they are moving back toward the midline. Timing of the open phase is marked by a slow opening, occupying about 50% of the cyclic period, and a more rapid closing phase, taking about 37% of the period. The glottis remains closed slightly for a slightly longer phase of the period as amplitude (vocal loudness) increases (Sulter and Wit, 1996).

Graphic projection of the two parameters of glottal area and air volume velocity over time might begin at the peak of glottal opening at the instant that subglottal force exerted over the surface of the glottal underside has blown the vocal fold tissues as far as their physical characteristics and the power that moves them permit. In the next instant, the vocal folds begin to return to the midline, and the *closing phase* of the glottal cycle begins. The beginning of the closing phase is convenient because it is there that glottal area and volume velocity begin to decrease, after a slower increase observed during the opening phase. Note that even though the vocal fold tissues are returning to the medial position, that is, closing, the glottis is still open until they make contact. There is still space between the vocal folds, and air is still flowing. Thus, both the opening phase and the closing phase of the glottal cycle take place during the open phase.

During the closed phase of a glottal cycle, the vocal folds are in contact. Air from the lower respiratory system is impounded just below the vocal folds in the *subglottal* area. This air is pressurized by either expiratory relaxation forces of the thorax or by active contraction of expiratory musculature. Even though it occupies a short phase in the entire glottal period, the closed phase is important because in this phase of the cycle, pulmonary air is compressed.

Glottal Flow Waveform

Study the two cycles of a typical glottal flow waveform above. The variables shown are as follows:

T_p the part of the glottal cycle in which airflow rate is increasing (thus, the slope of the curve is going upward during this time). Think of 'p' for positive.

T_n the part of the glottal cycle in which airflow rate is decreasing (downward slope in the curve). Here, 'n' represents negative.

T_o the length of time during each cycle in which air is flowing (i.e., the folds are open)

T the total duration of each vibrational cycle

u_o the average rate of airflow

u_{ac} maximum rate of flow

Glottal Cycle Frequency

Pitch is perceived to increase and decrease with changes in glottal frequency. The frequency of glottal cycles is the number of cycles completed in one second, expressed a *"Hertz"* (after Heinrich Hertz, 1857–1894). The glottal frequency is also the lowest, or fundamental, frequency of the complex, quasiperiodic glottal sound source. Its value depends upon factors such as the age and gender of the speaker, as well as upon situational or environmental factors, including the nature of the audience, the content of the message.

As a general rule, infants and young children have the highest glottal frequencies, whereas adult males have the lowest.

Since most speakers vary their vocal pitches during the course of an utterance, it is common to refer to a speaker's vocal pitch as being commensurate with a *habitual* or *modal* frequency, that is, that frequency to which the speaker returns most often.

Adult males' voices are generally of lower glottal frequency than those of females, owing to the greater masses of males' vocal folds. Increase in the masses of vocal folds is an accompanying change of puberty, affecting males to a greater extent than females. More massive folds naturally vibrate slower than less massive ones. Behrman (2006) reviewed several sources and reported approximate guidelines for typical glottal frequencies. For adults, modal frequencies of 115 Hz. for males and 215 Hz. for females were typical during reading aloud.

Interestingly, male and female children approximately five years of age have similar values for fundamental frequency, about 240 Hz., with an average speaking range of about 200 Hz. (Awan and Mueller, 1996). Children between one and two years of age had higher average glottal frequencies than the five-year-olds, around 357 Hz., with the same 200 Hz. average range, according to a study by Robb and Saxman (1985). In 1942, Grant Fairbanks, pioneer researcher in human communication, recorded his infant son's hunger wails monthly, during the first nine months of life. The junior Fairbanks' fundamental frequencies ranged from 63 Hz. to 2631 Hz.

Temporal parameters of the glottal cycle vary with glottal frequency. Specifically, the closed phase of the glottal cycle becomes shorter as frequency increases. The *closed quotient (CQ)* (Tarnoczy, 1951) of the glottal cycle is a frequently examined parameter of the glottal cycle because it expresses the relationship of open and closed phases. Closed quotient is the ratio of closed phase duration over the entire glottal cycle period. Thus, CQ becomes smaller as frequency increases. In high female voices, the CQ may be even be 0, indicating that the glottal aperture never fully closes. Higher CQs are observed in lower, male voices.

Every larynx is different, and each apparently has an *optimal* operating frequency, depending upon the masses of the vocal folds. This frequency purportedly gives the best voice with the least effort (Boone, 1971; Johnson, Darley and Spriestersbach, 1963). However, the persistent notion of an optimal vocal frequency is difficult to sustain with physical evidence (Thurman, 1958). Perhaps the most compelling argument for an optimal glottal frequency was posited by Zemlin (1998), referring to Kunze's (1962) unpublished doctoral research on glottal air resistance. Using hypodermic needles connected to manometers and inserted into the subglottal area, Kunze found glottal resistance to be lowest at about 30% of the fundamental speaking range. This is approximately the frequency recommended by Pronovost (1942) and Fairbanks (1944) as optimal.

More useful than any specific frequency or range of frequencies as being optimal for vocal health (Boone, 1971) is the idea that optimal speaking frequencies are near, but not at, the lower end of the natural frequency range, and that most effective voice comes with varying the frequency and accompanying suprasegmental pitch contours. A classic sign that a speaker is habitually speaking below the optimal range is a rather monotonous delivery and the occurrences of sudden high pitch breaks at the ends of declarative sentences, where vocal pitch usually drops off.

A speaker can change glottal frequency by altering the length and tension of the vocal folds. This is accomplished mechanically by exerting or relaxing transverse force on the vocal folds. When the vocal folds are stretched, their mass per unit of length is reduced, and they can move faster, creating more cycles per second. The mechanical characteristics of the folds are similar in this respect to those of a rubber band.

Tension of the vocal folds is increased by contractions of two *intrinsic laryngeal muscles*, the cricothyroid and the thyroarytenoid muscles. The cricothyroid muscles are sometimes called the "external tensors," since they are located outside the laryngeal cavity. The thyroarytenoid muscles are part of the mass of the vocal folds, and, being located inside the laryngeal cavity, are sometimes called the "internal tensors." Paradoxically, the thyroarytenoid muscles can also decrease vocal fold tension and lower vocal pitch. The intrinsic muscles and the extrinsic muscles will be covered in detail later in the chapter.

Glottal Cycle Amplitude

Loudness of the voice is a perceptual phenomenon associated with changes in the amplitude of the acoustic wave energy created during phonation. Increased amplitude is the result of increased subglottal pressure. While logic may suggest otherwise, increased vocal loudness may or may not be associated with increases in the distances vocal folds are displaced during phonation. Until the middle of the last century (Zemlin, 1998), it was thought that the louder voice was always associated with greater amplitude of vocal fold displacement and resulting larger glottal area, since more force was used to open the glottis. More recent research has reported, however, that this is only the case in some speakers. For others, there is no difference in the distance of maximum vocal fold displacement to accompany greater vocal loudness.

The mechanism for changing glottal cycle amplitude is the coordinated action of the muscles of respiration and the muscles of glottal adduction. To increase glottal wave amplitude, respiratory forces increase subglottic pressure. This increase can result from decreased inspiratory muscle opposition to thoracic relaxation forces or by active contractions of expiratory muscles. Coordinated with the increasing subglottal pressure, glottal adductors increase vocal fold resistance. Acting in concert, the two mechanisms create higher subglottal pressures and greater acoustic amplitudes.

The balance between respiratory and glottal muscle contractions varies throughout the duration of a prolonged utterance. The only constant is the speaker's intent to produce a voice at the desired loudness level, which requires a certain subjectively determined subglottal pressure.

Muscular and mechanical respiratory forces vary according to the amount of vital capacity being consumed at the moment, with relaxation forces being greatest at the peak of inspiration. As the thorax returns to its relaxed or neutral state, relaxation forces diminish. Thoracic inspiratory muscles that once resisted relaxation forces subside, and, eventually, expiratory muscles must contract to maintain the desired subglottal pressure. Not only do muscular and mechanical forces vary throughout the expulsion of pulmonic air, but the so-called desired loudness level can and does change from moment to moment, according to various prosodic requirements.

Glottal adductor muscles respond to variations in respiratory pressures. As subglottal pressure increases, and higher pressure volumes flow through the glottis, the aerodynamic forces that drive the Bernoulli Effect increase, as well. This results in greater forces to bring the vocal fold mucosa back to the midline. Increasing the force of pulmonic air by contracting the muscles of expiration increases subglottal pressure, but the loosely adducted folds will simply give away if there isn't also a countering increase in glottal adductive force. The medial forces that resist subglottal pressure are called *medial compression* forces.

Thus far, we have three phenomena associated with increased loudness: increased subglottal air pressure, increased Bernoulli-effected glottal closing forces, and increased medial compression. These combined increases have an expected effect: vocal fold mucosa return to the midline together with more force, because they are under more Bernoulli effect and under more muscular tension. They stay closed a little longer for the same reasons: inertia from the increased velocity of medial movement combined with increased muscular tension. The longer closed phase is needed to build the desired increased subglottal pressure for the next cycle. Of course, at a given fundamental frequency, the glottal cycle must take place within the corresponding period, and, since the closed phase is slightly longer in duration, vocal fold tissues must move faster through the open phase.

Alternative Glottal Configurations and Speech Functions

Alternative vocal fold articulatory configurations can be used to produce the plosive and fricative sources as well as changes in the phonatory source. Adduction and resistance to pulmonic air flow, followed by a sudden abduction and air release, produces the transient, aperiodic plosive source, creating the *glottal plosive* (/ʔ/), sometimes called the glottal *stop*: "Uh!" This phone is substituted for other plosives in some dialects of English, and is arguably a consonantal onset of syllables usually thought to begin with vowels. Loose vocal fold adduction produces a thin aperture, and forcing air through this aperture produces a continuous, a periodic source called the *glottal fricative*. There are two variants of the glottal fricative, one unvoiced, /h/ and the other voiced, /ɦ/. The voiced variant is commonly articulated in English when a glottal fricative is intervocalic, as in the word "ahead," with less medial compressive force than the unvoiced version.

Medial articulation of the vocal processes of the arytenoid cartilages combined with abduction of their bases produces a posterior aperture in the otherwise adducted vocal folds. Forcing pressurized air through this aperture produces the aperiodic, continuous, fricative source used for the whispered voice.

Another alternate glottal posture produces the high pitched *"falsetto"* voice. In this posture, the lateral aspects of the vocal folds are tensed to the point that they vibrate little or not at all. Only the inner or medial aspects of the two folds vibrate, and, since their masses are much less than those of the full folds, their vibratory frequency is much higher.

Gross Laryngeal Anatomy and Physiology

The form and function of the larynx has evolved to make it suited to its intended communicative and biological functions. Its location and configuration make it far more variable in function than the primitive muscular sphincter of the lungfish, adapted specifically to keep water out of the respiratory system (Negus, 1949).

Location of the Larynx

The larynx occupies the inferior part of the pharynx, giving that region its name, *laryngopharynx*. This same region is also sometimes called the *hypopharynx*. The larynx is a midline cervical structure, with a cartilaginous, multiply articulated skeleton, an intrinsic muscular system and covered with several types of epithelium, depending upon location. It is located in the middle cervical area, at the levels of vertebrae C-4 to C-6, or between the hyoid bone at C-3, superiorly, and the trachea, at C-6, inferiorly. The larynx is anterior to the spine and anterior to the esophagus. As the most inferior part of the pharynx, the larynx is part of the upper respiratory passages, and so forms a hollow tube to allow air to pass to and from the lungs.

Biological Laryngeal Functions

The primary function of the larynx is to protect the upper airway. The laryngeal mechanism is a gateway to the lower respiratory tract, with its three valves to keep unwanted material out, or to help expel unwanted material from within. Its complex articular arrangement makes the larynx surprisingly flexible and capable in its various functions.

Closing the laryngeal valves can block escape of pulmonic air when we want to, "Hold our breaths," or when thoracic rigidity is important. For example, when we want to lift something heavy, we inflate our chests and lift. The "grunt" often emitted as one lifts something heavy is the sound of sound of a little air escaping through tightly adducted vocal folds. Similar thoracic rigidity is also required for defecation, childbirth, and other strenuous activities.

Laryngeal Valves and Cavities

Three valves enable the larynx to perform its various functions. These valves form the borders of two laryngeal cavities or swellings in the laryngopharyngeal tube.

The uppermost of these valves is a sphincter at the entrance to the larynx, protecting a cavity called the laryngeal *vestibule* or *additus*, or simply the *superior division*. The vestibule is separated from the oropharynx by the *aryepiglottic fold*, extending around the entryway and forming a sphincter, covered by epithelial tissue and supported by the epiglottis, cuneiform, corniculate, and arytenoid cartilages. Muscular function of the sphincter is achieved by contraction of the thin *aryepiglottic muscle*, which cinches the cartilages and tissues tight. Opposing the aryepiglottic muscle is the *thyroepiglottic muscle*.

Anteriorly, the epiglottis provides a flexible cover of elastic cartilage that moves posteriorly during swallowing to protect the laryngeal entrance. Posterolaterally, the hyaline *arytenoid cartilage* apices and associated *corniculate* and *cuneiform cartilages* provide substance to the flexible aryepiglottic folds that cover the spaces between the cartilages.

The most prominent feature of the vestibule is the eminence in the aryepiglottic fold created by the epiglottis. Posterior to this eminence are two smaller elevations, the cuneiform and corniculate tubercles created by the cuneiform and corniculate cartilages. Two lateral *pharyngoepiglottic folds*, one on either side of the epiglottal eminence, connect the aryepiglottic fold to the lateral pharyngeal walls. These folds, combined with the median *glossoepiglottic fold* between the anterior surface of the epiglottal eminence and tongue's root, form the *glossoepiglottic valleculae*. Lateral to the aryepiglottic folds and the laryngeal vestibule are two *pyriform recesses (sinuses)*, bordered laterally by the superior cornua of the thyroid cartilage and its protective epithelium. The superior laryngeal nerve and laryngeal blood and lymph vessels traverse these pyriform recesses *en route* to piercing the cricothyroid membrane. Both the glossoepiglottic valleculae and the pyriform sinuses are notable for entrapping foreign material, usually food or drink.

At the inferior extent of the vestibule are two flaccid, *ventricular folds*, formed by a band of fibrous connective tissue, two weak *ventricular muscles*, and covered by stratified squamous epithelium.

The *ventricular folds* form the middle laryngeal valve as well as the entrance to the laryngeal *ventricle*, named *Ventricle of Morgagni*, for Italian anatomist and pathologist Giovanni Battista Morgagni (1682–1771). Soft and flaccid, the ventricular folds are composed of thin muscle covered with mucous membrane. They attach to the

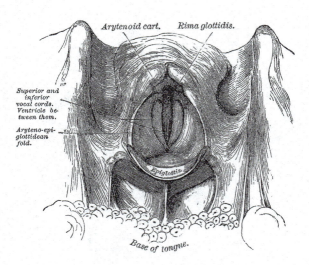

Arytenoid cart.   *Rima glottidis.*

Superior and inferior vocal cords. Ventricle between them.

Aryteno-epiglottidean fold.

Epiglottis.

Base of tongue.

anterolateral surface of arytenoid cartilages posteriorly, near the triangular fovea, a lateral recess, of each arytenoid cartilage. Anteriorly, they are more prominent than they are posteriorly, and attach at that end to the inner surface of the thyroid cartilage, about where the epiglottis attaches. The ventricular folds are sometimes called the *"false" vocal folds*, and the space between the ventricular folds is sometimes called the *"false glottis."* Ventricular folds adduct under strong laryngeal effort and can be contributors to the laryngeal tone in abnormal phonation.

The most inferior of the laryngeal valves is strong and flexible, and is formed by the two laterally opposed, horizontally oriented, medially projecting *vocal folds*. These folds not only form the strongest of the laryngeal valves, but are of primary interest to the study of speech because their adduction produces the voice. The vocal folds are highly mobile posteriorly, and like the ventricular folds above, attach to the paired arytenoid cartilages. Thus, they move in concert with the ventricular fords at that end. Anteriorly, the vocal folds attach at a fixed point on the inner surface of the thyroid cartilage, near its angle. The ligaments are connected at that point by an *anterior commissure*. Since the anterior ends of the vocal folds are fixed, abduction at their posterior ends forms a v-shaped opening, the glottis.

The vocal folds mark the inferior extent of the ventricle and are complex in structure. They have a superficial layer of epithelium, covering a fibrous tissue layer with a muscular and ligamentous infrastructure.

The surface of the vocal folds is a layer of stratified squamous epithelium, supported, as is typical of epithelial tissues, by a basement membrane. The reader will recall that stratified squamous epithelial tissue is designed to resist wear by virtue of its flat superficial cells that slough off when worn, to be replaced by fresh cells in the next deeper stratum. This is the perfect tissue to line the medial surface of the vocal folds, for that is where they make contact during phonation.

Deep to the surface epithelium, as is the case with many organs, is a *lamina propria* of loose connective tissue. The vocal fold lamina propria is composed of three layers of layers of elastic and collagen fibrous connective tissue: a *superficial* layer, an *intermediate* layer, and a *deep* layer. Each is composed of combined straight and coiled elastic and collagen fibers, ideal for maintaining the shape and organization of the vocal folds during the deformations of phonation (Ishii, Zhai, Akita, and Hirose, 1996).

The superficial lamina propria is mostly loose collagen and elastic fibers bound with matrix. This area is called *Reinke's Space*, and is technically not a space, but instead is filled with fibers and matrix in a gelatinous composition. Such composition permits the free mobility of the superficial layers of the vocal folds during phonation.

VENTRICULAR PHONATION

It is possible to adduct and excite the air in the pharynx with pulses generated by the ventricular folds. This is not a normal phonation, and sounds very rough. Some people phonate in this pathological way. The condition is called *dysphonia plica ventricularis* and is treated by voice therapy which focuses on generating and recognizing the normal voice as distinct from the from the ventricular voice, beginning with non-speech vocalizations such as sighing or yawning and progressing to speech activities.

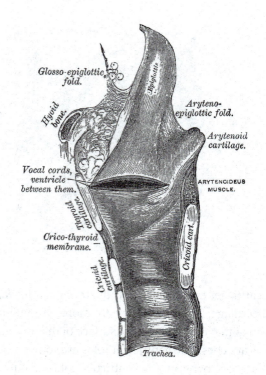

The mobility of the lamina propria with respect to the deeper parts of the vocal folds is extremely important, for without its flexibility, phase differentials along the medial vibrating parts of the folds would not be possible, and the human voice as we recognize it would not exist. The complex phase differentials between the leading and trailing edges of the vocal folds, considered caudally to rostrally, allow the superficial tissues to undulate as compressed pulmonary volumes emanate from the subglottal area and pulse through the glottis and vocal tract.

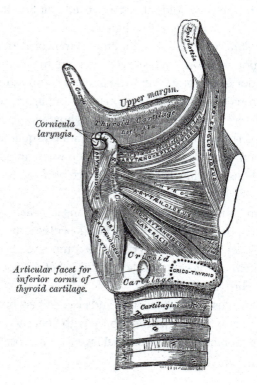

The vocal ligament, itself a strong connective tissue support for the vocal folds, may also be considered to form the substance of the intermediate and deep lamina propria (Hirano, 1975). It is composed of dense parallel collagen and elastin fibers running parallel to the planes of the folds, and is reported to have the tensile strength to resist tension forces ten times those that can be generated by the intrinsic laryngeal musculature. Rather than having the vibrating characteristics of a string, recent evidence suggests the vocal ligament follows a beam arrangement, fixed at each end, during vibration (Titze and Hunter, 2004).

At each end, the vocal ligament becomes broader and yellowish, forming *macula flavae* (Latin for "Yellow Spots"). At the anterior ends, the maculae flavae matrices are continuous with those of the connective tissue of the anterior commissure, while at their posterior ends the matrix material is continuous with that of the arytenoid cartilages.

The vocal ligament also forms the free superior border of the *cricovocal membrane*, a connective tissue sheet that attaches inferiorly to the superior border of the cricoid cartilage and posteriorly to the arytenoid bases. The cricovocal membrane is also known as the *conus elasticus*.

Deep to the connective tissues of the vocal folds are their muscular components. The most medial of these are the *vocalis* muscles, sometimes referred to as the *medial thyroarytenoid* muscles, bordered laterally by the thicker *(lateral) thyroarytenoid* muscles.

The vocal folds protect the *infraglottal (subglottal) division* of the larynx, the space immediately inferior to the vocal folds. They mark the border between the lower and upper divisions of the respiratory system. In the infraglottal division is the area where pulmonic air pressure is increased to drive the phonatory source. As we have seen, increases in subglottal air pressure attend increases in phonatory loudness. Compromises in the integrity of this division, such as those that accompany a tracheostomy, can prevent sufficient air pressure from building up to a magnitude that can drive the glottal source. The *conus elasticus* covers the epithelium of the subglottic division, and its shape contributes to the generation of Bernoulli Effect during the glottic cycle.

TRACHEOSTOMY

When patency of the upper airway is compromised, a surgical procedure called a *tracheostomy (tracheotomy)* is required to ensure airflow to and from the lungs. The procedure involves making an opening or *stoma*, in the skin of the anterior neck and in one or two tracheal rings.

Following the opening, a plastic tube called a *cannula* is inserted in the stoma to preserve patency of the opening. A tracheostomy may be temporary or permanent, as required.

A tracheostomy is required as postoperative treatment following *laryngectomy* or removal of the larynx, since the trachea is no longer protected by the laryngeal valves. Other conditions for which tracheostomy may be required are injury, congenital malformations of the upper airway, blockage of the upper airway by foreign objects, laryngeal paralysis, or chronic unconscious conditions such as coma.

Speech with a tracheostomy can be accomplished by restoring subglottal air pressure when the larynx is present or by creating an alternate phonatory source when the larynx is removed or non-functional.

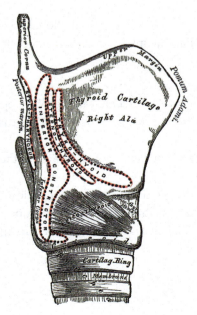

Cytology and Histology of the Laryngeal Skeleton and Cavities

Connective Tissues

Connective tissues form the skeleton, ligaments, and membranes of the larynx. The skeletal structure of the larynx is composed of hyaline and elastic cartilage, supplemented with dense and loose connective tissue ligaments and membranes. The thyroid, cricoid, and most of the arytenoid cartilages are hyaline cartilage. Elastic laryngeal cartilages include the epiglottis, corniculate, and cuneiform cartilages and the apices of the paired arytenoid cartilages.

Areolar tissues form the vocal fold lamina propria, turning to dense connective tissue at the vocal ligament. The superficial and intermediate layers of lamina propria consist of straight and coiled collagen and elastic fibers. This gives it resiliency during phonation and other glottal activities. The deep layer consists of dense collagen and coiled elastic fibers. Oxytalan fibers are distributed throughout (Ishii, Zhai, Akita, and Hirose, 1996). Cells of the maculae flavae appear to play an important role in the generation of extracellular matrix material to be infused into neighboring connective tissues of the lamina propria. This material is necessary for maintenance of the viscoelasticity of the lamina propria (Sato, Hirano, and Nakashima, 2003). Dense connective tissue forms the ligaments that bind laryngeal cartilages to each other and to the trachea and hyoid bone.

Epithelial Tissues

Most of the lining of the laryngeal cavities is pseudostratified, ciliated epithelium. Each of these cells is tipped with about 200 cilia, waving upward toward the oral cavity at a rate of about 10 to 20 Hz. to move debris toward the opening of the airway. As is common in respiratory epithelium, mucous-secreting goblet cells are interspersed among the columnar cells. Goblet cell secretions create a *mucous blanket* in the laryngeal cavity to moisten dry air and engulf small particles. Secretions also appear to lubricate the laryngeal tissues and contribute to phonatory efficiency (Fukuda, Kawaida, Tatehara, Ling, Kita, Ohki, et al, 1988). Stratified squamous epithelial tissue is found wherever wear is likely. This includes medial, contact surfaces of the vocal and ventricular folds and the pharyngeal surface of the epiglottis.

Skeletal Framework of the Larynx

The skeleton of the larynx is composed entirely of articulated cartilages. These are attached externally to the hyoid bone, superiorly, and to the trachea, inferiorly, by extrinsic ligaments. Cartilages are bound to one another by intrinsic ligaments. Hyaline and elastic cartilage, make the larynx it tough and flexible and keep the pharyngeal tube open for the passage of air.

The hyoid bone, properly a bone of the skull, has a functional relationship with the larynx. It suspends the larynx from above and provides a mechanical link between the oral cavity and the larynx.

Cartilages of the Larynx

Most of the laryngeal skeleton is composed of hyaline cartilage. The epiglottis is the only large elastic cartilage. The minor corniculate and cuneiform cartilages are formed of elastic cartilage, as are the apices of the arytenoid cartilages.

The unpaired laryngeal cartilages are single structures, located at the midline of the larynx. These include the *epiglottis*, *thyroid* and *cricoid* cartilages.

One each of the paired laryngeal cartilages is located on either side of the midline. All but one pair of the paired cartilages are associated with the vocal folds, the ventricular folds, and the aryepiglottic folds. These include the *arytenoid, corniculate,* and *cuneiform* cartilages. The paired *triticeal* cartilages are located in the posterior thyrohyoid ligament, superior to the laryngeal inlet.

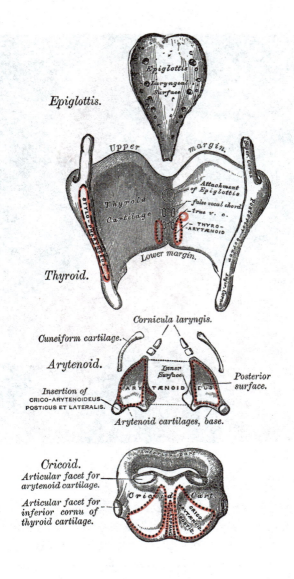

Unpaired Laryngeal Cartilages

Thyroid Cartilage

The thyroid cartilage is the largest laryngeal cartilage. It is the most visible and readily palpated cartilage in the neck. A large *thyroid eminence* is a secondary sexual characteristic in post pubescent males and is colloquially called the "Adam's Apple."

Following the pattern of branchial arches, also manifested in the maxilla, mandible, and hyoid bone, the thyroid cartilage is fused anteriorly and open posteriorly. It is suspended from the hyoid bone superiorly by ligaments and membranes, and it articulates with the cricoid cartilage inferiorly, surrounding it on three sides, anteriorly and laterally.

Internally, the thyroid cartilage provides the anterior attachment and support for the vocal ligaments, described earlier as the connective tissue infrastructure of the vocal folds. The vocal folds are attached to the interior aspects of the thyroid laminae from their anterior extent and all along their lateral margins, so that when the thyroid cartilage moves, the vocal folds move with it.

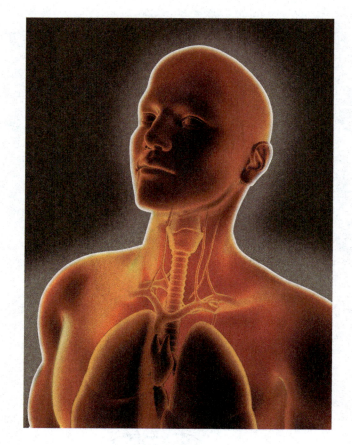

Thyroid Cartilage Structure

Thyroid Cartilage Laminae

The thyroid cartilage has two flat facets, *laminae* or *alae*, anteriorly, and two horns, or *cornua*, one superior and the other inferior on either side at the posterior part of each lamina. The laminae converge anteriorly in the center of the anterior cervical area where they form a prominent anatomical landmark. The fusion of the thyroid lamina forms a "V" shape creating a visible and palpable mid-cervical eminence known as the *laryngeal prominence*. Owing to the generally grater size of the larynx in adult males, the laryngeal prominence is particularly visible in post-pubescent members of that gender and is the anatomic land mark for the C-5 vertebra. There are descending angles in the medial edges of the superior margins of the thyroid laminae, forming the *thyroid notch* with its deepest part at the point of their fusion.

Descriptions of the thyroid laminae often include mention of an *oblique line*, beginning at the superior and posterior edge of a lamina, near the base of a superior cornu, and extending downward and anteriorly (Clemente, 1975; Yoffey, 1976; Palmer, 1993). Zemlin (1998) disputed the presence of an oblique line as part of the thyroid cartilage morphology. Instead, he reported that the oblique line was, in about 90% of his specimens, a tendon attaching the thyrohyoid and sternothyroid muscles to the cartilage.

Various authors have asserted significant sexual dimorphic differences in the angle of thyroid lamina fusion. The most common assertion is that in males the angle of laminar fusion is about 90 degrees, while in females the angle is closer to 120 degrees (Yoffey, 1976). Zemlin (1998) wrote that the angle for males was 80 degrees while that for females was 90 degrees. Although neither of these reports was based on cited empirical evidence, the supposition is that the larger, flatter angle of thyroid laminar fusion in females results in a shorter glottal length and, hence, a smaller, less massive vocal fold.

Among the very few empirical reports available, Kahane (1978) found no significant differences in angles of thyroid lamina fusion is between males and females in his small sample of ten Caucasian larynges ranging in age from 9 to 18 years of age. In 1982, with twenty specimens, Kahane did note a greater post-puberty increase in anterior/posterior thyroid cartilage length as compared to the distance between the posterior margins favoring males. However, more overall growth appeared to have occurred in the male larynx during puberty than had occurred in females. This created greater size and mass of laryngeal structures, including the vocal folds, so that post-pubescent male vocal folds were up to twice as long as female vocal folds. Eckel, Sittel, Zorowka, and Jerke (1994) examined in detail the measurements of 28 male and 25 female larynges, aged 25 to 88 years. They and found about ten to twenty degrees differences between the sexes. Among 95 measurements, most indicated an overall larger laryngeal framework for males as compared to females. They measured angles from the thyroid notch to the bases of the superior cornua relative to the mid sagittal plane on the right and left sides separately. Adding the right and left angles, male angles totaled 65.7 degrees and female angles totaled 80.7 degrees. These researchers noted that the elasticity of the vocal folds during life made post-mortem measurements unreliable. In summary, adult males appear to have larger larynges than adult females, and this may include a longer anterior-posterior glottal length created by a smaller thyroid laminar angle.

Thyroid Cartilage Cornua

The thyroid cartilage has two processes, the thyroid cornua ("horns"), extending superiorly and inferiorly, from the posterior margins of each lamina. The superior cornua are attachments for ligaments that connect to the posteriorly projecting greater cornua of the hyoid bone, and the inferior horns end in slightly bulbous swellings, which articulate with the cricoid cartilage, at lateral amphiarthrodial joints, on either side posterior portion of the cricoid arch. It is at this articulation that the thyroid and cricoid cartilages rotate and glide in relationship to one another, effecting changes in vocal ligament tension.

Thyroid Cartilage Attachments

The thyroid cartilage attaches to the hyoid bone superiorly by a complex extrinsic ligament, the *thyrohyoid membrane*. This connective tissue is thicker anteriorly, in its center, and at its two posterior ends. Posteriorly, the thickened areas are two *lateral thyrohyoid ligaments*, connecting the posterior ends of greater hyoid cornua to the superior thyroid cornua. In front and medially, a *median thyroid ligament* connects the thyroid notch to the hypoid corpus. Between the thickened thyrohyoid ligaments on either side are the thinner portions properly called *thyrohyoid membranes*. On either side of the membrane, roughly in their centers, are foramina, through which pass the superior laryngeal blood vessels, including the superior laryngeal arteries, accompanying veins and lymph ducts, and the internal branches of the superior laryngeal nerve.

The vocal ligaments attach to the thyroid cartilage's interior, at approximately the thyroid angle. At this anterior point, the vocal ligaments are fixed. Their movement is accomplished only by virtue of the freedom of their posterior ends. The vocal folds attach to the inside to this interior laterally.

The inferior thyroid cornua end in a slight bulbous shape, with smooth articular facets. These articulate in paired synovial joints formed with the lateral articular surfaces of the cricoid cartilage, with which it freely rotates within its anatomical limits, and glides antero-posteriorly (front-to-back). The angle of rotation is approximately 60E (Filho, Bohadana, Perazzio, Tsuji, and Sennes, 2005).

Cricoid Cartilage

The cricoid cartilage is the lowest cartilage of the larynx, and the only one with a complete ring shape, as its name implies (*krykos (Gr.): "ring"*). It articulates with the thyroid cartilage laterally, with the arytenoid cartilages superiorly and with the first tracheal ring inferiorly.

Cricoid Cartilage Structure

When viewed from the top, the cricoid cartilage is cylindrical, with its greatest height at its posterior aspect, tapering to a much shorter anterior part. The posterior part the cricoid ring is flattened *lamina*, while the circular anterior aspect is called the cricoid *arch*.

Cricoid Cartilage Lamina

The cricoid cartilage has a posterior lamina. The lamina is relatively flat posteriorly and, on its top margins, slopes gradually down in the anterior direction to the arch. The articular surface of the cricoid lamina is located on its superior surface, and extends down onto the arch a bit. This is the smooth surface upon which the arytenoid cartilages move during glottal maneuvers. The joint between the cricoid lamina and the paired arytenoid cartilages is a very flexible amphiarthrodial joint by virtue of which the vocal folds can assume a variety of postures.

On either side of the cricoid cartilage, just anterior to the lamina, are articular surfaces for the thyroid cartilage. At this point, the rounded ends of the inferior thyroid cornua articulate.

Cricoid Cartilage Arch

The superior surface of the cricoid arch slopes inferiorly from the lamina and becomes thinnest at its ventral surface. The slope of the arch is one of its notable aspects, since the paired arytenoid cartilages glide along this slope, creating a downward movement of the vocal folds upon adduction.

The smooth articular facet on the upper and lateral surface, near the lamina, is a distinguishing feature of the cricoid arch. This facet is convex and irregularly oval in shape, widening posteriorly (Wang, 1998). It slopes downward anteriorly and laterally at 30 to 60 degrees. Movement of the arytenoid bases along his slope results in a downward and forward movement of the vocal folds during adduction, and an upward and rearward movement during abduction.

Cricoid Cartilage Attachments

The cricoid cartilage articulates with the thyroid cartilage at the inferior thyroid cornua laterally. The thyroid extends superior to the cricoid, and the two are connected by a cricothyroid membrane. Below, the cricoid cartilage is attached to the first tracheal ring by means of a cricotracheal membrane.

The paired arytenoid cartilages articulate with the cricoid cartilage at the articular surfaces on the superior and lateral aspects of the arch, near the lamina. Here, the arytenoid cartilages can glide in a variety of directions.

Epiglottis

Epiglottic Structure

The epiglottis is a flat, leaf shaped elastic cartilage, located at the base of the tongue. It projects posteriorly and superiorly posterior away from the root of the tongue, to which it is connected by loose epithelial tissue called the *glossoepiglottic fold*. Its posterior and superior margin is free.

The epiglottis is the most prominent anatomic feature of the laryngeal inlet, marking the divergence of the respiratory system from the pathway it shares with the alimentary tract. When the airway is open, the epiglottis protrudes upward and back and is close to the root of the tongue.

The laryngeal inlet is composed of a cartilaginous skeleton covered with the mucous membrane described above. At the posterior end of the inlet, the cartilaginous framework is provided by the paired arytenoid, corniculate, and cuneiform cartilages. The epithelial membranes of the laryngeal inlet are very loosely attached to the sides of the pharyngeal walls. If irritated, they can become swollen and block the airway, causing a life-threatening emergency situation.

On either side of the laryngeal inlet are depressions in the epithelium called the *pyriform sinuses (fossae)*. Anterior to the epiglottis, between it and the tongue root, the tissue is elevated in three places by the presence of the *hyoepiglottic ligament* and the lateral pharyngeal walls, forming the median and the lateral *glossoepiglottic folds*. Depressions between the folds are the *glossoepiglottic valleculae*.

If the extrinsic laryngeal muscles fail to function properly during swallowing, material may collect or "pool" in the pyriform sinuses or glossoepiglottic valleculae. Later, the pooled material may enter the airway, causing coughing or choking in attempts to expel it, or aspiration if the material enters the lower airway (Tanner, 2007). Such paresis is not uncommon following a cerebral vascular accident.

The epithelial tissue continues inferiorly to cover the arytenoid, corniculate, and cuneiform cartilages. Since this membrane stretches from the epiglottis to the arytenoid cartilages, it is called the *aryepiglottic fold* once it passes to the posterior aspect of the epiglottis.

Epiglottic Attachments

The epiglottis is attached anteriorly to the inner face of the hyoid corpus by means of a hyoepiglottic ligament. It is also attached to the inner aspect of the thyroid cartilage by the thyroepiglottic ligament, just below the thyroid notch.

Paired Laryngeal Cartilages

One of each paired laryngeal cartilages is located on either side of the mid-sagittal plane. Although they appear to be mirror images of each other identical upon cursory examination, dimensional asymmetries are well-documented (Wang, 1998). The paired cartilages of the larynx are the *arytenoid, corniculate, cuneiform,* and *triticeal cartilages*. Of these, the most important are the arytenoid cartilages.

Arytenoid Cartilages

The arytenoid cartilages are positioned atop the cricoid lamina and are surrounded anteriorly and laterally by the thyroid cartilage.

> **EPIGLOTTITIS**
>
> If disease causes inflammation of the inlet mucosa around the epiglottis, the condition is called epiglottitis, and is a MEDICAL EMERGENCY. Epiglottitis is a life-threatening condition and is characterized by "THE THREE D'S": drooling, dysphasia, and dysphagia.
>
> Drooling is caused by the build up of saliva in the oral cavity. It is the obvious result of the patient's difficulty with swallowing, or dysphagia. Dysphasia is difficulty speaking. Such difficulty is the result of the swollen tissues limiting the passage of air ands mobility of vocal tract structures.

Arytenoid Structure

The arytenoid cartilages are shaped like roughly irregular pyramids, and the names of their parts reflect this analogy. The bases of these pyramids sit atop the upper edge of the cricoid cartilage, at about the site of the lamina, but these bases extend anteriorly and down along the upper edges of the arch. From each base, a *vocal process* extends anteriorly, and a *muscular process* extends laterally and inferiorly.

The vocal processes of the arytenoid cartilages project anteriorly, toward the inside of the thyroid laminae. The vocal processes serve as posterior attachments for the vocal ligaments. The joints between the arytenoid and cricoid cartilages are freely moveable, and the posterior ends of the vocal ligaments can assume various postures by virtue of the fact that their posterior ends move about with the arytenoid cartilages. The vocal ligament anterior attachment is a fixed point on the inner surface of the thyroid cartilage, near the fusion of the lamina.

The muscular processes of each arytenoid cartilage project laterally and inferiorly. They serve as the points of attachment of two intrinsic muscles, the *lateral* and *posterior cricoarytenoid* muscles.

The arytenoid bases have smooth facets on their inferior aspects for articulation with the corresponding facets on upper edge of the cricoid cartilage. The bases and, hence, the facets, are concave, to match the convex surface of the cricoid arch. They are also oblong and, but with their long axes perpendicular to those of the cricoid side of the articulation.

Minor Laryngeal Cartilages

There are several "minor cartilages" of the larynx. These are vestigial structures and probably serve only to give additional substance to the aryepiglottic folds and thyrohyoid ligaments. They are all paired and located between the arytenoid apices and the hyoid bone.

Corniculate Cartilages

The corniculate cartilages are located within the aryepiglottic folds, superior and posterior to the apices of each arytenoid cartilage. Since the arytenoid apices tilt posteriorly anyway, the corniculate cartilages look like continuations of the tips of the arytenoid cartilages.

Cuneiform Cartilages

A little further up in the aryepiglottic folds from where the corniculate cartilages are located are two more minor cartilages, the cuneiform cartilages. Cuneiform cartilages are not present in all larynges.

Triticeal Cartilages

Triticeal cartilages are not much more than cartilaginous lumps in the lateral thyrohyoid ligaments, between the superior horns of the thyroid cartilage and the greater horns of the hyoid bone.

Laryngeal Support

The Hyoid Bone

The hyoid bone is considered a bone of the skull, and, as such, it is part of the axial skeleton. It not an atomically a part of the larynx, but is considered here because it plays an important supporting role in laryngeal function. Situated just below the mandible, the hyoid bone can be palpated (felt) by moving a finger posteriorly beneath the mandibular corpus to the point at which the neck begins. The hyoid bone provides the superior support for the larynx. In other words, the larynx "hangs" from the hyoid bone by muscular tethers.

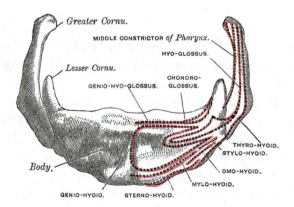

Structure of the Hyoid Bone

The hyoid bone is shaped like a "U" or a horse shoe. It is open posteriorly and curved in the front. This form gives the hyoid bone an anterior mass, called the *corpus* or body and two large, posteriorly projecting *greater cornua* or horns. On the lateral and superior margins of the corpus are two smaller superior projections, the *lesser cornua*. The hyoid bone is unique in that it has no articulations with other bones.

Function of the Hyoid Bone

Muscle Attachments

The hyoid bone serves as an attachment for many of the muscles of the head and neck, including the thyrohyoid muscle, extrinsic to the larynx. Other muscle with hyoid bone attachments include the *digastric, sternhyoid, omohyoid, stylohyoid, hyoglossus, genioglossus, mylohyoid, geniohyoid,* and *middle pharyngeal constrictor*. Through these muscle attachments, the hyoid bone serves as an intermediate between the tongue and the larynx, adding patency to the oropharynx.

Relationship to Tongue and Larynx

The hyoid bone provides inferior support to the tongue and superior support to the larynx. Since the tongue and the larynx are freely moveable, this relationship allows the muscles between them to function in various ways. For example, they can pull the larynx up, toward the tongue, or they can pull the tongue down, toward the larynx.

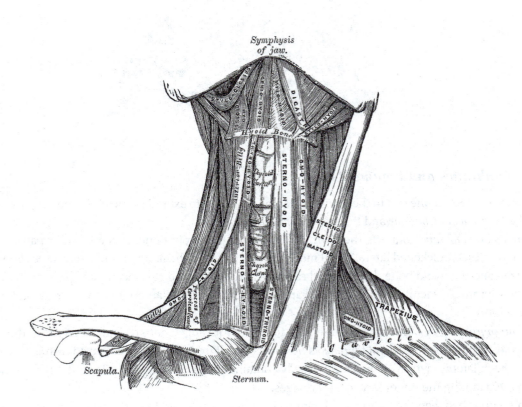

Sternum

Muscular attachments connect the sternum directly and indirectly to the larynx. The sternothyroid and sternohyoid muscles provide inferior support and movement.

Laryngeal Articulations

Skeletal components are linked to each other and to structures outside the larynx with a set of extrinsic and intrinsic ligaments and membranes. These ligaments and membranes have a direct effect on the acoustics of the glottal source.

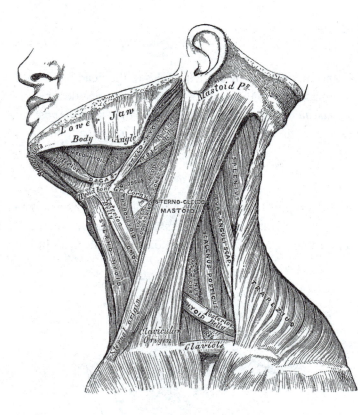

Extrinsic Membranes and Ligaments

Extrinsic ligaments attach at one end to the larynx and at the other end to external structures. There are two extrinsic ligaments: the *thyrohyoid membrane* and the *cricotracheal ligament.*

The *thyrohyoid membrane* connects the thyroid cartilage to the hyoid bone. It is a broad expanse of dense, elastic connective tissue, thickened laterally and medially, and covering the anterior and lateral areas between the thyroid cartilage and the hyoid bone. Like thyroid cartilage and hyoid bone, it is incomplete posteriorly. The thick sections of the thyrohyoid membrane are the median thyrohyoid ligament, anteriorly, and the lateral thyrohyoid ligaments posteriorly.

The *median thyrohyoid ligament* attaches inferiorly to the thyroid angle margins and superiorly to the posterior part of the hyoid corpus. The two *lateral thyrohyoid ligaments* extent from the superior thyroid cornua to the greater hyoid cornua. Each lateral thyrohyoid ligament usually contains a small kernel of cartilage, the *triticeal cartilages.* These are counted among the minor laryngeal cartilages.

The thin sections between the thyrohyoid ligaments are usually referred to as *thyrohyoid membranes.* Bilateral foramina in the thyrohyoid membranes allow the superior laryngeal nerve and the laryngeal blood and lymph vessels to transit.

The *cricotracheal ligament* connects the inferior border of the cricoid cartilage to the superior border of the first tracheal ring. The flexibility of the cricotracheal ligament and of the other respiratory cartilages allows movement of these structures during breathing and as the individual assumes various postures.

Intrinsic Membranes and Ligaments

Intrinsic membranes ligaments attach to larynx at both ends. To a certain extent, the membranes extending within the larynx from inferior to superior are all parts of one expanse of dense connective tissue, with a few gaps in the ventricular region. However, parts of this expanse of connective tissue have been named to facilitate study.

Cricothyroid Membrane

The *cricothyroid membrane* is, by far, the most important intrinsic laryngeal ligament, as far as the study of speech is concerned. As its name suggests, its main attachments are the cricoid cartilage, below, and the thyroid cartilage, above, but its complex morphology and additional attachments make it difficult to apprehend. For example, the cricothyroid membrane includes the *conus elasticus* and the *vocal ligament*, making it essential to normal phonation.

The *conus elasticus* is a complex formation of connective tissue, serving dual roles as a connector of cartilages and as the vocal ligament, itself. Its great importance to speech results not only from the fact that its upper, free border is the *vocal ligament*, but also that its funnel-like shape facilitates the aerodynamics of the phonatory mechanism. It is, perhaps, most properly conceived as a membrane, because it is thin over most of its area.

Conus elasticus might be best visualized as having three parts, one anterior and two lateral. Anteriorly, it is thickened. Here, it is referred to as the *anterior* or *middle cricothyroid ligament*. This portion is readily visible in dissection in the anterior space between the two cartilages, and forms the anterior portion of the conus elasticus. Superiorly, it attaches to the inner part of the thyroid cartilage near the angle of the laminae. Two lateral parts of the conus elasticus, left and right, extend back into the inside of the arch formed by the thyroid cartilage laminae, but are not attached to it. These lateral aspects of the conus elasticus have also been called the *cricovocal membrane*, implying study independent of the rest of the cricothyroid membrane. Below, the conus elasticus attaches to the inner aspect of the cricoid arch. It extends upward as the cricovocal membrane, inside the arch formed by the thyroid cartilage laminae, and attaches posteriorly to the vocal processes and bases of the arytenoid cartilages and anteriorly to the anterior cricothyroid ligament. The superior border is free, and, as stated earlier, forms the paired vocal ligaments.

The *vocal ligaments* form the inner connective tissue framework of the vocal folds. In front, their two ends are attached to the inner aspect of the thyroid cartilage, near the point at which the laminae fuse. There, they have a mutual point of attachment, at the anterior commissure. The posterior attachments of the vocal ligaments are to the vocal processes of the two arytenoid cartilages. Since the articulation between the cricoid and arytenoid cartilages is diarthrodial, the vocal folds can move freely at their posterior ends. Anteriorly, they are fixed to the inner thyroid angle.

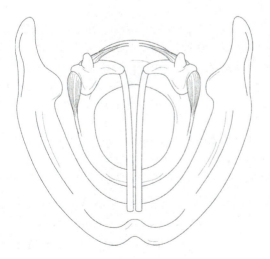

Attached to thyroid cartilage anteriorly, and to the arytenoid apices posteriorly, the *vestibular ligament* may be considered the superior analog to the vocal ligament, except there is nothing analogous to the conus elasticus at its level. The vestibular ligament forms the infrastructure of two weaker ventricular (vestibular) folds.

The sheet of inner laryngeal connective tissue continues upward to form the thyrohyoid membranes discussed earlier. When the connective tissue lining reaches the upper levels, it is often referred to as the *quadrangular membrane*.

Other intrinsic ligaments support articulations between the thyroid and cricoid cartilages. These ligaments are called the *ceratocricoid ligaments* (anterior, lateral, and posterior). There are also other ligaments that support the articulation between cricoid and arytenoid cartilages. These are called the anterior and the posterior *cricoarytenoid ligaments*.

Laryngeal Cartilage Articulation

An understanding of how the laryngeal cartilages move in relationship to one another is fundamental to an appreciation of how the larynx performs its biological and speech functions. These cartilage articulations create changes in the tension and length of the vocal ligaments, in the area of the glottal aperture, and modifying the opening of the laryngeal inlet.

In particular, the relationships of the thyroid, cricoid, and arytenoid cartilages can be confusing to the student. Understanding that the arytenoid cartilages are attached to the cricoid cartilage posteriorly and that the thyroid cartilage partially surrounds the cricoid and arytenoid cartilages is the first step in appreciating those relationships. Students must then realize that one vocal ligament is attached posteriorly to the vocal process of each arytenoid cartilage in the rear and to the inside of the thyroid cartilage angle in front. Thus any movements of thyroid, cricoid, or arytenoid cartilages must have effects on the vocal ligaments, vocal folds and, hence, the voice.

Cricothyroid Articulation: Two-Way Movements

Rotation about the horizontal axis changes tension of vocal ligaments, assuming the arytenoid cartilages are fixed by action of intrinsic muscles. Modifying vocal ligament tension changes the perceived pitch of the voice by altering the mass per unit length of the vibrating bodies. If there is greater tension on the vocal ligament, it will vibrate faster and increase the pitch in a manner similar to that of increasing the tension on a guitar string by cranking the tuning pegs at the head. What actually happens is the ligament becomes thinner at any given portion of its extent. Thinner means less mass and increased velocity, given a constant driving force.

Gliding anteriorly and posteriorly also changes vocal ligament tension. There is a little room at the cricothyroid joint for such movement as a means of changing glottal frequency.

Cricoarytenoid Articulation: Three-Way Movement

The arytenoid cartilages articulate in a surprising variety of directions with respect to the cricoid cartilage. This smooth and free movement allows the posterior ends of the vocal folds to assume a corresponding variety of postures to perform their physiological and vocal functions. The physiological function of the vocal folds is opening and closing of the airway. To accomplish the physiological function of opening and closing the airway, the arytenoid cartilages bring the posterior ends of the folds to and from the midline through contractions of the appropriate muscles.

Varied arytenoid movements also bring about the more subtle modifications of the glottal aperture required to vary phonatory source production as the speaker desires.

Arytenoid movements with respect to each other and with the cricoid cartilage include:

Rotation: when the bases of the arytenoid cartilages rotate about an eccentric axis from their bases to their apices they cause the vocal processes to move to and from the midline.

Gliding or Sliding: The arytenoid cartilages can slide laterally/medially and anteriorly/posteriorly across the superior surface of the cricoid lamina. These movements would necessarily include some lateral rocking as the arytenoid cartilages move along the convex cricoid facet and also some superior/inferior movement as the arytenoid cartilages glide along the sloping cricoid arch surface.

Tilting: The cartilages can tilt medially at the tops (apices) bringing the tips of the pyramids together or apart.

These movements can occur in concert, to varying degrees, and allow for the great flexibility of the human glottis. Wang (1998) found that vocal process movement during adduction and abduction was mainly lateral and vertical, whereas muscular process movement was anterior and posterior with a slight degree of superior/inferior and medial/lateral excursion varying among his sample of five excised larynges.

The cricoarytenoid joint is a true synovial joint with an articular capsule and ligament support. Movement at the cricoarytenoid articulation is limited by a complex system of connective tissue. The vocal ligament and the conus elasticus limit posterior arytenoid displacement and abduction. A cricoarytenoid ligament also limits abduction along with anterior and superior displacement. Anterior and posterior capsular ligaments limit movements in the anterior/posterior and medial/lateral dimensions (Wang, 1998).

Glottis

The glottis is the space between the vocal folds. Sounds created through actions of the vocal folds are called *glottal*, and the sound of the voice is called the *glottal* or *phonatory* source. Since a space may be defined by its boundaries, the vocal folds, themselves, are regarded when considering the anatomy and physiology of the voice producing mechanism.

The glottis is considered to have two distinct regions determined by the underlying skeletal structure. The skeleton of the *cartilaginous glottis* is formed by the vocal processes of each arytenoid cartilage. This part of the glottis occupies about one-third of its length at the posterior end. Compared to the remaining, anterior part of the glottis, the cartilaginous glottis is relatively stiff, especially when the folds are adducted. There is very little movement of the cartilaginous glottis during phonation in normal voice registers. The anterior two-thirds of the glottis is the *membranous glottis*. Its skeletal infrastructure is the vocal ligament. Normal glottal opening and closing during phonation takes place in the membranous glottis.

Point of Maximum Displacement

During phonation, one might expect the glottis to open widest at the midpoint of the membranous glottis, since the vocal folds have the least structural support there. Considering the vocal ligaments as single mass beams, they are supported at their anterior ends by the thyroid cartilage and at their posterior ends by the arytenoid cartilages, but they have no support in the middles. Thus, they will naturally be displaced farthest during phonation in their middle, all things being normal. This widest opening is called the *point of maximum displacement* and ranges in width from .15 to .8 mm during the phonatory cycle (Zemlin, 1998; Gunther, 2003).

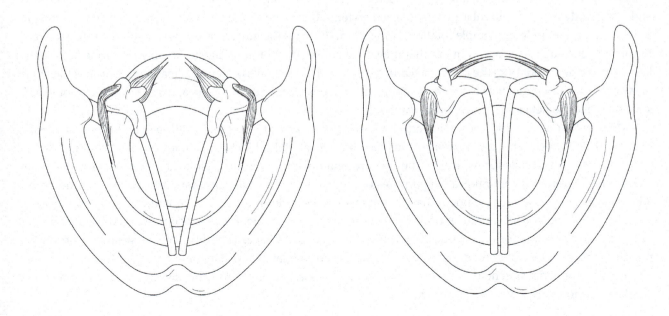

Studies of vocal fold dynamics during phonation reveal that the vocal fold membranes are far more complex than a single mass approach would explain. Rather, as many as sixteen interacting or "lumped" masses making up the membranes have been considered as forming the vibrating bodies of the vocal folds (Gunther, 2003).

Aryepiglottic Articulation

The opening of the laryngeal inlet (additus) is modified when the epiglottis and arytenoid apices are brought closer together. During this articulation, the epiglottis is drawn posteriorly and inferiorly, assuming a horizontal attitude. Simultaneously, the arytenoid apices are drawn medially and anteriorly, so that the entire laryngeal inlet is drawn around the epiglottis, sealing the vestibule from unwanted intrusion. Aryepiglottic articulation occurs most commonly during deglutition (swallowing) when it is necessary to pass food or liquid into the esophagus.

Muscles of the Larynx

Laryngeal muscles may be considered as intrinsic or extrinsic. *Intrinsic* laryngeal muscles have origins and insertions within the body of the larynx, whereas *extrinsic* muscles have one end attached to structures outside the larynx.

Intrinsic Muscles

Intrinsic laryngeal muscles are attached to laryngeal structures at both ends. These muscles perform the internal laryngeal functions, including abduction and adduction of the vocal folds, modulation of vocal ligament tension, and modification of laryngeal inlet dimensions. There is some functional overlap between muscles in the first two categories. All are paired except one, the transverse arytenoid muscle.

Muscles that Abduct and Adduct the Vocal Folds

Vocal folds abduct to open the airway during inspiration and expiration and adduct for physiological and communicative purposes. Abduction and adduction is accomplished by moving the arytenoid cartilages, attached to the posterior parts of the vocal folds, either toward or away from the midline.

The only muscles that abduct the vocal folds are the paired *posterior cricoarytenoid* muscles. These originate at a relatively broad area on either side of the cricoid lamina. Their fibers course superiorly and laterally to converge and insert at the narrow muscular process of each arytenoid cartilage. Contraction of the posterior cricoarytenoid muscles creates more than a simple rotational movement. It draws the muscular process backward and downward toward the cricoid lamina, a movement that draws the entire arytenoid cartilage up the incline formed along the lateral cricoid arch and rocks the arytenoid cartilage outward to the limits of the joint ligaments. This has the effect of swinging the arytenoid vocal processes, along with the posterior ends of the vocal folds, upward and outward, spreading like the petals of a flower on inspiration.

Muscles that adduct the vocal folds are the *lateral cricoarytenoid, transverse arytenoid,* and *oblique arytenoid* muscles. Additionally, the *thyroarytenoid* muscles assist in glottal adduction. Employed in unison and at full strength, they provide an effective seal for the lower respiratory tract. Of course, glottal adductors also approximate the vocal folds for their communicative purposes of phonation and glottal consonant articulation. The degree of adductive tension for speech purposes changes with variations in desired acoustic amplitude.

The *lateral cricoarytenoid* muscles oppose the posterior cricoarytenoid muscles. They arise on the lateral border of the cricoid arch, anterior to the articular facet for the thyroid cartilage. Their fibers converge and insert on the muscular process of each arytenoid cartilage. Contraction of the lateral cricoarytenoid muscles has the opposite effect of posterior cricoarytenoid contraction. Muscular processes are drawn forward, swinging the vocal processes medially and inferiorly, effecting glottal closure.

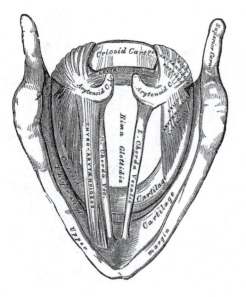

The *transverse arytenoid* muscle is the only unpaired muscle in the intrinsic group. Since its ends are attached to moveable structures at both ends, the distinction of origin and insertion does not apply. Fibers attach at either end to the posterior surfaces of the muscular process of an arytenoid cartilage. Contraction of transverse arytenoid muscle fibers draws the arytenoid cartilages medially, across the superior surface of the cricoid lamina.

The paired *oblique arytenoid* muscle fibers form a thin muscle that stretches from the base of one arytenoid cartilage to the apex of the other. They cross each other and are superficial to the transverse arytenoid muscles. Contraction of oblique arytenoid fibers tilts the apices of the arytenoid cartilages together.

Muscles that Modulate Tension/Length of the Vocal Ligament

Changes in tension and length of the vocal ligaments and, hence the vocal folds, bring about changes in glottal cyclic frequency and amplitude. Tension is increased when muscular forces stretch the ligament and when increased tonus stiffens glottal tissues. Increasing loosely adducted vocal fold length and tension increases glottal cyclic frequency. Vocal fold tension may also increase to resist rising subglottal pressures that accompany phonatory amplitude increases, this in accompaniment with increasing medial compressive force. The increased tension gives rise to the pitch increases that sometimes accompany amplitude increases. Muscles that modulate tension and length of the vocal ligaments are the *cricothyroid* and *thyroarytenoid* muscles.

As their name indicates, the *cricothyroid* muscles originate on either side of the cricoid arch and insert on the thyroid laminae on each side. It should be easy to understand that such attachments mean that shortening the muscles must being their points of attachments together. What may not be easily as appreciated it that contracting the cricothyroid muscles actually moves the cricoid and thyroid cartilages apart.

The cricothyroid muscle fibers insert on the thyroid cartilage in two sections. One section, called the *"Pars Recta"* has a nearly vertical course and inserts on the lower part of the posterior thyroid lamina. The other fiber group has an oblique course and inserts on the inferior thyroid cornu. This group is called the *"Pars Oblique."*

The cricothyroid muscles act on both the cricoid and the thyroid cartilages, moving each in relationship to the other, depending upon which one of the two is more moveable, by virtue of contractions of other muscles at the time. When the pars recta contracts, it rotates the thyroid and/or cricoid cartilages about a transverse axis through the cricothyroid joint. When the pars oblique contracts, it pulls the thyroid cartilage forward in relationship to the cricoid. Both actions move the inside of the thyroid cartilage away from the cricoid lamina.

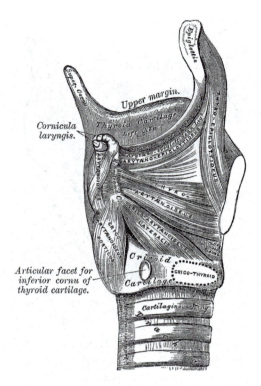

Since the vocal ligament is firmly attached the inside of the thyroid cartilage at one end, and to the arytenoid cartilages at the other end, and the arytenoid cartilages are attached to the cricoid cartilages, it should be apparent that both actions, assuming a fixed relationship between the arytenoid and cricoid cartilages, increase vocal ligament tension. For this reason, and because it is located outside the laryngeal cavity, the cricothyroid muscle is sometimes called the "external tensor."

Thyroarytenoid muscle fibers originate at several sites: anteriorly and laterally, along the inner aspect of the thyroid laminae, from their fusion and along the sides, also from the cricovocal membrane, medially. Insertion is on the antero-lateral aspect of the arytenoid cartilage bases.

The thyroarytenoid muscles oppose the cricothyroid muscles by drawing the thyroid cartilages closer to the arytenoid cartilages and relaxing tension on the vocal ligament. Paradoxically, they can also stiffen and thus increase vocal fold tension, if the arytenoid cartilages are held in place atop the cricoid lamina and if there is equal tension imposed by the contracted cricothyroid muscles. For this reason, and because they lie inside the laryngeal cavity, they are sometimes called the "internal tensors."

The thyroarytenoid muscles have a distinct medial bundle of fibers at their medial aspects, where they attach to the cricovocal membrane. These medial fibers form muscle masses that are sometimes called the *"vocalis muscles"* or the *"medial (or internal) thyroarytenoid muscle."* They can be contracted or relaxed independent of the lateral thyroarytenoid muscles.

Muscles that Modify Laryngeal Inlet Dimensions

The laryngeal inlet, also called the *vestibule* or *additus*, is the entrance to the lower airway. It serves as the "fork in the road" to separate the digestive tract from the respiratory tract. The airway must be protected from intrusion of food, liquid and the like, for such intrusion could, at worst, be fatal. Around the inlet opening is a loose fold of epithelial tissue covering an infrastructure of cartilage and thin muscles, the *aryepiglottic, thyroepiglottic,* and *vestibular* muscles. The fold is called the aryepiglottic fold, since it covers arytenoid cartilages (along with the corniculate and cuneiform cartilages) and epiglottis.

During regular breathing, the opening is almost vertical, sloping steeply back from the higher epiglottic tip to the lower arytenoid apices. During the passage of food into the esophagus, the larynx is drawn superiorly, while the epiglottis is pulled inferiorly into a horizontal attitude. Simultaneously, the arytenoid cartilages are drawn together and approximate the epiglottis. This closes the laryngeal inlet in a manner somewhat like cinching the opening of a sack.

The *aryepiglottic* muscles, as their name indicates, are embedded within the aryepiglottic fold. They are actually continuations of the oblique arytenoid muscles and draw the aryepiglottic fold together in a sphincteric action to close the laryngeal inlet.

The *thyroepiglottic* muscles are actually continuations of the thyroarytenoid muscles, and extend from the thyroid cartilage to the epiglottis. They oppose the aryepiglottic muscles and widen the laryngeal inlet.

Within the flaccid ventricular folds are thin *ventricular* muscles. These can stiffen and adduct the ventricular folds slightly.

Extrinsic Laryngeal Muscles

Extrinsic muscles of the larynx attach at one end to laryngeal structures and at the other end to the soft palate, temporal bone, hyoid bone, or sternum. These muscles adjust the position of the larynx in relationship to the skull and sternum. The extrinsic laryngeal muscles are the *palatopharyngeus, stylopharyngeus, thyrohyoid,* and *sternothyroid* muscles and the *inferior pharyngeal constrictor.*

The *palatopharyngeus* muscles originate on the posterior border of the hard palate and the palatine aponeurosis within the substance of the velum. Most of its fibers pass inferiorly to insert into the posterior margin of the thyroid cartilage, while others insert in the pharyngeal wall. Contraction of its fibers, assuming a fixed velum, draws the larynx and pharyngeal tube superiorly. Alternately, with the larynx fixed by contraction of other extrinsic muscles, the palatopharyngeus muscle draws the velum inferiorly, coupling the nasal cavity to the vocal tract. The palatopharyngeus forms the muscular infrastructure of the *posterior faucial arch,* a prominent landmark of the posterior oral cavity.

The *stylopharyngeus* muscle originates at the styloid process of the temporal bone. Its fibers course downward and anteriorly to insert in the thyroid cartilage with fibers of the palatopharyngeus muscle. Some fibers insert into the pharyngeal wall. Contraction of stylopharyngeus fibers pulls the thyroid cartilage upward and posteriorly, toward the styloid process.

The *thyrohyoid* muscles attach at one end to a ligamentous oblique line sloping downward and anteriorly on the posterior part of each thyroid lamina and at the other end to the hyoid bone corpus. Since both the thyroid cartilage and the hyoid bone are moveable, the muscle can either depress the hyoid bone or elevate the larynx, depending upon which structure is fixed by contraction of other muscles.

The *sternothyroid* muscles originate on the manubrium of the sternum, on its posterior surface, and insert on the thyroid cartilage, just below and sloping downward along the same oblique line that serves as the inferior attachment of the thyrohyoid muscle. Contraction of sternothyroid muscles draws the thyroid cartilage downward in the neck.

The *inferior pharyngeal constrictor* is a dual band of muscle fibers connecting the larynx and the rest of the lower pharyngeal tube. Its upper fibers attach at one end to the thyroid cartilage, near the oblique line, and to a connective tissue band over the cricothyroid muscle. Lower fibers attach to the sides of the cricoid cartilage. The other ends of the fibers attach to the median pharyngeal raphe, with the upper fibers ascending posteriorly at an angle, and the lower fibers passing straight back.

Extra-Laryngeal Muscles

Extra-laryngeal muscles do not attach directly to the larynx, but control the posture of the larynx in relationship to the other structures of the head and neck by displacing the hyoid bone or pharyngeal constrictors. These muscles are properly grouped as muscles of the anterior neck. They are relevant to study of the larynx because of their roles

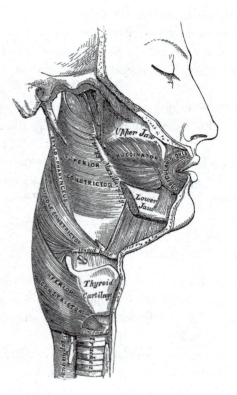

in swallowing and in pharyngeal resonance, accomplished by moving the larynx up and down, allowing swallowed material to pass harmlessly over the epiglottis and changing the dimensions of the vocal tract resonator.

Extra-laryngeal muscles include a *suprahyoid* group, with attachments to structures above the hyoid bone, and an *infrahyoid* group, with attachments below the hyoid bone.

Suprahyoid Muscles

The *suprahyoid* muscles are so-named because they are positioned superior to the hyoid properly considered intrinsic muscles of the neck, since their attachments are to skull bones. They are important to laryngeal functioning because their contractions draw the hyoid bone superiorly toward the rest of the skull. When the hyoid bone is elevated, and the muscles that attach the thyroid cartilage to the hyoid bone are fixed, the larynx will elevate with the hyoid bone. The suprahyoid muscles are the *digastric, stylohyoid, mylohyoid, geniohyoid,* and *hyoglossus* muscles.

The *stylohyoid* muscles originate at the styloid process of the temporal bone and insert into the hyoid corpus at its posterior edge. The styloid process is a thin, pointed, bony projection beginning just medial to the external auditory meatus and pointing down and forward, toward the hyoid bone. Contraction of stylohyoid fibers draws the hyoid bone posteriorly and superiorly during swallowing, moving the larynx with it.

The *digastric* muscles, as their name implies, have two bellies. The anterior belly of each digastric muscle originates at the inner surface of the mandible` on either side of the mental symphysis (chin). Its fibers course posteriorly in relatively narrow bundle to an intermediate tendon, at which point the posterior belly inserts. The posterior belly originates on the temporal bone, deep to the mastoid process. A loop of connective tissue connects the intermediate digastric tendon to the posterior end of the hyoid corpus, near the insertion of the stylohyoid muscle. Simultaneous contraction of the sling formed by the two digastric bellies lifts the hyoid bone and depresses the mandible against resistance.

The *mylohyoid* muscles are flat and triangular, forming a muscular floor to the oral cavity. Their name, like the name "molar," is derived from the Greek word, *myle,* meaning "mill," and refers to the mouth, where food is ground. The two mylohyoid muscle origins occupy the anterior four-fifths of a horizontal ridge, the *mylohyoid ridge,* on the inner surface of the mandibular body. They insert on the anterior surface of the hyoid body and on a median raphe, which unites the two muscles in the midline. Contraction of the mylohyoid muscles draws the hyoid bone up toward the mandible or draws the mandible down toward the hyoid bone. They can also elevate the oral cavity floor and its contents during swallowing.

The *geniohyoid* muscles originate at the inferior mental spine, or *geneion*, on the inner surface of the mandible opposite the chin. Their fibers insert on the anterolateral surface of the hyoid corpus. Contraction of the geniohyoid muscles draws the hyoid forward and up or depresses the mandible.

The *hyoglossus* muscles connect the hyoid bone and tongue. Their fibers form a flat quadrilateral, and originate on the hyoid bone, where the greater cornua and corpus meet. Hyoglossus fibers insert on the lateral aspects of the tongue and draw the two structures together when contracted. The hyoglossus muscle is an important anatomical landmark, for the lingual and hypoglossal nerves, as well as the hypoglossal salivary duct, pass across it superficially, and the glossopharyngeal nerves, lingual arteries and veins pass deep to its posterior edge.

Infrahyoid Muscles

The *infrahyoid* extra-laryngeal muscles are the *sternohyoid* and *omohyoid* muscles. So-named for their attachments inferior to the hyoid bone, they displace the larynx by drawing the hyoid bone inferiorly.

The *sternohyoid* muscles have their origins on the posterior surfaces of the manubrium of the sternum and on the medial ends of the clavicles. They pass upward to insert on the inferior border of the hyoid body. Contraction of sternothyroid fibers depresses the hyoid bone.

The *omohyoid* muscles, whose name comes from the Greek word *omos* (shoulder), have two bellies, one superior and the other inferior. The shoulder reference applies the attachment of the inferior belly. The superior omohyoid belly attaches at its superior end to the hyoid bone, low on the corpus. The other end of the superior belly is a connective tissue intermediate tendon. This tendon attaches to the medial end of the clavicle and first rib with a band of connective tissue. The inferior omohyoid belly's fibers course laterally and posteriorly, creating an obtuse angle between the two sections of the muscle. The inferior belly attaches at its posterior end to the scapula, in its inner surface, near the scapular notch. Contraction of the omohyoid fibers depresses the hyoid bone.

Suprahyoid and infrahyoid musculature acts on the hyoid bone and thyroid cartilage in a reciprocal manner. Muscles that attach to the hyoid bone can either elevate or depress it, and if the thyrohyoid muscles fix the thyroid cartilage in relationship to the hyoid bone, move the larynx with it. Alternatively, the same muscles can fix the hyoid bone, while the extrinsic laryngeal muscles move the larynx in relationship to the hyoid bone.

Neurology of the Larynx

The larynx receives its innervation from the *superior* and *recurrent laryngeal nerves*, both branches of *vagus nerve*, with assistance by the *accessory nerve*. The superior laryngeal nerve also carries fibers originating in the superior cervical sympathetic ganglion.

Two *superior laryngeal nerves* emerge from the inferior ganglia of the vagus nerve at the medulla oblongata and course inferiorly on either side and external to the pharynx. At the level of the greater hyoid cornua, the superior laryngeal nerves split into external and internal branches. They enter the laryngeal cavity through the thyrohyoid membrane's foramina, along with laryngeal blood and lymph vessels.

The *recurrent laryngeal nerves* branch from the vagus nerves in the thorax, as they descend to the levels of the subclavian artery on the right, and the aortic arch on the left. They loop posteriorly, below and around these arteries, and then ascend between the trachea and the esophagus to enter the laryngeal cavity near the cricothyroid joint. As they ascend, the recurrent nerves are in close proximity to the inferior thyroid arteries.

The relationship between the recurrent laryngeal nerves and neighboring arteries is of significance to practitioners. Abnormalities, such as aneurysms, can impinge upon these nerves and cause paresis of the intrinsic muscles. Thyroid surgery can also affect recurrent nerve functions. Thus, persistent dysphonia may be a sign of arterial anomalies, and patients having this complaint should be seen by a physician.

Motor Innervation

Laryngeal motor innervation has both autonomic and voluntary function. The distinction important, as most speech is a voluntary function. Autonomic functions such as coughing or involuntary airway closure is a function of the efferent vagus nerve fibers (cranial nerve X), combined, as mentioned above, with fibers from the cervical sympathetic ganglion on each side. Voluntary muscle function is accomplished through the action of fibers of the internal branches of the bilateral spinal accessory nerves (cranial nerve XI). These fibers emerge from the accessory trunks of the spinal accessory nerves and join the vagus nerves at the jugular foramina, traveling with them to be distributed with the superior and recurrent laryngeal nerves.

Most of the intrinsic laryngeal muscles receive their motor innervation from the recurrent laryngeal nerves. Only the cricothyroid muscles receive their motor innervation from the external laryngeal nerves, branches of the superior laryngeal nerves.

Sensory Innervation

Action potentials to be interpreted as somesthesis (touch) are conveyed from the larynx by the superior and recurrent laryngeal nerves. Those originating in laryngeal cavity from the vestibule to the level of the vocal folds are conveyed the internal divisions of the superior laryngeal nerves. These also convey sensations from the dorsum of the posterior tongue. Those originating in the inferior laryngeal division, below the vocal folds, are conveyed by the recurrent laryngeal nerves.

Developmental Anatomy of the Larynx

The endodermal primitive foregut, also known as the *laryngotracheal groove*, appears at three to four weeks of gestation as a ventral depression at the rostral end of the embryo and develops caudally. It will become the lower respiratory system, including the inferior part of the larynx, and the rostral part of the alimentary system. Laryngeal structures appear as rostral respiratory structures a short time later in the forms of two lateral *arytenoid swellings*, followed shortly thereafter by an *epiglottic swelling*. These form the primitive laryngeal additus. As development continues, these swellings and their adjacent mesenchyme will form, as their names suggest, the arytenoid, corniculate, and cuneiform cartilages and the epiglottis.

Tissue folding and fusion from the laryngotracheal groove forms a separation along the foregut, the *tracheo-esophageal septum*, which separates the respiratory and alimentary systems during the fourth and fifth weeks. Incomplete morphology of this septum, forming a *laryngotracheoesophageal cleft*, results in aspiration of digestive materials.

As the embryo develops, epithelial tissue proliferates and completely blocks the laryngeal opening. At about the tenth gestational week, this tissue disappears in a process called *recanalization*, and in so doing, forms the laryngeal cavities. Failure of the epithelial tissue to recanalize is the cause of *laryngeal stenosis* and *laryngeal webs*.

Branchial Arches, Clefts, and Pouches

Laryngeal structures develop from the second through sixth branchial arches. Skeletal support for the larynx, including the styloid process of the temporal bone, origin of the stylohyoid muscles, the stylohyoid muscle, itself, the stylohyoid ligament, which attaches the stylohyoid muscle to the lesser hyoid cornua, and the superior part of the hyoid corpus are derivatives of the second (hyoid) branchial arch.

The third branchial arch gives rise to the remainder of the hyoid bone, including its lower corpus and greater cornua.

The fourth branchial arch will develop into supraglottal laryngeal structures. The thyroid cartilage develops from the fourth arch. There remains some controversy about the embryonic source of the epiglottis, but most

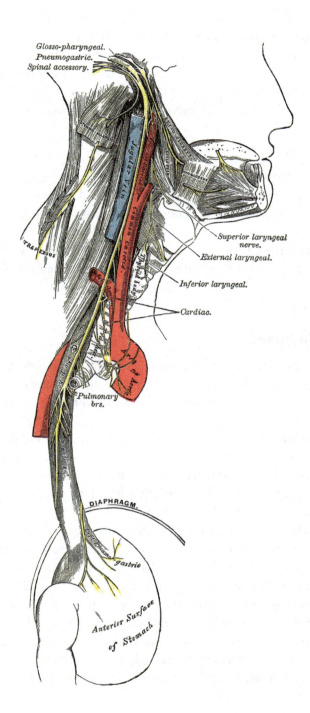

authorities hold that the epiglottis comes from the fourth arch (i.e., Brookes and Zietman, 1998). Some evidence exists that the epiglottis forms from mesenchyme not originally part of that arch (Larsen, 2001). The three pharyngeal constrictors, levator palatini, and cricothyroid muscles also develop from the fourth branchial arch.

From the sixth branchial arch develop glottal and subglottal structures. These include the cricoid, arytenoid, corniculate, and cuneiform cartilages. Muscles derived from the sixth branchial arch originate from mesenchyme of the occipital somites, and include the lateral and posterior cricoarytenoid muscles, the thyroarytenoid and vocalis muscles.

The nerves of branchial arch laryngeal sources are the glossopharyngeal nerve (cranial nerve IX) and the vagus nerve (cranial nerve X). The glossopharyngeal nerve is the nerve of the third branchial arch and supplies sensory innervation to pharyngeal structures along with the vagus nerve. Structures deriving from the fourth and sixth branchial arches share innervation from the vagus and spinal accessory nerves. Fourth arch structures receive fibers from the superior laryngeal nerves and sixth arch structures receive fibers from the recurrent laryngeal nerves.

Ontogeny of the Larynx

The larynx undergoes dimensional and structural changes during life, resulting in changes in the sound of the phonatory source and in its function as a respiratory structure. These changes result from several factors, including normal growth patterns, hormonal influences, environmental stresses, decreases in tissue plasticity, and weakening of the musculature.

During the embryonic stage, the larynx grows rapidly, occupying a relatively large space at the rostral end of the developing respiratory system by the time of birth. At birth, the configuration of the skull and neck are, in some ways, similar to those of a chimpanzee. The larynx is situated at approximately the level the C-4 vertebra, and the hyoid bone overlaps the thyroid cartilage.

Laryngeal Differences In Infancy

The human infant has the ability to juxtapose the epiglottis and the velum, effectively separating the oral cavity from the respiratory tract. This enables the infant to nurse and breathe simultaneously, as the larynx separates the oropharynx into two channels, allowing milk to pass around it and protecting the respiratory tract from aspiration of liquids (Crelin, 1987).

The larynx gradually increases in size and descends in the cervical region. At birth, the cricoid cartilage is located at the level of vertebrae C-3 and C-4, whereas it is located at C-4 and C-5 in an adult. The laryngeal cartilages and hyoid bone are much closer approximated in the infant than they are in the adult, the thyroid and cricoid cartilages nearly overlapping. The result is an angled relationship of the vocal folds to the horizontal plane of the thyroid cartilage and a ventral curve to the laryngeal cavity.

Being composed mostly of hyaline cartilage, the laryngeal skeleton can be quite plastic at birth. During the early months of life, the laryngeal framework may be so flexible as to incompletely support the patency of the vestibule or ventricle. This condition, called *laryngomalacia*, or *soft larynx*, is fairly common. In some cases, the result is a disturbing but benign *laryngeal stridor*, a "wheeze" on inspiration, most audible during inhalation and while supine, and lasting until the framework becomes more rigid with age, usually by the second year.

Growth rate decreases after birth until puberty, when hormonal influences that cause growth changes increase again. By age six, the larynx has moved so far inferiorly that the epiglottis can no longer articulate with the velum. The result is a more flexible vocal tract, enabling the individual to produce a wide variety of phones. The inferior laryngeal migration also creates a longer common passageway for respiration and deglutition, enabling food and liquid greater opportunity to enter the airway.

Puberty

Dimensional changes during puberty include increases in the mass and length of the vocal folds and decreases in the angles of the thyroid cartilage laminae. Changes in skeletal dimensions and increases in the overall size of the larynx make the laryngeal prominence more noticeable in the anterior cervical region, especially in males. With the dimensional increase in the length of the vocal folds comes an increase in their mass and with that, an attendant reduction in velocity of the folds during phonation and lowering of glottal frequency.

Pubescent laryngeal growth is especially marked in males, and is associated with growth and development of the testes. In males castrated during childhood, dimensional changes in the larynx do not occur, and there is much less decrease in glottal frequency during phonation. After middle age, decreases in hormonal influence may cause some reversal of the pubescent morphogenesis.

Environmental factors that change laryngeal morphology include airborne pollutants, lack of sufficient hydration and ingestion of certain medications. The same environmental stresses associated with aging of lower respiratory organs may also be associated with changes in the structure of laryngeal tissues. These stresses include inhalation of airborne substances, including urban fumes and pollutants, tobacco smoke, as well as indoor and outdoor air pollutants. The effects of these substances vary with the individual, but generally, and over time, compromise cellular structure and function. In particular, cilia at the tips of columnar epithelial cells may be paralyzed by tobacco smoke rendering them ineffective for conveying particles suspended in mucosal secretions toward the pharynx. This causes frequent coughing, as the smoker reflexively tries to move the irritants from the respiratory passages, including the larynx, through explosive pulses of air from the lungs. Of course, other airborne substances may cause the same changes in laryngeal function, depending upon their composition. Chronic exposure to irritants also results in inflammation of laryngeal tissues and the increased vocal fold mass causes changes in the frequency of vocal fold vibration with the attendant lowering in perceived voice pitch. Another environmental factor influencing laryngeal function is hydration. In some patients, failure to drink sufficient water results in changes to tissues in general, including those of the larynx. Changes associated with dehydration include loss of flexibility and atrophy.

Long-Lived Larynx

Laryngeal cartilages undergo calcification during life, resulting in a gradual stiffening of the structure. In the long-lived individual, hardening of laryngeal cartilages can reduce the plasticity of the structure to the extent that articular mobility is restricted. This reduced mobility, coupled with a senescent nervous system, creates incomplete vocal fold adduction and an attendant breathy voice quality known as *presbyphonia*.

Aging also brings gradual weakening of skeletal muscles. Muscle mass and tone decrease with varying results between individuals. Expected changes associated with chronic muscular weakness include incomplete and slowed vocal fold adduction, decreased glottal frequency range, increased shimmer and jitter, and possible inferior displacement of the larynx and hyoid bone in the cervical region. Such inferior carriage results in dimensional changes in the vocal tract and accompanying resonance variations.

CHAPTER SUMMARY

This chapter has described the form and function of the larynx, a complex, highly moveable structure which serves biological and communicative functions. As a series of three rostral respiratory valves, the larynx protects the airway, helps expel unwanted material from the respiratory passages, and helps increase stiffness of the thoracic cavity for such actions as lifting, defecation and childbirth. The strongest, most flexible, and most important of these valves for speech is the glottis, formed by two vocal folds at the bottom of the laryngeal ventricle.

Laryngeal speech functions include generation of the quasiperiodic phonatory, or glottal, source for vowels, approximants and nasal phonemes, plus generation of two glottal fricatives and a plosive. Other communicative, non-speech functions, such as grunts and sighs are common and originate in the larynx. Speakers can alter the frequency and amplitude characteristics of the phonatory source to modify suprasegmental features of speech and to sing. Frequency modifications come through mainly through actions of intrinsic laryngeal muscles on the length and tension of the vocal ligament, whereas amplitude variations come through variations of respiratory muscles, including those of the larynx, on air flow.

Two types of epithelial tissue line the laryngeal cavity. Pseudostratified, ciliated columnar cells, interspersed with mucous-secreting goblet cells, form most of the lining. Stratified squamous epithelial tissue is located in locations subject to mechanical wear, such as the medial surfaces of the vocal folds.

The skeleton of the larynx is formed of hyaline and elastic cartilage. Three of these, the epiglottis, thyroid, and cricoid cartilages, are single, midline structures, and the rest are paired. The hyoid bone, a bone of the skull, provides a rigid suspension from above. The mechanical relationships of the thyroid, cricoid, and arytenoid cartilages are critical to laryngeal functioning.

Skeletal components are linked to each other and to structures outside the larynx with a set of extrinsic and intrinsic ligaments and membranes. These ligaments and membranes have a direct effect on the acoustics of the glottal source. One such membrane is the conus elasticus. The thickened upper free border of this membrane is the vocal ligament.

The most prominent laryngeal cartilage is the thyroid cartilage, whose presence as the "Adam's Apple" is usually quite visible in the middle of the neck. The thyroid cartilage articulates with and nearly surrounds the cricoid cartilage. Paired arytenoid cartilages articulate with the top and rear of the cricoid cartilage within, the arch of the thyroid cartilage, to open and close the posterior of the glottal aperture. The vocal ligaments, skeletal infrastructure of the vocal folds, attach to the arytenoid cartilages posteriorly and to the inside of the thyroid cartilage anteriorly. Minor laryngeal cartilages add rigidity to the aryepiglottic folds and lateral thyrohyoid ligaments.

Muscles that create laryngeal movements may be grouped into intrinsic and extrinsic categories. Intrinsic laryngeal muscles abduct and adduct of the vocal folds, modulate vocal ligament tension, and modify laryngeal inlet dimensions. Extrinsic laryngeal muscles adjust the position of the larynx in relationship to the skull and sternum, particularly for swallowing. A set of extra-laryngeal muscles also effects laryngeal movements within the neck.

Motor and sensory nervous supply to the larynx comes from superior and recurrent laryngeal nerves, both of which have fibers originating in the vagus and spinal accessory cranial nerves. The superior laryngeal nerve also carries fibers originating in the superior cervical sympathetic ganglion.

Laryngeal structures develop from the second through sixth branchial arches. Approximately four weeks of embryonic development. During life, the larynx undergoes dimensional and structural changes, resulting in changes in the sound of the phonatory source and in its function as a respiratory structure. These changes result from several factors, including normal growth patterns, hormonal influences, environmental stresses, decreases in tissue plasticity, and weakening of the musculature.

REFERENCES AND SUGGESTED READING

Anatomy: A Review. San Diego, Ca.: Singular.

Awan, S.N., and Mueller, P.B. (1996). Speaking fundamental frequency characteristics of white, African American and Hispanic kindergarteners. *Journal of Speech and Hearing Research, 39*, 573–577.

Boone, D.R. (1971). *The Voice and Voice Therapy*. Englewood Cliffs, N.J.: Prentice-Hall.

Clemente, C.D. (1975). *Anatomy: A Regional Atlas of the Human Body*. Philadelphia: Lea and Febiger.

Crelin, E.S. (1987). *The Human Vocal Tract*. New York: Vantage.

Darley, F., Aronson, A., and Brown, J. (1975). *Motor Speech Disorders*. Philadelphia: W.B. Sanders.

Eckel, H.E., Sittel, C., Zorowka, P., and Jerke, A. (1993). Dimensions of the laryngeal framework in adults. *Surgical and Radiologic Anatomy, 16*, 31–36.

Filho, J.A., Bohadana, S.C., Perazzio, A.F., Tsuji, D.H., and Sennes, L.U. (2005). Anatomy of the cricothyroid articulation: differences between men and women. *Annals of Otology, Rhinology and Laryngology, 114*, 250–252.

Fink, B. R. (1975). *The Human Larynx: A Functional Study.* New York: Raven Press.

Flanagan, J.L. (1958). Some properties of the glottal sound source. *Journal of Speech and Hearing Research, 1,* 99–116.

Fukuda, H., Kawaida, M., Tatehara, T., Ling, E., Kita, K., Ohki, Y., Kawasaki, Y., and Saito, (1988). A new concept of lubricating mechanisms of the larynx. In Fujimura, O. (Ed.). *Vocal Fold Physiology: Vocal Physiology Voice Production, Mechanisms, and Functions.* (Volume 2, pp. 83–92) New York: Ravens.

Gray, S.D., Chan, K.J., and Turner, B. (2000). Dissection plane of the human vocal fold lamina propria and elastin fiber concentration. *Acta-Otolaryngologica, 120,* 89–91.

Gunther, H.E. (2003). A mechanical model of vocal-fold collision with high spatial and temporal resolution. *Journal of the Acoustical Society of America, 113,* 994–1000.

Haji T., Mori K., Omori K., Isshiki N. (1992). Experimental studies on the viscoelasticity of the vocal fold. *Acta Otolaryngologica, 112(1).* 151–159.

Hirano, M. (1975). Phonosurgery: basic and clinical investigations [in Japanese]. *Otologica (Fukuoka), 21,* 129–440.

Hirano, M., and Sato., K. (1993). *Histological Color Atlas of the Human Larynx.* San Diego, CA: Singular.

Hiroto, P. (1966). Patho-physiology of the larynx from the standpoint of vocal mechanism [in Japanese]. *Practa Otologica Kyoto, 59,* 229–292.

Ishii, K., Zhai, W.G., Akita, M., and Hirose, H. (1996). Ultrastructure of the lamina propria of the human vocal fold. *Acta Otololaryngologica, 116,* 778–782.

Johnson, W., Darley, F.L., and Spriestersbach, D.C. (1963). *Diagnostic Methods in Speech Pathology.* New York: Harper and Row.

Kahane, J. C. (1978). A morphological study of the human prepubertal and pubertal larynx. *American Journal of Anatomy, 151,* 11–9.

Kahane, J. C. (1982). Growth of the human prepubertal and pubertal larynx. *Journal of Speech and Hearing Research,* 25, 446–455. Retrieved June 5, 2011 from EBSCO*host*.

Kent, R. (1976). Tutorial: Anatomical and neuromuscular maturation of the speech mechanism: evidence from the acoustic studies. *Journal of Speech and Hearing Research. 19,* 421–427.

Kent, R. D., and Vorperian, H.K. (1995). *Development of the craniofacial-oral-laryngeal anatomy: a review. Journal of Medical Speech-Language Pathology, 3 (3),* 145–190.

Linville, S. E. (2001). *Vocal Aging.* San Diego, Ca.: Singular.

Liss, J., Weisemer, G., Rosenbek, J. (1990). Selected acoustic characteristics of speech production in very old males. *Journal of Gerontology, 45,* 35–45.

Minifie, F.D. (1973). Speech Acoustics. In Minifie, F.D., Hixon, T.J. and Williams, F. (1973). *Normal Aspects of Speech, Hearing, and Language,* pp. 235–284. Englewood Cliffs, N.J.: Prentice-Hall.

Nair, G. (2006). *The Craft of Singing.* San Diego, CA: Plural.

Negus V.E. *The Comparative Anatomy and Physiology of the Larynx.* London: Heinemann, 1949.

Palmer, J.M. (1993). *Anatomy for Speech and Hearing* (4th ed.). Baltimore: Williams and Wilkins.

Robb, M.P., and Saxman, J.H. (1985). Developmental trends in vocal fundamental frequency of young children. *Journal of Speech and Hearing Research, 28,* 421–427.

Sato, K., Hirano, M., and Nakashima, T. (2003). 3D structure of the macula flava in the human vocal fold. *Acta Otolaryngologica, 123,* 269–273.

Sellars, I.E., and Keen, E.N. (1978). The anatomy and movements of the cricoarytenoid joint. *Laryngoscope, 88,* 667–674.

Stevens, K.N., and House, A.S. (1961). An acoustical theory of vowel production and some of its implications. *Journal of Speech and Hearing Research, 4,* 303–320.

Sulter, A.M., and Wit, H.P. (1996). Glottal volume velocity waveform characteristics in subjects with and without vocal training, related to gender, sound intensity, fundamental frequency and age. *Journal of the Acoustical Society of America, 100*, 3360–3373.

Tanner, D.C. (2007). Medical-Legal and forensic Aspects of Communication disorders, Voice Prints and Speaker Profiling. Tucson, Az.: Lawyers and Judges.

Tarnoczy, T. (1951). The opening time and opening-quotient of the vocal folds during phonation. *Journal of the Acoustical Society of America, 23*, 42–44.

Titze, I.R., and Hunter, E.J. (2004). Normal vibration frequencies of the vocal ligament. *Journal of the Acoustical Society of America, 115*, 2264–2269.

Titze, I.R. and Hunter, E.J. (2007). A two-dimensional biomechanical model of vocal fold posturing. *Journal of the Acoustical Society of America, 121*, 2254–2260.

Van den Berg, J. (1958). Myoelastic-aerodynamic theory of voice production. *Journal of Speech and Hearing Research, 1*, 227–244.

Wang, R.C. (1998). Three-dimensional analysis of cricoarytenoid joint articulation. *Laryngoscope, 108, April 1998 Supplement*, 1–17.

Yoffey, J.M. (1976). Respiratory System. In Hamilton, W. (1976). *Textbook of Human Anatomy.* St. Louis, MO: C.V. Mosby.

Yumoto, E., and Kadota, Y. (1998). Pliability of the Vocal Fold Mucosa in Relation to the Mucosal Upheaval during Phonation. *Archives of Otolaryngology, Head and Neck Surgery, 124*, 897–902.

Zemlin, W.R. (1998). *Speech and Hearing Science: Anatomy and Physiology (4th. Ed.).* Englewood Cliffs, N.J.: Prentice-Hall.

IMPORTANT TERMS

Bernoulli Effect: Fluid dynamic principle that dictates that the internal pressure of a liquid medium, such as air, will decrease when pressure imbalances increase its volume velocity. Named for French mathematician Daniel Bernouilli, 1700–1782.

Cannula: A tube that can be inserted in to a body cavity or duct to administer drugs, withdraw fluids, or ensure adequate respiratory ventilation. A laryngectomy tube is a type of cannula inserted after surgery to remove the larynx to maintain an open airway. Another frequently encountered cannula is a nasal cannula, a soft plastic tube connecting a supplemental oxygen supply to a patient's nose.

Cartilaginous Glottis: The portion of the glottis supported by the bases and vocal processes of the arytenoid cartilages. The cartilaginous glottis occupies the approximate posterior one-third of the entire glottis.

Closed Quotient: The duration of the glottal cyclic phase during which the glottis is closed, expressed as a fraction of the entire period.

Conus Elasticus: The part of the cricothyroid membrane that is attached at its base to the upper border of the cricoid arch and that has a free superior border in the thickened vocal ligament. It extends upward inside the arch formed by the thyroid cartilage laminae, attaching posteriorly to the vocal processes and bases of the arytenoid cartilages, and to the median cricothyroid ligament, anteriorly.

Dysphonia Plica Ventricularis: Phonatory disorder wherein the speaker approximates the ventricular (vestibular or "false") folds instead of the vocal folds to create a rough, breathy sound source.

Extra-laryngeal Muscles: Anterior neck muscles that do not attach directly to the larynx, but control the posture of the larynx in relationship to the other structures of the head and neck by displacing the hyoid bone or pharyngeal constrictors, which, in turn, are attached to and displace the larynx.

Extrinsic Laryngeal Muscles: Extrinsic muscles of the larynx attach at one end to laryngeal structures and at the other end to the soft palate, temporal bone, hyoid bone, or sternum. The extrinsic laryngeal muscles are the *palato-pharyngeus, stylopharyngeus, thyrohyoid* and *sternothyroid* muscles, and the *inferior pharyngeal constrictor.*

Falsetto Voice: An abnormally high-pitched voice produced by males by fixing the lateral vocal fold tissues and allowing only the medial tissues to vibrate.

"False" Vocal Folds: Colloquial name for the ventricular folds.

Glossoepiglottic Valleculae and the Pyriform Sinuses: Recesses anterior and lateral to the laryngeal inlet in which solid or liquid material can become impounded during swallowing.

Glottal Pulsing ("Fry"): Mode of phonation in which the glottal frequency is very low, around 20–50 Hz., and listeners can identify individual pulses of air. Known by several names, including vocal fry, creaky voice, laryngealization, pulse register, or combinations of these terms, glottal pulsing is often heard at the end of a declarative utterance in English.

Glottal Fricative: Continuous aperiodic speech sound source used in whispering and produced by forcing pressurized subglottal air through a small posterior aperture in the glottis. A looser glottal posture produces accompanying phonation, producing a voiced cognate, especially in an intervocalic phonetic environment. Phonemic in English, its International Phonetic symbols are /h/ (unvoiced) and /σ/ (voiced).

Glottal Plosive: Transient aperiodic speech sound source produced by sudden release of impounded subglottal air pressure. Glottal plosives, sometimes called "glottal stops," are often observed in English as vowel releasers or as non-speech vocal gestures. The International Phonetic Alphabet symbol is /σ/.

Glottis: The space between the vocal folds; alternatively includes the tissues surrounding the space. The term "glottal" refers to the glottis.

Habitual or Modal Frequency: Glottal frequency to which the speaker returns most often.

"Hertz": Unit of periodicity denoting cycles per second. Named for German mathematician and physicist, Heinrich Hertz, 1857–1894.

Infraglottal (Subglottal) Division: Region of the laryngeal cavity inferior to the vocal folds.

Intrinsic Laryngeal Muscles: Muscles that attach to laryngeal structures only.

Jitter: Variations in cyclic energy frequency.

Laryngeal (Thyroid) Eminence: Anterior protrusion in the mid cervical area created by the fusion of thyroid laminae; most obvious in post-pubescent males.

Laryngectomy: Surgical procedure by which part of, or the entire, larynx is removed.

Laryngomalacia: Softness of laryngeal cartilages, usually resolved spontaneously during the first two years of life.

Laryngopharynx: The most inferior division of the upper airway, also called the hypopharynx.

Macula Flavae: Yellowish regions at either end of the vocal ligaments.

Medial Compression: The muscular force with which a speaker adducts the vocal folds for phonation.

Membranous Glottis: Region of the glottis supported by the vocal processes at the posterior ends and the thyroid cartilage at the anterior end, and with no cartilaginous support in the middle. The membranous glottis occupies about two-thirds of the entire glottis.

Mucous Blanket: Protective and cleansing coat of mucous secretions produced by goblet cells in the respiratory epithelium.

Mucoviscoelastic Principle: Principle of vocal fold function that accounts for vertical phase differences of proximal and distal vocal fold tissues during phonation. The lower, or leading edges of the vocal folds, being affected by air flow first, are slightly out of phase with the upper, trailing, edges (Hiroto, 1966; Haji, Mori, Omori, and Isshiki, 1992; Yumoto, Kadota, 1998).

Myoelastic and Aerodynamic Principle: Principle of vocal fold phonatory function whereby muscle tonus of loosely adducted vocal folds and the Bernouilli effect create quasiperiodic openings and closings of the glottis as long as air flows from the lungs (Van den Berg, 1958).

Optimal Pitch: That glottal frequency that provides the best sustained voice with the least muscular effort.

Phonatory or Glottal Source: Speech sound source created by sustained quasiperiodic variations in subglottal air pressure. The phonatory source is used in vowels, approximants, nasals, and voiced obstruent consonants.

Point of Maximum Displacement: The point along the horizontal plane of the glottis at which vocal fold excursion during phonation is greatest; this point is usually midway along the membranous glottis.

Presbyphonia: Changes in voice accompanying advanced age.

Reinke's Space: superficial lamina propria of vocal folds, mostly loose collagen and elastic fibers bound with matrix.

Shimmer: Variations in cyclic amplitude.

Subglottal: Below the level of the vocal folds.

Supraglottal: Above the level of the vocal folds.

Syllabic Stress: A suprasegmental feature of increased amplitude, duration, and often frequency of a syllable.

Tracheostomy (Tracheotomy): Surgical procedure whereby an opening is created to connect the upper end of the trachea to the outside environment.

Ventricle of Morgagni: Middle division of the laryngeal cavity, bounded superiorly by the ventricular folds and inferiorly by the vocal folds. Named for Italian anatomist Giovanni Battista Morgagni (1682–1771).

Vestibule, (Additus; Superior Division): Superior division of and entrance to the laryngeal cavity.

Lamina Propria: Superficial tissue of the vocal folds (or any organ). Vocal fold lamina propria are composed of three layers of elastic and collagen fibrous connective tissue.

Vocal Folds: Horizontal tissue projections of epithelium, muscle, and connective tissue projecting horizontally into the laryngeal cavity at the inferior extent of the ventricle. Vocal folds are highly mobile posteriorly but fixed anteriorly.

Volume Velocity: Air volume displacement per unit of time.

Whispering: Mode of speech in which the phonatory source is replaced by glottal frication.

CHAPTER 6
Speech Articulation

CHAPTER PREVIEW

The chapter on speech articulation covers what is probably regarded by most as the focus of the speech-language pathologist. This chapter begins with a discussion of articulatory (physiological) phonetics and the acoustic aspects of the speech wave. It progresses to examine the vocal tract in successively more detailed sections, using the skeletal structure of the skull as a foundation, and moving on to cover the form and function of vocal tract tissues, their skeletal foundations, musculature, innervation, and soft tissues. One distinguishing feature of this text is the special examination given to dental anatomy. This feature is important because of the special role dentition plays in speech development and disorders. The articulators are examined in two main groups: those outside the oral cavity and those within the oral cavity. Articulators outside the oral cavity include those on the face and those in the larynx. Facial articulators will be considered as extending beyond labial phoneme articulation and include muscles of facial expression. The peripheral visual system will not be covered in detail beyond the general contents of the orbits and the external muscles of facial expression surrounding their openings. Developmental anatomy and growth of the articulatory mechanism is the topic of later chapter sections.

Sidebars include the chart of the International Phonetic Association, a list of muscles of mastication, maneuvers to clear the Eustachian tubes, a list of muscles of speech articulation, and descriptions of disorders of articulation in children and adults.

CHAPTER OUTLINE

Articulatory Physiology: The Syllable, Articulation, and Coarticulation

The sounds we employ for the purpose of speech are made, in the simplest of terms, by closing and opening the rostral end of the respiratory tract and disturbing the free flow of air from the lungs. Moving the structures of this part of the respiratory system in relationship with one another is called *speech articulation*, or simply, *articulation*. It is not difficult to appreciate that the more open system allows air to flow freely, whereas the more closed system slows the egress of air. Opening and constricting the vocal tract while air is flowing creates the sound pulses we recognize as *syllables*.

Because of its great flexibility, attributable in no small part because it is also the rostral end of the digestive tract, the mechanism we use for creating the sounds of speech can shape or modify the flow of air from the lungs in a wide variety of ways and at a broad range of sites from the vocal folds to the lips or nostrils. Since it is the very mechanism for creating the connected syllables we call speech, the rostral end of the vocal end is known as the *vocal tract*.

The Vocal Tract

The vocal tract is an anatomical tube, about 17 cm in length in adults, extending from the glottis to the lips or, as articulatory needs present, the nares. This tract is, for acoustical analysis, considered to be closed at its proximal end, that is, the glottis. It is a resonating system, with the driving source produced at the glottis or elsewhere and the resonating elastic medium being the air molecules in the vocal tract's resonating cavities.

Speech Sound Classification

The science that examines speech sounds is *phonetics*. It is a lively science and fascinating to almost everyone. Phoneticians have spent centuries studying ways to analyze and classify speech sounds, and the study continues today.

Any sound one can make with one's vocal tract is called a *phone*. Shriberg and Kent (1995) estimated that there are approximately 100 different noises or phones one can make with one's vocal tract. English speakers, depending upon the dialect, use about forty-four. When a phone is associated with meaning in a spoken word, it becomes a member of a *phoneme* group and is termed and *allophone*. Minor variations in allophone articulation may noticeable, but not sufficient to change the meanings of words in which they are spoken. When articulatory variations are sufficient to change (or eliminate) meaning, then the phone becomes an allophone of another phoneme group. Acoustic differences sufficient to signal a change of meaning from one phoneme to another are termed *distinctive, contrastive*, or simply: *phonemic*.

One may classify phonemes in several ways, from simple to complex. The reason for any classification, besides a natural desire to arrange complex maters, is to better understand the dynamics of syllable articulation, with its sequential opening, closing, and conjoining features.

Since phones are produced by opening and closing the vocal tract, the simplest phoneme classification system would be by degree of vocal tract openness, represented by a continuum ranging from "wide open," with minimal vocal tract restrictions, to completely closed, allowing no air passage. At the most open end of the open-closed continuum are *open vowels*, created with a lowered mandible and with the tongue held low in the mouth. At the closed end are the *plosives*, created by completely obstructing pulmonary airflow.

Of course, to create 100 phone possibilities, there must be degrees of open-ness and closed-ness among these categories. For example, vowels can be articulated with infinite variation in mandibular elevation and subsequent degree of vocal tract openness, combined with equally infinite possibilities of tongue positioning. This leads to the *International Phonetic Association's* vowel classifications of *open, mid-open, mid, mid-close*, and *close* combined with *front, mid*, and *back* tongue postures. The closest of the close vowels are most obstructive in their postures, and are nearly as obstructive as approximant consonants. These close postured phonemes can form syllable boundaries as the pace of speech increases.

Variation in open and closed postures is essential to the recognition of acoustic markers that distinguish and signify meaning in the syllables of speech. Therefore, boundaries between syllabic pulses must be created with closed postures called *consonants*.

OPEN AND CLOSED VOCAL TRACT POSTURES	
OPEN------------------------CLOSED	
Open Vowels--------------Close Vowels	
Approximants------------------Nasals	
Fricatives----------------------Plosives	

We can readily see that a simple "open-closed" system for organizing speech sounds is not detailed enough to accommodate even the forty-four phonemes of English. One major fault of such a system it that it doesn't account for the wide range of articulatory sites available along the proximal-distal extent of vocal tract length.

A more complex and far more widely used phoneme classification system is that of *"Place-Manner-Voicing."* This system classifies phonemes according to the vocal tract location of maximum constriction, how the breath stream is managed, and whether or not there is accompanying phonation in obstruent consonants. This system is used by the *International Phonetic Association (IPA)* to arrange their charts of vowels and pulmonic consonants.

Still more complex are systems of *distinctive features*, first proposed by Jakobsen in the 1940s (Jakobson, 1940; 1972) and refined to their most useful extent by Edwards in 2003. Distinctive feature systems are arrays of binary attributes felt to be contrastive "distinctive" in their presence of absence. Distinctive differences in phonemes are those that change meanings in spoken words. The reality of the situation is that not all "distinctive features" in distinctive feature systems are always distinctive, but the usefulness of the systems in further understanding articulatory dynamics is undeniable.

Pulmonic consonant and vowel charts of the International Phonetic Association (2005)

	Bilabial	Labiodental	Dental	Alveolar	Post alveolar	Retroflex	Palatal	Velar	Uvular	Pharyngeal	Glottal
Plosive	p b			t d		ʈ ɖ	c ɟ	k g	q ɢ		ʔ
Nasal	m	ɱ		n		ɳ	ɲ	ŋ	N		
Trill	ʙ			r					R		
Tap or flap		ⱱ		ɾ		ɽ					
Fricative	ɸ β	f v	θ ð	s z	ʃ ʒ	ʂ ʐ	ç ʝ	x ɣ	χ ʁ	ħ ʕ	h ɦ
Lateral fricative				ɬ ɮ							
Approximant		ʋ		ɹ		ɻ	j	ɰ			
Lateral approximant				l		ɭ	ʎ	ʟ			

Where symbols appear in pairs, the one to the right represents a voiced consonant. Shaded areas denote articulations judged impossible.

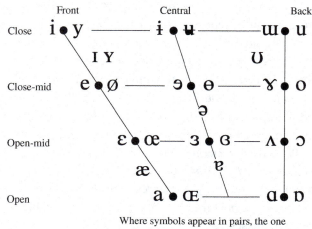

Where symbols appear in pairs, the one to the right represents a rounded vowel.

Distinctive feature systems and "Place-Manner-Voicing" classifications are used to help describe the motor strategies used in single syllable articulation and in running speech articulation by children and adults. Such strategies are called *phonotactics* and depend upon the central linguistic constraints of phoneme combinations inherent in the individual's *phonological system*.

Vocal Tract Acoustic Driving Sources

All acoustic resonating systems must have a driving source, which transfers energy to the system, and an elastic medium to vibrate in harmony with the driving source. The vocal tract driving sources, as we have learned, are of three main types. The phonatory or glottal source is a quasiperiodic series of air pulses that pass between the vocal folds and up through the vocal tract during the open phase of the glottal cycle. The fricative source is a random, aperiodic continuous source, created by forcing a relatively laminar air flow through a tight constriction, disturbing the orderly flow of molecules and creating turbulence. A third source is the plosive source, a transient burst of energy created by trapping air behind a complete vocal tract closure and releasing it as articulatory demands require. The sources of all normal speech are one of these three sources or a combination of the phonatory source and one of the other two.

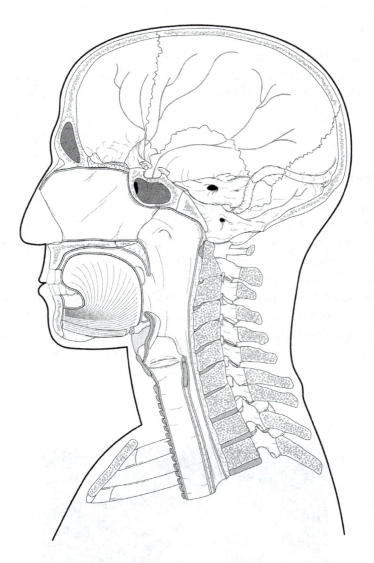

The vocal tract's major resonating cavities are all located superior (rostral or distal) to the glottis, since for acoustic purposes, the glottis represents the closed end of the vocal tract tube. The major resonating cavities include the pharynx, oral, and nasal cavities, and are considered major not only because of their sizes, but because the speaker consciously varies their dimensions and hence, their resonating characteristics in the production of phonemes. Smaller resonating cavities still pay a resonating role and include the spaces between the teeth and outer tissues of the oral cavity, the small cavities of the larynx, and the cavity of the trachea. These cavities are important inasmuch as they contribute to the unique, personal characteristics of an individual's speech.

The resonating cavities are said to *filter* the driving sound source. In contrast to the usual concept of a filter, these can either amplify or dampen components of the complex driving source, depending upon the momentary shape of the vocal tract. It is patterns of amplification and dampening that, along with the nature of the driving source, are key perceptual features in the decoding of speech.

Sound Sources and Phonemes

As we have learned, one of the keys to phoneme perception and production is the nature of the driving source or sources. These sources, *phonatory, fricative,* and *plosive,* are sequenced and combined in various ways to produce the sounds of our language or any language. We will limit our discussion to American English, but the same sources can be described for all languages.

The Phonatory Source

The phonatory source, described in detail in the previous chapter as a quasiperiodic, continuous source, is produced only at the glottis. It is essential for the production of all vowels, approximants, and nasals, as well as for the distinctive voicing feature of voiced obstruent consonants. The obstruent consonants are the plosives and fricatives. When an obstruent consonant is to be perceived as voiced, the phonatory source must begin before the consonant is released.

Obstruent Consonants

Obstruent consonants are plosives and fricatives. They represent the most closed phonemes in the phonology of any language because they are formed by placing a significant obstruction in the vocal tract. It is easiest to appreciate how obstruent consonants form the boundaries of syllables by imagining a speaker moving from their closed postures to the more open ones of a vowel (or syllabic consonant) nucleus.

Obstruent consonants are also the only ones for which the speaker can combine the phonatory source with one of the other sources, either the fricative or the plosive one. Most of the obstruent consonants exist in pairs of *homophones,* or *cognates*. Homophonous cognates are consonant pairs distinguished only by the voicing feature. The member of the pair articulated with accompanying vocal fold vibration is called the "voiced" or "sonant" member of the pair. Voiced cognates are articulated with the phonatory source in combination with the fricative or plosive source.

The Plosive Source

The acoustic source for plosives, as is implicit in the name, is called the *plosive source*. The plosive source is an aperiodic, transient burst of acoustic energy, created by completely blocking the egress of pulmonic air and releasing it when the speaker perceives that sufficient pressure has been created behind the blockage. More pressure is derived from longer hold duration and the consequences are greater acoustic intensity.

Plosives

Plosives are articulated with complete vocal tract occlusion and sometimes, a sudden release of pulmonic air. The release causes turbulence in the same way the fricative does, with major variations resulting from variations in hold duration and hence, pressure accumulation. Since driving the vocal folds adds impedance to the vocal tract, voiceless plosives generate greater oropharyngeal air pressure and greater acoustic energy than their voiced cognates.

The plosive release is the major source of allophonic variation within this category. The release may be weak (*lenis*) or strong (*fortis*). Affricated plosives generate release turbulence at the place of articulation, while aspirated plosives add glottal frication to the output. If a plosive is unreleased, the acoustic cue is transition from the preceding sonant phoneme to the closure posture.

Plosive Spectra

The spectra of plosives are of short duration and very wide bandwidth, and contrasts are based on similar resonating cavities as with fricatives. In other words, the more anterior is the place of articulation, the higher are the frequencies of major resonant energies. The exception to this found in the case of the bilabial plosive, with which the major resonating energies are low (500–1500 Hz.). A plausible explanation for this phenomenon is that the labial place of articulation frees the vocal tract posterior to the place of articulation to resonate as a whole tube (Minifie, 1973).

Voice onset time (VOT) for plosives distinguishes members of plosive cognate pairs. In the case of plosives, VOT may be negative, occurring as much as 0.195 sec. before the release (Lisker, & Abramson,1964). This, of course, results in a *sonant* or *voiced* plosive.

Affricates

The English affricates are phonemes associated with the "Ch" and "soft G" (tʃ/ and /dʒ/) sounds. These are combinations of postalveolar plosive and fricative articulations, with spectral characteristics similar to both. Timing of affricate articulation is crucial to contrastive discrimination between them and the corresponding plosive or fricative phonemes.

Locations of plosive blockages, or places of articulation, range from the glottis, proximally, to the lips, distally, with several sites between created by juxtaposition of the tongue with various parts of the oral cavity. Variations in the location of the blockage changes the relative sizes of the cavities proximal and distal, and along with that, distinctive resonating characteristics of the plosive source.

Small variations in the manner of plosive articulations can create perceptible differences within a phoneme group. Such small differences are not sufficient to change the associated meanings of words in which the plosives are used, and are called allophonic differences. For example, the release of a plosive may be forceful (*fortis*) or weak (*lenis*). It may, in fact, be inaudible (*unreleased*). If the sound of an audible release is created by turbulence at the glottis, the release is said to be *aspirated*. If the sound of an audible release is created by turbulence at the place of articulation, the release is said to be *affricated*. Although any plosive can be affricated, two notable affricated plosives are the English affricates. These are exemplified as the syllabic boundaries of "church" and "George."

The Fricative Source

Fricatives are articulated by forcing a stream of pressurized air through a small vocal tract constriction. Constrictions can be created by loosely juxtaposing structures of the face, oral cavity, or pharynx. These include the two lips, the tongue and the superior or posterior oral cavity, supraglottal, and glottal structures. The air pressure differential across this constriction is created by a relatively high pressure region caudal to the constriction and a relatively low pressure region between the constriction and the opening to the atmosphere. Air particles are forced through the constrictive resistance and in crease their velocity as they do so. This creates random scattering of air molecules and the continuous hissing turbulent sound that goes with it. The fricative source is a continuous, aperiodic sound. Creating a fricative requires more precision and coordination than what is required for plosive articulation.

There is a critical fluid dynamic parameter that must be exceeded before audible turbulence is generated. This parameter is *Reynold's Number*, and specifies the relationship of air viscosity to its inertia, taking into account the contributions of aperture width at the articulatory constriction (Minifie, 1973). Speaker respiratory effort sustains pressure and velocity as long as speaker wants during what phoneticians call the fricative "hold phase."

As is the case with plosives, acoustic markers for distinctive fricative perceptions are created by varying the locations of vocal tract constrictions. Among all the other pulmonic consonant phonemes, only the fricatives are distinctively different when articulated at the dental, alveolar, or postalveolar sites. Allophonic variations within fricative phoneme groups can be created by altering the duration of the sound or by minute variations in the place at which the articulatory constrictions is located.

Fricative Spectra

Fricative acoustic spectra depend on the place and shape of constriction, as well as the degree of pressure differential across constriction. The ideal constriction shape for generation of the fricative source appears to be the cylinder. This is the shape that generates the best sound with the least effort, resulting in maximum acoustic amplitude of the signal. Consider the apparent diffee /s/, with its narrow constriction, and /ʃ/, with its broad one.

Since the constriction shape is almost identical for both members of a cognate pair, the spectral characteristics of the fricative source will be the same for both members. There will, however, be differences in output amplitudes.

Voiced fricative articulation includes the glottal or phonatory source accompanying the fricative source. The act of phonation, that is, adduction of the vocal folds, creates additional impedance to air flow at the glottis. The result is decreased air flow and pressure at the constriction point, and a consequent reduction in output amplitude for voiced fricatives as compared with their voiceless cognates.

We rarely articulate any phoneme in isolation, and obstruent consonants are no exception. When a speaker articulates fricatives or plosives before vowels or approximants or nasals, the difference between a voiced or voiceless fricative depends upon the timing of phonation. This timing is called "Voice Onset Time" or "VOT." An earlier VOT, relative to fricative source onset, means the fricative will be perceived as sonant, while a later one means the fricative will be perceived as surd.

Constriction Location and Spectrum

The location of vocal tract constriction in fricative articulation is essential to its contrastive perception. Readers will recall that resonance occurs between any driving source and elastic medium, and the random fricative sound is just such a driving source. The spectral characteristics of acoustic output in such combinations will also be random but acoustic energy will be amplified or dampened in certain frequency bands depending upon the volumes of the resonating chambers in front of and behind the source. A more anterior position means less volume in front of constriction and higher resonant frequencies. Thus, the "s" spectrum is higher than the "sh" spectrum, and "h" is lowest of all.

Peter Strevens (1960) studied the fricative spectra of nine voiceless fricatives. He sorted these into three groups: front, middle, and back. The front group included dental and labiodental fricatives. Their spectra were of low intensity and were 5 KHz to 6 KHz wide, with major energy in the band between 7 to 8 KHz. In the middle group were the alveolar, postalveolar and palatal fricatives. These generated higher output intensity and narrow spectra of 3 to 4 KHz. width. Major energy was on the 6 to 7 KHz. range. The only English fricative in the back group was the glottal fricative. It was of medium intensity and spectrum width, (4 to 5.5 KHz.), with highest amplitudes at 300 Hz to 4 KHz.

Approximants and Nasals

Approximants and nasals are resonance phenomena in speech and are created by filtering the quasiperiodic phonatory source through various vocal tract postures. While the vocal tract is relatively closed, compared to its status when vowels are articulated, it is still open in comparison to its status when articulating plosives and fricatives. These phonemes occupy an indeterminate zone between open and closed and were once called *semivowels* for that reason. Small changes in their articulation or phonetic environment can create a vowel from an approximant or vice versa. Approximants and nasals regularly begin, end, and form the nuclei of syllables.

Vocal Tract Dynamics

Up to now, we have considered phonemes as singular phenomena. Of course, this is far from the reality of normal speech. Instead, the vocal tract is in almost constant motion, its internal spaces quickly changing shape as the mobile articulators glide from one posture to another posture, creating a tract that is closed, then open, and then closed again. Vocal tract movement, that is, changes occurring over time, is also essential to the normal perception and articulation of speech. The movement begins as the speaker attacks the releaser or onset of the first syllable, and only stops when the speaker senses a need for a pause, or a breath intake, or has finished speaking (for the time being). Such movement is programmed in the central nervous system through learning and requires precise coordinated contractions and relaxations of muscle groups in the face, mouth, pharynx, and thorax.

Vocal tract movement, considered over time, is called "transition." Transition is most readily observable in diphthong articulation, during which the vocal tract posture and resultant resonance shifts from that of one vowel to that of another. It may also be observed in the articulation of certain approximants, including those we associate with the letters "Y" "R" and "W" (/j/, /r/ and /w/). In the case of either the diphthongs or the approximants, transition is an essential perceptual key. Less obvious is the importance of transition in the perception of all other phoneme combinations.

The transition from closed to open, and back to closed again, in varying degrees creates the syllables of speech. Syllables are the minimal motor segment of speech, and can generally be observed as variations in airflow through the vocal tract or as variations in the acoustic amplitude of the speech wave.

Syllables

Syllables can be examined in terms of their parts. Each syllable in normal speech has a beginning, a middle, and, depending upon the intentions of the speaker, an end. The beginning of a syllable may be termed its *onset* or *release*, and the end, if one exists, is termed its *offset*, *arrest*, or *coda*. Onset and offset of a syllable are said to be its boundaries. The middle of a syllable is its *nucleus*.

Syllable Boundaries

Syllable boundaries are formed with the more closed vocal tract postures called consonants. The more closed vocal tract postures produce the least acoustic amplitude. In contrast, syllable boundaries offer the most linguistic information.

The most closed consonant boundaries are the plosives and fricatives, called, collectively, *obstruent* consonants. *Approximants* and *nasals* also form boundaries, but are more open and can even form syllabic nuclei under the right circumstances. Even syllables usually thought to begin with vowels, such as the one that forms the word "ask," actually begin with a vocal tract closed at the glottis, and arguably begin with glottal plosives or fricatives. In any case, syllable articulation is dynamic. Whatever the nature of the syllabic beginning, the speech articulators dwell for only a brief time, if at all, and move on to the syllabic nucleus.

Syllable boundaries can be complex, that is, formed by sequencing several closed articulatory postures before the nucleus. Such boundaries are said to be formed by *consonant blends*. Syllable boundaries produce the lowest acoustic amplitude in a syllable, but provide the most linguistic information.

Syllable Nucleus

The nucleus of a syllable produces the most acoustic energy. Nuclei are formed by vowels, approximants, or nasals, and can be simple or complex. A syllabic nucleus formed by gliding from one vowel posture to another without interruption is called a *diphthong*, or in some dialectical variations, a *triphthong*. When an approximant or nasal forms a syllabic nucleus it is said to be *syllabic*. Acoustically, a nucleus formed by a syllabic consonant is similar to a diphthong.

The last part of a syllable, that is, its nucleus and arresting boundary, is sometimes called its *rime*.

Articulatory Juncture

Most syllables are articulated in sequence, and the transition from one to another is called *articulatory juncture*. Articulatory juncture varies according with intent of the speaker and can be open or closed.

If juncture is *open*, there is an observable temporal gap between the syllables. Open juncture occurs when a speaker articulates *one-word-at-a-time*. With open juncture, each syllable has all of its parts: an onset to begin, a nucleus in the middle, and an offset at the end.

Most running speech is by connecting syllables in *closed juncture*. Here, the offset of a previous syllable serves as part of the onset of the following one, and there is no observable break in the sequence.

Coarticulation

Movement from one articulatory posture to another, gliding from one phoneme to the next, creates inertial and positional changes in phoneme articulations. In these cases, the next phoneme articulated depends upon the articulatory posture created at the end of the previous one. Such inertial and positional effects are termed *coarticulation* and are quite variable, even within a single speaker.

Examples of coarticulatory effects include the inclusion of a bilabial plosive in the word "something," (/sʌmpθɪŋ/), created as a labial pressure release by sequencing of a bilabial nasal immediately before the dental fricative, and a change from an alveolar to a velar nasal in the first syllable of "income," (/ɪŋkʌm/) created by anticipation of the velar plosive while articulating the nasal.

Gross Anatomy of the Vocal Tract

Location, Extent, and Dimensions of the Vocal Tract

The vocal tract is a muscular tube, about 17 cm in length, extending from the vocal folds in the larynx to the openings of the oral or nasal cavities. It is curved anteriorly at its rostral extent, and has a valve separating the nasal part from the rest of the system.

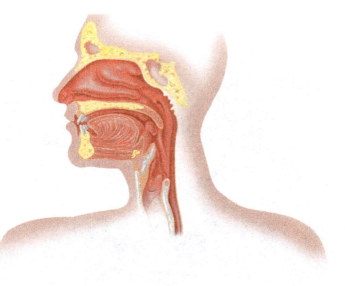

At its rostral end, the vocal tract includes the oral and nasal cavities. These are connected by the pharynx. Adjustments of the sphincter separating the oropharynx and nasopharynx increase or reduce the effects of nasal resonation. The laryngopharynx is part of the vocal tract. Adjustments of structures in the oral cavity and larynx produce the sound sources for speech.

Because of the multitude of movements that occur during the articulation of speech, the tube of the vocal tract has constantly varying cross sections in all planes. Gunnar Fant (1970), in a landmark study done for Bell Laboratories, described the relative resonances of the oral and pharyngeal cavities during vowel productions. His findings suggested that the vocal tract is more accurately called a "Double Hemholtz Resonator," since it has two resonating cavities. The oral and pharyngeal cavities, affected most by changes in muscular forces and configurations, are the primary resonating cavities for the first three harmonics of the phonatory source, or *formants*, and labeled F1, F2, F3. Formants F4, F5, F6 are resonated more in the laryngeal cavities, not subject to nearly as much shape change.

When aperiodic sources are considered, the higher frequencies of the source spectra are amplified when the resonating chamber in front of the tongue is small. As the tongue moves posteriorly in the oral cavity, this anterior chamber becomes larger, and the amplified band of the source spectrum becomes lower in frequency.

The Vocal Tract and Its Relationship to Other Systems

The vocal tract is primarily part of the respiratory system, although it shares structures with the digestive system. The oral cavity and oropharynx are both parts of the respiratory and digestive systems.

The close relationship of the digestive and respiratory (speech) systems gives rise to the delegation of the treatment of swallowing disorders to the speech-language pathologist. While swallowing may seem distinctly different from speech and language functions, the shared musculature and structures of the oral and pharyngeal cavities makes treatment plan development in cohesive. Even in light of the fact that nervous system control of either system has, in most manifestations, different origins, efforts to increase voluntary control, strength, coordination, and flexibility of oral cavity and oropharyngeal structures may be seen, subjectively to be as beneficial to speaking as it is to swallowing.

The Skull

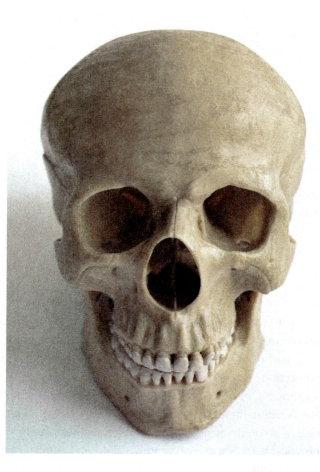

The skeletal foundation of the articulatory mechanism is, as we know, formed by the bones of the head and neck. In the neck, the bones of the cervical spine and the cartilages of the larynx have been described in Chapter Four, Respiration, and Chapter Five, Phonation.

In this chapter, we will focus on the skeleton of the head, the skull, in order to provide a foundation for attachments of speech articulatory structures.

Gross Structure of the Skull

Most beginning anatomy students are surprised to discover that the skull is formed by twenty-nine separate bones. Of these, eight form the cranium, and fourteen form the face. The remaining bones are the tiny *ossicles*, three each in the middle ears, and the independent hyoid bone inferior to the mandible. Most of the skull's bones articulate at fibrous (synarthrodial) joints called "sutures." The dense connective tissue forming the interstices of these articulations helps them perform two important skull functions: to protect the brain and to keep the airway open. Some skull bones articulate at synovial joints. Synovial skull articulations exist between the temporal bones and the mandible and between the ossicles.

The skull has had diverse and complex origins. It may be generally divided into two main sections: a *neurocranium*, which houses the brain, auditory, and olfactory end organs, and a *viscerocranium* that consists of the outer and middle ears and the structures of the distal alimentary and respiratory tracts, including the jaws. The

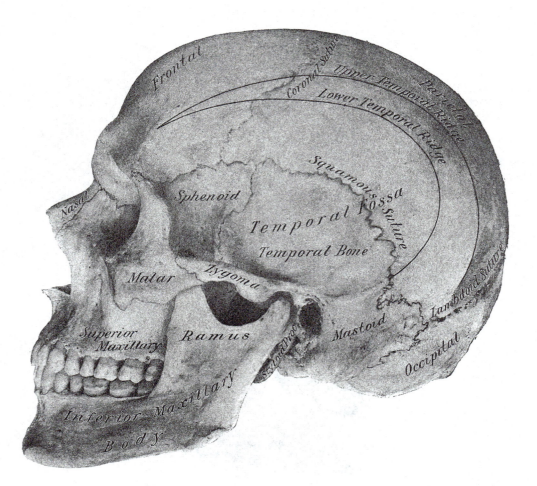

viscerocranium develops from the pharyngeal (branchial) arches and are felt to be derivatives of primitive gill structures in chordates.

Cranium

The cranium is the skull without the mandible and hyoid bone. Most of the bones of the skull are contained in the cranium. The bones of the cranium include the *frontal, parietal, occipital temporal sphenoid, ethmoid, lacrimal, inferior nasal concha, palatine, vomer, maxilla, zygomatic,* and *nasal.* The cranium has two subdivisions: the facial skeleton and the calvaria.

Facial Skeleton

The lower, anterior part of the skull is the facial skeleton. In the newborn, the facial skeleton is relatively flat compared to that of the adult skull. As the individual grows, the facial skeleton grows anteriorly and inferiorly. This change in geometry has some effect on the acoustics and movements of speech by changing the dimensions of the vocal tract and the resonating characteristics of the facial bones. The changing dimensions of the facial skeleton also create spaces between the layers of skull bones. These spaces are the facial sinuses. The facial skeleton part of the cranium overlaps with the rest of the skull, because the face is not complete without the mandible.

Ten bones form the facial skeleton in addition to the mandible. These include the frontal, sphenoid, maxilla, lacrimal, ethmoid, inferior nasal concha, nasal, palatine, vomer, and zygomatic bones. Some of the bones of the facial skeleton form the structure underlying what we think of as the "face."

Calvaria

The bones of the superior skull form the calvaria. These bones form the bowl that protects the brain, and are sometimes referred to as the "skull cap." The bones of the calvaria are the frontal, parietal, and the squamous parts of the temporal and occipital bones. They articulate with fibrous sutures to maintain rigidity, but they may give a little in the event of impact or "sudden deceleration" meeting with objects in the environment. This giving or flexing is beneficial because it spreads the force of an impact over a greater area and time, allowing a more gradual dissipation of its energy.

Mandible

The mandible is what we know casually as the "lower jaw." It articulates with the temporal bone at the synovial *temporomandibular joint*. The mandible may also be considered part of the facial skeleton.

Hyoid Bone

Although it is most commonly associated with the larynx, the hyoid is a skull bone. It is the only bone in the body that does not articulate with another bone.

Cavities of the Skull

There are four large cavities of the skull: the *nasal cavity*, *oral cavity*, *orbits*, and *cranial cavity*. Smaller cavities include the two *tympanic cavities*, which house the soft tissues of the peripheral hearing mechanism, deep in the petrous portion of the temporal bones, and the six *paranasal sinuses*, surrounding the nasal cavity and created

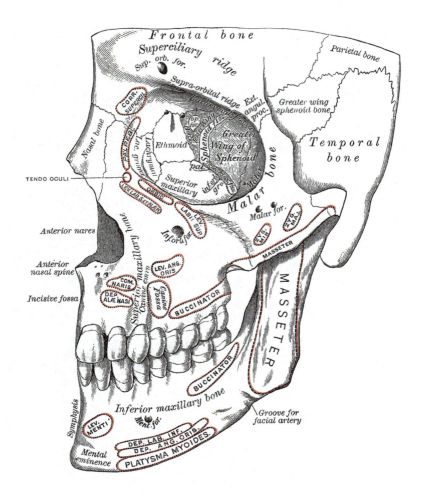

within the maxillae and the frontal, sphenoid, and ethmoid bones by the anterior and inferior growth of the facial skeleton. All the skull's cavities contain the soft tissues and functional organs of the cranial region, and provide space for the resonating bodies of the vocal tract.

The nasal cavity is situated in the approximate center of the facial skeleton. In this position, its bony infrastructure is shared by all the other skull cavities. The nasal cavity is the primary inlet for pulmonary air, and it provides an alternate resonating body for the vocal tract.

The oral cavity is the inlet for the digestive system and provides an alternate vent for the respiratory tract. It is also the flexible rostral end of the vocal tract, allowing almost limitless variations in the size and shape of its resonating body and great variation in the modification of pulmonary egress.

The orbits house the eyes, their extrinsic musculature, and soft tissue support structures. The importance of theses in communication is to receive graphic and gestural input, including written and pictorial signals, as well as hand and facial gestures.

The cranial cavity contains the brain and its support structures. The most complex structure on the nervous system, the brain is the mediator of input from and output to the peripheral nervous system.

Mechanical Stresses of the Skull

The bones of the skull are specialized to help perform their protective function. Most of them, particularly the ones of the calvaria, are laminated, like plywood. The outer layer is called the outer lamina. The inner layer is the inner lamina. The two laminae are separated by a spongy layer called *diploe*.

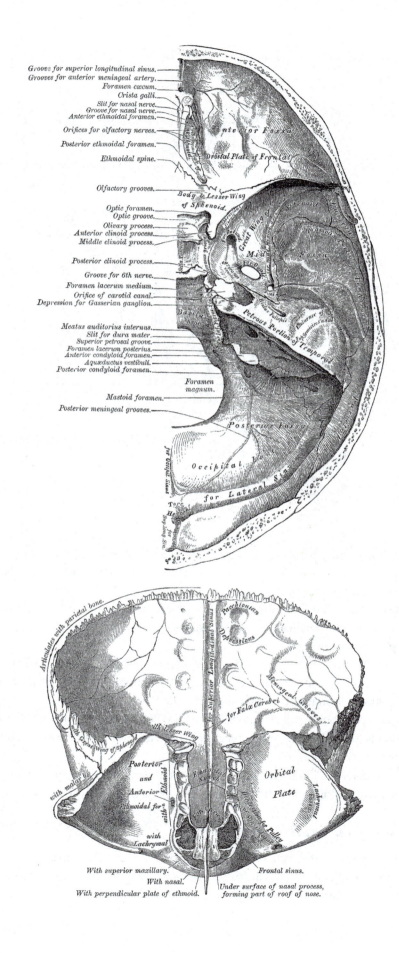

Groove for superior longitudinal sinus.
Grooves for anterior meningeal artery.
Foramen cæcum.
Crista galli.
Slit for nasal nerve.
Groove for nasal nerve.
Anterior ethmoidal foramen.
Orifices for olfactory nerves.
Posterior ethmoidal foramen.
Ethmoidal spine.
Olfactory grooves.
Optic foramen.
Optic groove.
Olivary process.
Anterior clinoid process.
Middle clinoid process.
Posterior clinoid process.
Groove for 6th nerve.
Foramen lacerum medium.
Orifice of carotid canal.
Depression for Gasserian ganglion.
Meatus auditorius internus.
Slit for dura mater.
Superior petrosal groove.
Foramen lacerum posterius.
Anterior condyloid foramen.
Aquæductus vestibuli.
Posterior condyloid foramen.
Mastoid foramen.
Posterior meningeal grooves.

Anterior Fossa
Orbital Plate of Frontal
Body & Lesser Wing of Sphenoid.
Great Wing of Sphenoid
Middle Fossa
Petrous Portion of Temporal
Foramen magnum.
Posterior Fossa
Occipital bone
for Lateral Sinus

Articulates with parietal bone.
Pacchionian Depressions
for Superior Longitudinal Sinus
Meningeal Grooves
for Falx Cerebri
with Great Wing of Sphenoid
with Lesser Wing
Posterior and Anterior Ethmoidal for
with Ethmoid
Orbital Plate
Lachrymal
with Lachrymal
With superior maxillary.
With nasal.
With perpendicular plate of ethmoid.
Frontal sinus.
Under surface of nasal process, forming part of roof of nose.

The skull is exposed to a variety of mechanical stresses. Extrinsic forces are generated by outside pressures and constitute the majority of forces acting upon the skull. Intrinsic forces are generated by the skull itself.

Extrinsic forces can be applied through sudden deceleration from outside the calvaria and along the vertebral column into the basal area of the skull through the basal portion of the occipital bone. An important intrinsic force is that generated by exertion of the muscles of mastication on the mandible against the maxilla. This force is usually applied along the line formed by the second upper molar, the zygomatic bone lateral margin of the orbit and into the anterior calvaria. This force can be the source of chronic pain and dental damage.

Individual Bones of the Skull

We now turn our attention to the morphology and function of the separate skull bones. These may be grouped not only by location, but according to those that are paired and those of which there is only one.

We will examine first the bones that form the cranial cavity. These bones include the calvaria, the *frontal, parietal, occipital, temporal, ethmoid,* and *sphenoid* bones. The sphenoid bone, interposed between the cranium and facial skeleton, serving the structure of both groups, will be described as part of the facial skeleton, as will the ethmoid bone, only a small part of which, the cribriform plate, presents a surface in the anterior cranial fossa.

Cranium

The bones of the cranial cavity include the *frontal, parietal, occipital, temporal, sphenoid,* and *ethmoid* bones. Of these, the parietal and temporal bones are paired. The bones that form the cap of the cranium are sometimes referred to as the *calvaria,* although that term is becoming rather old-fashioned. Only the squamous parts of the temporal and occipital bones are said to be parts of the calvaria.

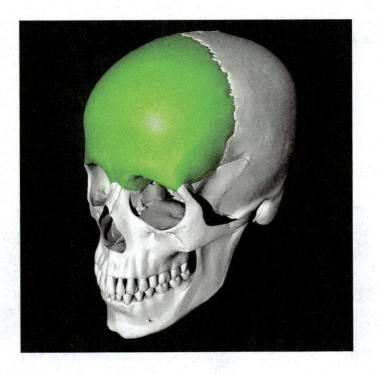

Frontal Bone

The frontal bone forms the anterior part of the calvaria and the superior part of the facial skeleton. Forming the anterior part of the skull cap, the frontal bone articulates posteriorly with the paired parietal bones at the *coronal suture*. The coronal suture continues laterally and inferiorly to the line of articulation between the frontal bone and the sphenoid bone.

During gestation, the frontal bone develops as two halves. These subsequently fuse at the *metopic suture,* an articulation that almost completely disappears in later life. Its only remnants appear between the two supraorbital eminences above the orbits.

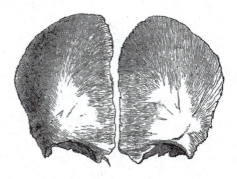

As part of the facial skeleton, the frontal bone forms the entire forehead of the facial skeleton with its *squamous portion.* Inferiorly, the frontal bone forms the superior borders of the two orbits. Between these is a lightly elevated area of the forehead called the *glabella.* The *orbital part* of the frontal bone extends into the orbital cavities and forms the superior part of each orbit. The superior orbital borders are usually marked by distinct eminences, just above each of which can be found a *supraorbital foramen.* These foramina lie immediately below the eyebrows and allow passage of a *supraorbital nerves* and blood vessels. A *supercilliary arch* may be seen anterior and medial to the orbits, whereas lateral and superior to these may be found the *frontal eminences.* Deep to these arches, two spaces, the frontal *sinuses* develop as the facial skeleton grows. They reach their full volumes after puberty and communicate with the middle nasal meatus.

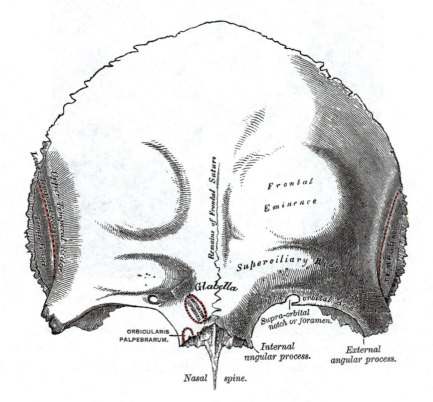

Lateral to the forehead, around the left and right sides, the squamous part of the frontal bone continues, marked by a curved temporal line. This line forms from the lateral supraorbital eminence and continues superiorly and posteriorly along the area of the skull popularly known as the *temple*. As it continues backward, the temporal line soon divides into superior and inferior temporal lines. The superior temporal line is the attachment for the temporalis fascia, whereas the inferior line is the superior limit of the broad temporalis muscle origin.

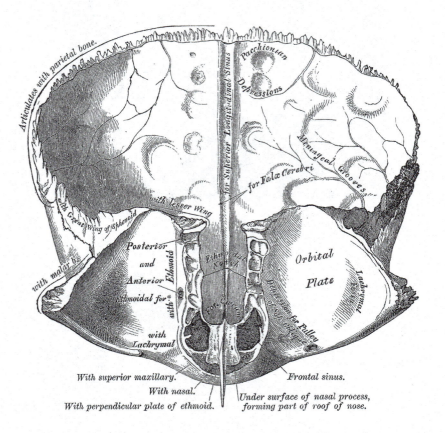

With superior maxillary.
With nasal.
With perpendicular plate of ethmoid.
Frontal sinus.
Under surface of nasal process, forming part of roof of nose.

Lateral to the orbital margins, the frontal bone forms two *zygomatic processes*. The paired zygomatic bones articulate here, at the *frontozygomatic suture*. Together, the frontal, zygomatic, and maxillary bones form the bony foundations of the cheeks.

Parietal Bones

The parietal bones are the only paired bones entirely of the calvaria. Together, they form most of the skull's dome and most of its sides, and are thus convex externally and concave internally. They articulate anteriorly with the frontal bone, at the coronal suture, inferiorly with the greater wing of the sphenoid bone and with the squamous and mastoid parts of the temporal bones, and posteriorly with the squamous part of the occipital bone. The anterior union of frontal and parietal bones is called the *bregma*. It is the location of the *anterior fontanel*, a soft spot where the bones have not fully ossified in early life. The articulation of both parietal bones with the occipital bone is called the *lambdoid suture*, because of its vague resemblance to the small case Greek letter. The central meeting point of all three bones of the lambdoid suture is called the *lambda* and is the site of the *posterior fontanel* in prenatal and postnatal stages of life.

Medially, the parietal bones articulate with one another at the *sagittal suture*. Along this suture, on the inner surface, is a groove. This groove forms half of a channel for the *superior sagittal dural sinus* and completes the channel when combined with the opposite parietal bone. The ridges on either side of the groove form the attachment for the *falx cerebri*, which widens at its superior extent to form two sides of the superior sagittal sinus.

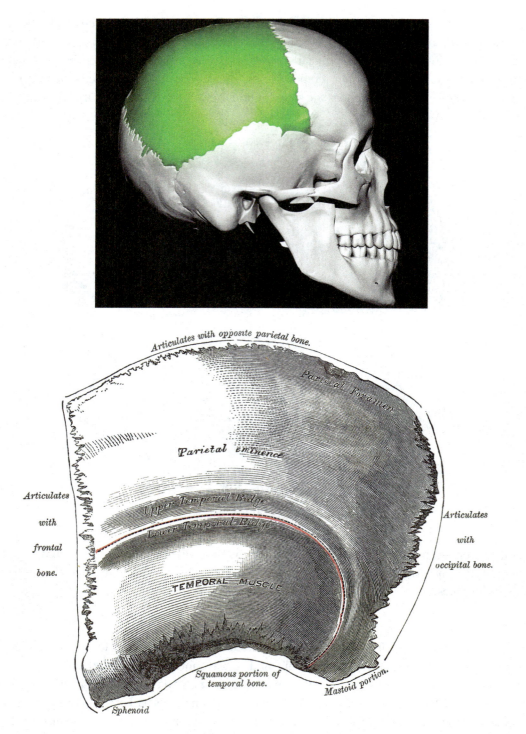

The outer surfaces of the parietal bones are marked by the continuations of the superior and inferior temporal lines. These began at the lateral aspect of the frontal bone, on either side, just lateral to the orbits, and serve as attachments for the temporal fascia and temporalis muscles.

Occipital Bone

The occipital bone forms the posterior wall of the cranial cavity and the floor of its posterior fossa. It is a complex bone, with its most visible portion at the back of the skull and wrapping around to form most of the base of the cranial cavity, including the foramen magnum.

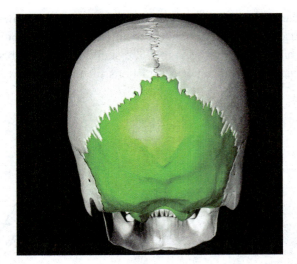

Since the inferior aspect of the brain, including the cerebellum and brain stem, is situated within, the occipital bone is distinguished by the foramen magnum in its base, through which the central nervous system passes. On either side of the foramen magnum are two smooth flattened areas, the *occipital condyles*, at which the first cervical vertebra articulates. Superiorly, the occipital bone has a flattened or *squamous* part, which articulates with the paired parietal bones at the lambdoid suture, forming the posterior part of the cranial cavity. Laterally and inferiorly, the occipital bone articulates with the paired temporal bones. The inferior portion the occipital bone is its *basilar* part. A narrow part of this, the *pharyngeal spine*, extends forward from the foramen magnum and articulates with the sphenoid bone and the vomer. An elevated line in the middle of this part is the *pharyngeal tubercle*, the superior point of attachment for the pharyngeal raphe, the tendonous origin of the pharyngeal constrictor muscles.

An *external occipital crest* extends posteriorly from the edge of the foramen magnum to the center of the squamous part, ending in the *occipital protuberance*, a raised knot at which tough dorsal ligament of the spinal column, the *nuchal ligament*, has its superior attachment.

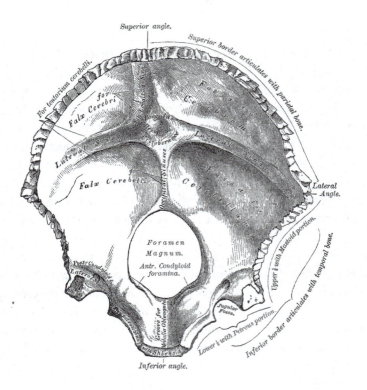

On its squamous and basilar surfaces, the occipital bone serves as an attachment point for the several cervical spinal erector muscles. The muscles have attachments both anterior and posterior to the occipital bone's articulation with the skull, and their bilateral contractions rock the skull forward and backward, "nodding," about the atlanto-occipital and atlantoaxial joint. Unilateral contractions tilt the skull toward the side of the contracting muscles.

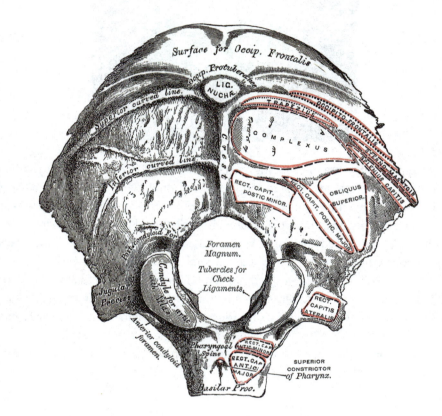

Anterior and lateral to the occipital condyles, in the *cervical prevertebral* region, the *longus capitis* and two *rectus capitis* (anterior and lateral) muscles insert from their origins on the upper cervical vertebrae.

Posterior to the occipital condyles, the musculature is much more complex, deeper, and stronger. Deep muscles are *rectus capitis posterior* (major and minor) and *obliquus capitis* (superior and inferior). These deep muscles originate on either the atlas (obliquus capitis and rectus capitis posterior minor) or axis (rectus capitis posterior major) and insert near the foramen magnum. More superficial posterior occipital muscles attach to the *nuchal lines*, a pair of transverse eminences, arranged one below the other, crossing the rear of the skull. *Semispinalis capitis* originates at the cervical and upper thoracic vertebrae and inserts just inferior to the superior nuchal line, whereas *splenus capitis* originates in the nuchal ligament of the lower cervical and upper thoracic vertebrae to insert partially into the lateral occipital area (as well as into the mastoid area of the temporal bone).

The most superficial of the muscles that insert into the posterior part occipital bone are the *occipitalis, trapezius,* and *sternocleidomastoid* muscles. The occipitalis muscle is the posterior member of several epicranial muscles attached at least partially to a loosely attached connective tissue aponeurosis, the *galea aponeurotica,* which covers the superior surface of the cranium. Included among these are the *frontalis* and *auricular* muscles. Fibers of the strong and bulky trapezius muscles originate in the scapulae and vertebrae, and can pull the skull and shoulders closer to the skull or can extend the skull, whereas those of the sternocleidomastoid originate at the sternum and clavicles and rotate the skull toward the opposite side. Most of the sternocleidomastoid muscle inserts into the mastoid process of the temporal bone.

Temporal Bones

The paired temporal bones are located on either side of the lateral and lower skull. They are complex in shape, having several parts and processes. The *squamous* parts of the temporal bones form the anterolateral parts of the calvaria, being located approximately where the "temple" is. The temporal bones continue posteriorly and inferiorly to contribute to the basilar part of the skull, where their medial, dense *petrous* parts articulate with the occipital and sphenoid bones and where the posterior and lateral *mastoid processes* form the lateral aspects of the skull's base.

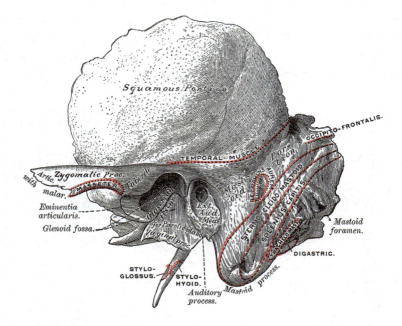

Inside the cranial cavity, the petrous parts form the boundaries that separate the middle and posterior cranial fossae, and the mastoid parts form the lateral walls of the posterior cranial fossae. A thin *zygomatic process* juts forward in a straight line, toward the face, to articulate with the zygomatic bones underlying the cheeks.

The temporal bones are distinguished by the fact that they house the organs for hearing. These are contained in hollow tunnels developed within the petrous parts of each bone. The hollow passages in the petrous part are collectively called the *bony labyrinth* in reference to an ancient Greek mythological tale of a cunningly designed

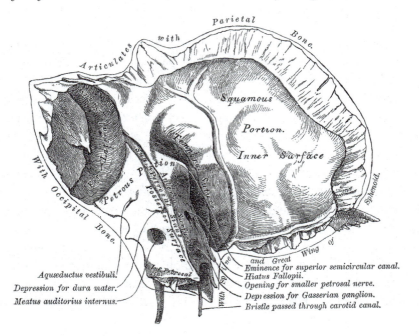

winding passage. Similarly, the delicate tissues that convert acoustic energy into auditory action potentials, along with those that convert changes in head position into action potentials associated with balance are collectively called the *membranous labyrinth*.

An *external auditory meatus* opens to the outside on the middle and lateral aspect of the bone. It is formed by a bony cylinder, the *tympanic part* of the temporal bone, and is continuous with the cartilaginous distal two-thirds of the external ear canal. Proximally, the external auditory meatus leads to a *tympanic cavity* first and then to the bony labyrinth hollowed out of the dense bone of the petrous part.

The tympanic cavities and taller than they are wide and open anteriorly to form the tympanic end of the Eustachian tubes, connecting the cavities to the nasopharynx. Inside both cavities, three *ossicles* articulate with each other, with *tympanic membranes* distally, and with the *oval windows* proximally, to conduct and amplify acoustic energy as it propagates to the membranes and fluids of the inner ear.

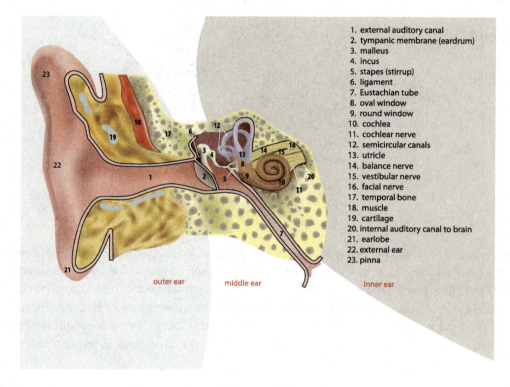

1. external auditory canal
2. tympanic membrane (eardrum)
3. malleus
4. incus
5. stapes (stirrup)
6. ligament
7. Eustachian tube
8. oval window
9. round window
10. cochlea
11. cochlear nerve
12. semicircular canals
13. utricle
14. balance nerve
15. vestibular nerve
16. facial nerve
17. temporal bone
18. muscle
19. cartilage
20. internal auditory canal to brain
21. earlobe
22. external ear
23. pinna

outer ear   middle ear   inner ear

Just inferior to the external auditory meatus, a *styloid process* projects inferiorly, medially, and anteriorly. The name derives from an archaic term for a thin pointed tool, particularly used for writing. It develops separately from the petrous, mastoid, and squamous parts of the bone and is sandwiched between the petrous and mastoid parts. This thin, pointed projection is the origin for the "stylo-" muscles, including the *styloglossus, stylopharyngeus,* and *stylohyoid* muscles. A *stylohyoid ligament* bridges the gap between the styloid process and the hyoid bones by attaching anteriorly to the lesser hyoid cornua.

Anterior to the tympanic portion is a deep *mandibular fossa*. Here, a smooth articular facet is offered for the condylar process of the mandible.

Posterior to the styloid process and external auditory meatus, the bulbous *mastoid process* project inferiorly. It can be palpated just behind the external ear and is the usual site for placement of the bone vibrating transducer for the testing of inner ear acoustic sensitivity. Irregular spaces in the bony tissue of the upper mastoid process are often referred to as "air cells." Air cells become smaller inferiorly and are ultimately replaced by marrow.

The deep groove, parallel to the protruding mastoid process, the *mastoid notch,* is the origin for the posterior belly of the *diagastric* muscle. This muscle, when contracted, draws the hyoid bone posteriorly and superiorly. Further out on the lateral surface of the mastoid process is the insertion for most of the *sternocleidomastoid* muscle, a large flexor and rotator of the neck and skull. Fibers of the *longissimus capitis* and *splenus capitis* also insert on the mastoid process to rotate and extend the skull in relation to the cervical and thoracic spine.

Facial Skeleton

The facial skeleton contains the orbits, the nasal cavity, and the oral cavity. The bones that form the facial skeleton can be grouped into those that form the deep facial infrastructure, interposed between the cranium and the outer facial bones, and those that form the superficial bones of the face. Bones of the deep facial infrastructure are not readily visible from the outside, but can be seen within the orbits and nasal cavities. The greater wing of the sphenoid bone is an exception, as it can be readily seen from the lateral aspect of the skull. Bones of the superficial face are those that give foundation to the facial features and provide attachments to the muscles of expression.

Deep Facial Bones

Bones of the deep facial infrastructure are the sphenoid, ethmoid, and palatine bones, the inferior nasal concha and the vomer. These bones are situated between the cranial cavity and the outer face, forming the inner structure of the orbits nasal cavity and nasopharynx.

Sphenoid Bone

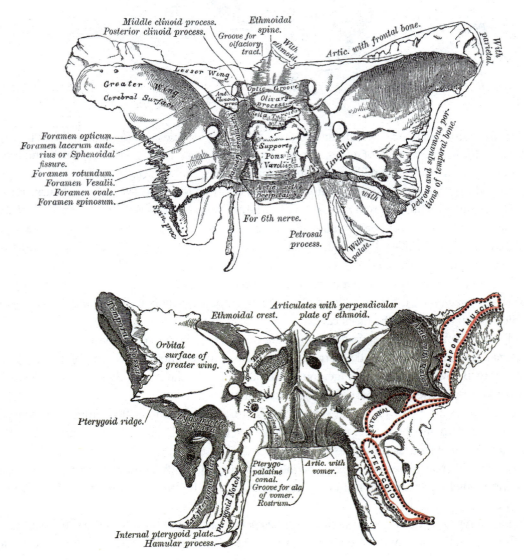

The complex sphenoid bone forms the core of the skull. Being situated roughly in the middle of the cranium, the sphenoid bone articulates with all the other bones in that cavity, and is the foundation of most of the middle cranial fossa. Its superior surface serves as the cranial seat for the middle and upper parts of the brainstem, and its inferior surface forms the bony infrastructure of the nasopharynx. It forms parts of the orbits, and can be seen superficially from the lateral and inferior aspects of the skull. Anteriorly, the sphenoid bone is an integral part of the facial skeleton, as it articulates with the frontal, ethmoid and zygomatic bones, the vomer, and the maxillae.

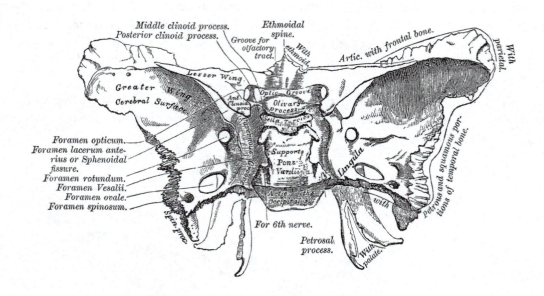

In shape, the sphenoid bone's resemblance to a winged creature gives rise to the naming of its main parts as a medial *body* with two lateral pairs of *wings*, one *greater* and the other *lesser*. Projecting below the greater wings, like four rudders, are two pairs of smaller wing-shaped projections, the *pterygoid plates*.

The sphenoid body projects superiorly into the cranial cavity to form the *sella turcica* ("Turkish saddle," also called the *pituitary fossa*), a prominent feature of the middle cranial fossa. This is the site of the *optic chiasma*, where the twin optic nerves have a mixed decussating after entering the cranial cavity through two anterior perforations in the body, the *optic canals*. The pituitary gland is situated in the "seat" of the "saddle." Surrounding the saddle are anterior, middle, and posterior *clinoid (Greek: "like a bed") processes*.

Within the body of the sphenoid bone are two *sphenoidal* sinuses, separated by a midline septum. Small at birth, these spaces reach their greatest volumes after puberty. Openings high in the sphenoidal sinuses allow them to communicate with the roof of the nasal cavity.

Inferiorly and externally, the sphenoid body and the vomer form the roof of the nasopharynx. Anteriorly to the body, an *ethmoid crest* articulates with the perpendicular plate of the ethmoid bone, forming part of the foundation of the nasal cavity. Below this is a small ridge, the rostrum, which articulates with the ala of the vomer. Posteriorly, the body articulates with the basilar part of the occipital bone.

The greater wings are relatively broad expanses of bone forming parts of the middle cranial fossa on their superior surfaces. Externally, the greater wings articulate with the frontal, temporal, and parietal bones to form an H-shaped intersection of sutures called the *pterion*.

A pair of pterygoid plates, one lateral and one medial, project inferiorly from each greater wing, near the body. The lateral and medial plates diverge from one another toward their free inferior surfaces. Between the two is a *pterygoid* fossa.

The lateral pterygoid plates form most of the *infratemporal fossa* on the outside of the skull, beneath the arches formed by the zygomatic bones and the zygomatic processes of the temporal bones. The lateral and medial surfaces of the lateral pterygoid plates are, respectively, the origins of the lateral and medial pterygoid muscles, both of which insert on the mandible and cause it to elevate and move laterally, forward and backward in relation to the cranium.

The medial pterygoid plates form the posterior and lateral walls of the nasal cavity. A *hammular process (hammulus)* projects posteriorly from each medial pterygoid plate, serving as articular surfaces for the ligaments of the *tensor velipalatini* muscle. Both ends of this muscle originate partially within the pterygoid fossae. The *superior pharyngeal constrictor* attaches to the medial pterygoid plate at its inferior edge.

The greater wings also have *orbital surfaces* that form parts of the posterior orbits. Within the orbits, the sphenoid bone articulates with the maxillae, vomer and the frontal, zygomatic, lacrimal, and ethmoid bones.

The lesser sphenoid wings are only visible within the cranial cavity, where they form part of its floor. Their posterior edges are the borders demarking the anterior and middle cranial fossae. The optic canals enter the cranial cavity at the point where the lesser wings project from the sphenoid body.

Ethmoid Bone

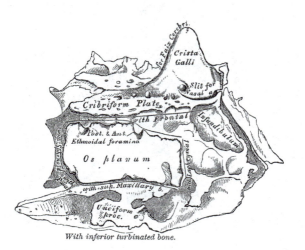

With inferior turbinated bone.

The convoluted, delicate ethmoid bone is best known as the foundation of the nasal cavity. It is located anterior to and slightly lower than the sphenoid bone. The vomer articulates posteriorly and below. Anterior and lateral to the ethmoid bone, the diminutive lacrimal bones and the maxillae articulate on either side, forming with it the lower nasal portion of each orbit.

In shape, the ethmoid bone is in the rough form of the letter "M," with a flat top. Two thick, bony complexities, or *labyrinths*, form the right and left sides, and a midline *perpendicular plate*, part of the nasal septum, forms its middle. At its most superior extent, a horizontal *cribriform plate* forms the roof of the nasal cavity and part of the anterior cranial fossa.

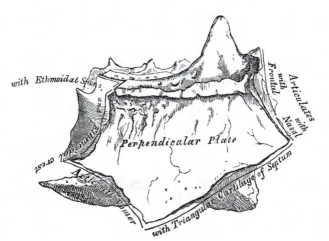

The lateral labyrinths form much of the nasal cavity's infrastructure. They consist of the superior and middle nasal *conchae*, or *turbinates*, delicate bony sheets projecting into each nasal passage and narrowing their openings at the middle and top. The small *uncinate processes* projects inferiorly from these to articulate with the *inferior nasal conchae*. The three paired nasal turbinates are formed by the uncinate process and inferior concha articulations, below, combined with the middle conchae and superior conchae of the ethmoid labyrinth. These are covered with mucosal epithelium to trap atmospheric pollutants during inspiration, and contribute to the resonating character-istics of the nasal cavities for speech. On the lateral surface of each labyrinth are smooth *orbital surfaces* that form parts of each orbit's interior.

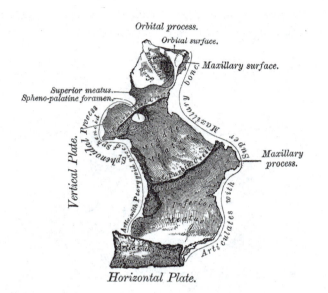

The superior aspect of the ethmoid bone is formed by the *cribriform plate* (from Latin, "in the form of a sieve"). This plate articulates around its margins with the frontal bone and forms the middle of the anterior cranial fossa. Its name derives from the perforations of its surface, through which penetrate the end organs for the sense of smell. These communicate with the olfactory nerves ion either side. A midline process, the *crista galli*, is the anterior attachment for the falx cerebri. It continues posteriorly and inferiorly to form the perpendicular plate, part of the nasal septum.

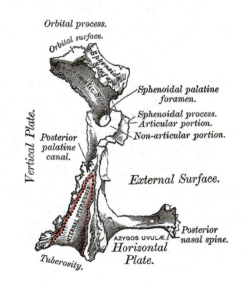

Palatine Bones

As their name suggests, the paired palatine bones form the foundations for parts of the roof of the oral cavity. They only form the posterior quarter of that roof, however, with the largest part being formed by the maxillae. The palatine bones are situated between the maxillae in front and the sphenoid bone's pterygoid processes posteriorly.

The paired palatine bones consist mainly of two plates, a *horizontal plate* and a larger *perpendicular (or vertical)* plate, meeting laterally to form the shape a rough "L." The horizontal plates of these mirror image bones project medially and fuse in the *median palatine suture,* thus forming the bony foundation of the posterior one-quarter of the palate. The perpendicular plates, situated at the lateral ends of the horizontal plates, project superiorly and medially to form the lateral foundations of the posterior oral and nasal cavities.

The horizontal plates articulate anteriorly with the palatal processes of the two maxillae at the *transverse palatine suture.* Their posterior borders are smooth and free. A medial *nasal crest,* formed by the fusion of the two horizontal plates, articulates with the vomer, superiorly. The aponeurosis for the *tensor velipalatini* muscle, a muscle of the soft palate, attaches along the posterior border of the two fused horizontal processes, and the uvular muscle originates at its anterior edge.

The perpendicular plates are complex, presenting internal and external surfaces and several angles. Viewed from the side, the perpendicular plates narrow as they project superiorly, curving backward then forward and ending in thickened *orbital processes*. The orbital processes form most of the posterior orbital cavities. Along their ways superiorly, the perpendicular plates form the posterior part of the nasal cavity's lateral wall. Their posterior edges articulate with the pterygoid processes of the sphenoid bone along the upper part of their borders, but present free edges lower, between the pterygoid plates as they diverge. In front, the perpendicular plates articulate with the medial surfaces of the maxillae.

Inferior Nasal Conchae

The thin, flat inferior nasal conchae form the inferior nasal turbinates within the deep parts of the nasal passages. These delicate scrolling bones articulate with the ethmoid bone, superiorly and with the maxillae, laterally. They project into the nasal cavity medially, curling downward to narrow the inferior openings of each nasal passage.

With the maxillae and the palatine bones, the inferior nasal conchae form the medial walls of the maxillary sinuses, one pair of the six chambers surrounding and communicating with the nasal cavity.

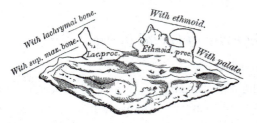

Vomer

The single, midline vomer occupies the space just posterior and superior to the hard palate. It forms the posterior part of the nasal septum. Superiorly, it has a flattened surface, forming two *alae* (wings), one on either side, that articulate with the inferior sphenoid body. From that point, a long, thin bony process extends inferiorly and anteriorly at an oblique angle, reminiscent of a plowshare, from which its (Latin) name was derived.

Along the vomer's inferior border, its long process articulates with the crest formed by the median palatine suture. This crest is formed by the fusion of the palatal processes of the paired maxillae (anteriorly) and the horizontal processes of the palatine bones (posteriorly). The anterior surface of the long process articulates with the perpendicular plate of the ethmoid bone at its upper border and with the septal cartilage of the nasal cavity at its lower end, forming, with the ethmoid bone, the bony part of the nasal septum. Posteriorly, the edge of the vomer is free and forms the septal edge of the choanae.

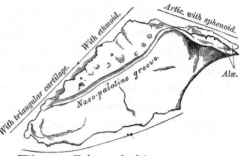

With sup. maxill. bones and palate.

THE LACRIMAL APPARATUS

Tears are secreted by the *lacrimal glands*, located at the upper, lateral edges of the orbits, by *tarsal (or Meibomian) glands* within the eyelids, and by the goblet cells of the conjunctiva. These combined fluids, consisting of water, proteins, and mucin, protect, clean, and lubricate the outer surfaces of the eyes. The purpose of tears in emotion is not clear, but may be a facial gesture exhibited only by human beings.

Tear fluids flow across the surfaces of the eyes, spread by blinking. They collect at the medial edges of the orbits, and are gathered into the orbital *puncta*, tiny openings in the medial edges of the upper and lower eyelids. Capillary action draws the fluids through the *lacrimal ducts* and into the *lacrimal sack*, the expanded superior end of the *naslolacimal duct*. Tears are then conducted through the duct into the nasal cavity.

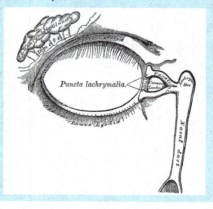

Puncta lachrymalia.

Lacrimal Bones

Two thin, flat lacrimal bones form the anterior medial walls of the orbits and the lateral parts of the nasal cavity's middle meatuses on the right and left. These delicate bones articulate anteriorly with the maxillae, posteriorly with the uncinate processes of the ethmoid bone on either side, superiorly with the frontal bone and inferiorly with the inferior nasal concha. The lacrimal bones have distinct *lacrimal crests*, running vertically nearly the vertical dimensions, which serve as attachments for the *medial palpebral ligaments* of the eyelids and for the palpebral parts of the *orbicularis oculi* muscles. Anterior to the crests are shallow *lacrimal fossae* in which the *lacrimal sacks* are situated.

Superficial Facial Bones

Maxillae

The term *"maxilla"* may refer to either of the two maxilla bones or to both of them when they are fused medially to form the "upper jaw." The paired maxillae form the upper jaw and the outer facial foundation just below the orbits and down to the mandible, or lower jaw. Internally, the fused maxillary bones contain the upper dental arch and form most of the hard palate in the oral cavity. They contribute to the three main cavities of the facial skeleton, including the orbits, forming most of their floors, the nasal cavity, forming parts of its lateral and inferior walls and, of course, the oral cavity in which they form the foundation for most of the oral roof or *"hard palate."*

Each maxillary bone presents a body with a facial surface, a posterior or infratemporal surface and an orbital surface, from which project a frontal process, superiorly, a palatal (also called palatine) process, medially, an alveolar process inferiorly, and a zygomatic process laterally.

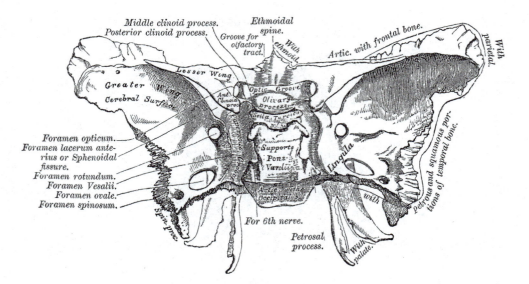

The body of each maxillary bone is roughly pyramidal in shape, and each is hollow, having a large *maxillary sinus* within. The maxillary sinuses communicate with the nasal cavity through ostia located in their superior areas. Laterally and posteriorly, the *infratemporal surface* lies medial to the *zygomatic arch* created by the fusion of the maxillae, the zygomatic bones and the zygomatic processes of the temporal bones. The anterior part of the maxillary bodies presents a *facial surface* at which originate several muscles of facial expression. These muscles are elevators of the upper lip, *levator labii superioris* and *levator anguli* oris, on either side of the nasal cavity and below each orbit. A prominent landmark, the *infraorbital foramen* allows passage of *infraorbital artery*, *vein*, and *nerve*. The facial surface curves into the orbit, becoming the *orbital surface*, which forms the floor of each orbit.

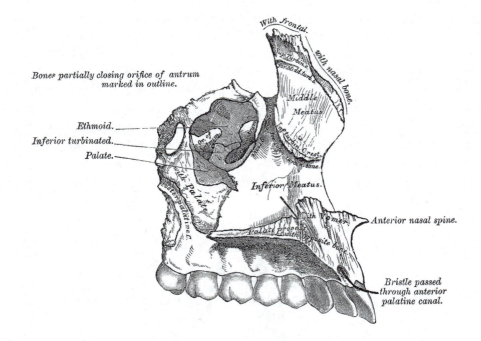

The frontal processes extend superiorly from their bodies to articulate with the frontal bone on their superior surfaces and with the lacrimal bones on their posterior edges, thus contributing to the bony foundation of the orbits' floors and nasal parts. *Orbicularis oculi*, the muscle which closes the eyelid, attached partially to the frontal process on the nasal side of the orbit. Outside the orbit, the frontal process is the origin of the *levator labii superioris alaquae nasi* muscle, another upper lip elevator. More posteriorly, along the orbital surface is an almost horizontal articulation with the ethmoid bone, giving way at its posterior end to a more perpendicular articulation with the perpendicular plate of the palatine bone.

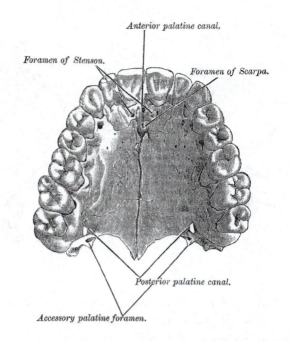

Anterior palatine canal.

Foramen of Stenson.

Foramen of Scarpa.

Posterior palatine canal.

Accessory palatine foramen.

The *palatal processes* of each maxilla extend medially within the oral cavity and fuse with one another in the midline to form the anterior three quarters of the oral cavity roof or "hard palate" skeleton. The term "hard palate" is most often simplified to "palate." Phoneticians classify a *palatal* place of articulation for several consonant phonemes. At their anterior ends, one or more incisive foramina allow passage of the sphenopalatine arteries and nasopalatine nerves. The incisive foramina also mark the posterior extent of the so-called *prepalate*, which develop separately form the posterior part of the palate, fusing with it at about eight weeks of gestation.

The *alveolar processes* extend inferiorly from the bodies in an arched shape below and around the palatal processes, forming their anterior and lateral borders. They are so-named because they contain the alveoli, from which the upper dentition erupts in the *maxillary dental arch*. The posterior part of the maxillary alveolar process, just behind the most anterior of the teeth serves as an important fixed articulatory surface for speech, the *superior alveolar ridge*. Covered by epithelium in life, it is referred to simply as the *alveolar ridge* by phoneticians. The tongue tip and blade are juxtaposed against this and immediately adjacent *dental* and *postalveolar* surfaces for production of various obstruent, approximant, and nasal consonants. The word "superior" is dropped from its name by phoneticians because the inferior alveolar ridge is not used as a speech articular surface in any known language. On its external surface, the alveolar processes are the origins for the upper part of the *buccinator*, a muscle of mastication.

The zygomatic processes extend laterally like bony wings from the body of each maxilla bone to articulate with the zygomatic bones on either side.

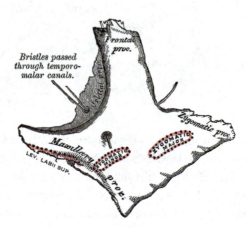

Frontal proc.

Bristles passed through temporo-malar canals.

Zygomatic proc.

Maxillary proc.

ZYGOMATIC MAJOR.

LEV. LABII SUP.

Zygomatic Bones

The zygomatic bones form the middle section of the arches covering the infratemporal fossa of the skull. Medially and anteriorly, they have a curved surface that forms the lower, lateral portion of each orbit, articulating with the greater sphenoid wings. From the orbits, the *temporal process* extends laterally and posteriorly to articulate with the zygomatic process of the temporal bone. The maxillae, zygomatic, and temporal bones give form to the cheeks. The zygomatic bones are the origins of the *zygomaticus major* and *zygomaticus minor* muscles, muscles of facial expression, which draw the upper lip superiorly and laterally toward their origins.

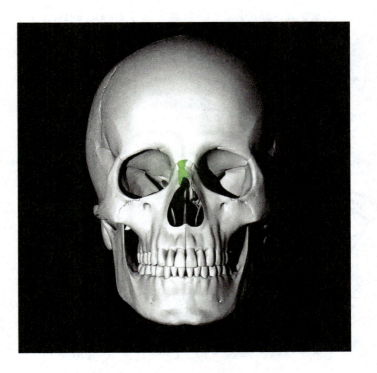

Nasal Bones

The paired nasal bones fuse in the midline to form the bony *dorsum* ("bridge") of the nose. Superiorly, they articulate with the frontal bone and laterally with the frontal processes of the maxillae. The lateral and septal nasal cartilages are attached below.

Mandible

The mandible is the "lower jaw," and, as such, it forms the floor of the oral cavity. It plays an important role in configuring the oral cavity for speech, swallowing, and respiration. The mandible articulates with the temporal bone by means of the synovial joint, the *temporomandibular joint ("TMJ")*. This freely moveable articulation allows the mandible to elevate and depress, move laterally, rotate and advance, and retract in relationship to the cranium.

The mandible moves very little in normal speech. Mandibular elevation is required to bring the tongue within reach of the oral roof to form labial, dental, palatal, alveolar, and postalveolar phonemes. Mandibular elevation is also required to create the vocal tract configuration necessary for close vowels, such as /i/ and /u/, and the rhotic vowels, /ɜ/ and /ə/.

Depression of the mandible is necessary from some aspects of speech. For example, open vowels, such as /ɑ/ and /ɑ/, are often produced with a more open oral aperture, created by lowering the tongue within its mandibular bed. Greater loudness can be achieved with a more open cavity. This is most important for public speakers and singers.

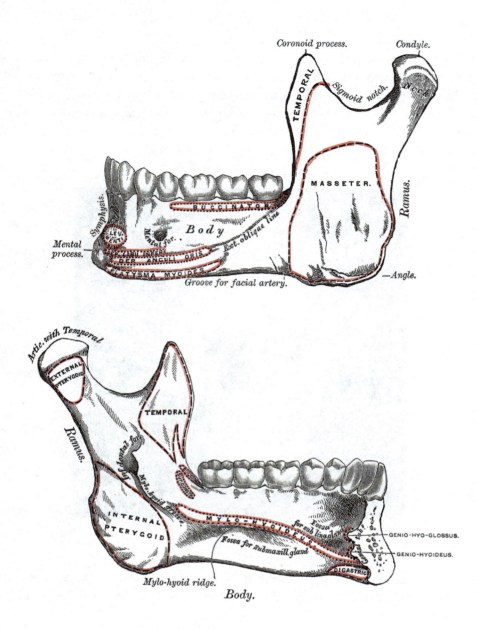

Mastication requires initial vertical movement to admit material to the oral cavity, or to slice it from a larger portion with the anterior teeth, followed by rotary movement of the mandible, to grind material between the posterior teeth.

The mandible is U-shaped, like the hyoid bone, following is brachial arch origins. The anterior part of the arch is the mandible's *corpus*, and the more vertical posterior parts, at the open ends of the arch, are called *rami*.

At the anterior extent of the corpus is a prominence, the *mental symphysis*, forming the underlying structure of the chin. In this case, the word "mental" comes from the Latin *mentum*, referring to the chin. Here the two sides of the mandible begin to fuse *in utero*, in the same manner as the maxilla bones of the upper jaw. The fusion continues post partum, as cartilage at this site turns into bone. On the external surface, a *mental eminence* forms the underlying bony structure of the chin. Its most anterior extent is sometimes called the *pogonion*. The inner surface of the mental symphysis is marked by a *mental spine*, or *genial tubercle*. This small posterior projection provides origins for the *genioglossus* and *geniohyoid* muscles.

Projecting superiorly around the arch of the mandibular corpus is the *alveolar process*, the inferior counterpart of the maxillary alveolar process. From here, the inferior dentition erupts. The inferior teeth are of little importance to speech. Their absence may make the speaker sound a bit unusual, but intelligibility should not suffer. They are, of course, of great importance to mastication.

The *rami* are the more vertical, flat portions of the mandible that projects superiorly to articulate with the temporal bone at the temporomandibular joint. Each ramus ends in a smooth articular knob, the *conylar process*. This articulates with the temporal bone in a similarly smooth depression, the *mandibular fossa*. Just anterior to the condylar processes the *coronoid processes* project superiorly toward the zygomatic arches like two knife blades. They serve as insertions for the *temporalis* muscles, whereas the bulky *masseter* muscles insert along a broader extent of the rami, from the coronoid processes down to their inferior margins.

The *angle of the mandible* is defined by the geometric relationship of the ramus to the mandibular body. It decreases with the development of dentition, then increases with the loss of dentition. Thus, the mandibles of infants and older adults are relatively flat when compared to those of individuals having complete sets of teeth.

The mandible provides numerous attachments for muscles of speech articulation and deglutition. Some of these attach to its interior surface and others attach to its external surface. Since both speech articulation and mastication involve elevation and depression of the mandible, there is a good deal of functional overlap. Muscles that attach to the mandible are described briefly below for convenience, and described again fully in the section on muscles of articulation.

Muscles That Attach to the Interior Aspect of the Mandible

The *genioglossus* is the largest of the tongue muscles. It is fan shaped and forms most of the body of the tongue. This muscle originates from mental spine of mandible (interior, at the chin) and inserts into the tongue from tip to root. Some of the fibers insert into the hyoid bone. Contraction of the genioglossus muscle protrudes the tongue.

The paired *mylohyoid* muscle forms the muscular floor of the anterior mouth. It attaches laterally to the anterior four-fifths of the *mylohyoid line* a slight, narrow bony elevation on the interior aspect of the mandible. In its posterior and middle section, the mylohyoid attaches to the hyoid corpus. It has a median raphe, running right along its antero/posterior length, to which the opposing pairs are attached along the remainder of its middle. Contraction of the mylohyoid muscles draws the hyoid bone upward and anteriorly.

The *anterior belly* of the *digastric muscle* originates at the internal surface of the mandible, near the symphysis. Its fibers extend posteriorly to a middle tendon, connected to the grater cornua of the hyoid bone by a fibrous connective tissue link. Anterior belly fiber contraction draws the hyoid bone toward the mandible or draws the mandible toward the hyoid bone.

The *superior pharyngeal constrictor* originates at several locations. From the most superior downward, these attachments are the most lower part of inferior border of medial pterygoid plate, the pterygomandibular raphe, and the mandible, at a location posterior to the last molar. The superior pharyngeal constrictor inserts into the pharyngeal tubercle of occipital bone and a median pharyngeal raphe.

The *lateral (or external) pterygoid* muscles have two heads. One originates from the lateral aspect of lateral pterygoid plate, and the other originates from infratemporal surface of greater sphenoid wing. The two bellies insert into the upper portions of the mandibular rami, near the condyles, and into the capsule of the temporomandibular joint. Contraction of the lateral pterygoid muscles protrudes the mandible and moves it from side to side.

The *medial pterygoid* muscles originate at the medial aspect of the lateral pterygoid plate and at the posterior surface of maxilla and palatine bones and inserts into the medial aspect of mandible near its angle. Medial pterygoid contraction moves the mandible from side to side and under can also produce mandibular protrusion. Both sets of pterygoid muscles draw the mandible forward and side-to-side to produce the grinding action of the so-called "rotary chew."

Muscles That Attach to the External Aspect of the Mandible

The thick, short *masseter* muscles originate at lower border and deep aspect of zygomatic arch insert into wide area of lateral aspect of ramus and coronoid process of mandible. Contraction of the masseters elevates the mandible.

The *temporalis* muscles originate on the lateral skull surfaces (parietal and temporal bones) and insert into the tip and medial aspect of coronoid processes, as well as the anterior margin and medial aspects of mandibular rami.

The buccinators originate from the alveolar processes of both the mandible and the maxilla and also from the *pterygomandibular raphe*. Buccinator fibers interdigitate with fibers of the *orbicularis oris* and *depressor anguli oris* muscles of the face. Contraction of buccinator fibers draws the oral margins posteriorly, or, with resistance from the facial muscles, stiffens the buccal interior against pressures produced by sucking.

Depressor labii inferioris and *depressor labi anguli oris* lower the inferior lip. They originate on a line at the inferior border of the mandibular body, on either side of the mental symphysis, and insert into the fibers of the orbicularis oris and superficial fascia.

The thin, flat *platysma* arises in fascia inferior to clavicles and inserts at lower border of mandible and facial muscles. Its contraction stiffens the anterior surface of the neck and assists in drawing the mandible inferiorly.

Importance in Deglutition

Mandibular movement is essential to normal mastication and, normal deglutition. Articulation with the temporal bone produced by paired or alternating contractions of the muscles of mastication elevates, depresses, or rotates the mandible from vertical and rotary chewing action.

The mandible is depressed to admit food into the oral cavity. Gravity is sufficient to depress the mandible in most circumstances, but in cases requiring extreme oral opening or when the mandible must be depressed against resistance, the *suprahyoid* and *infrahyoid* muscles (see Chapter 5) as well as the *platysma* may be brought in to play to overcome the resistance.

After the food or other material is admitted into the oral cavity, the mandible is elevated to help keep the food in. The lips seal the oral cavity, and the buccinator and tongue muscles push food material into the appropriate positions within the oral cavity for processing.

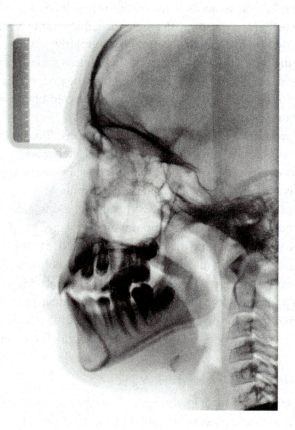

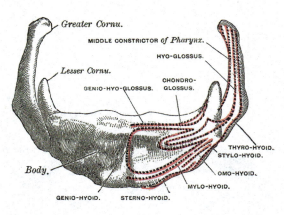

Mastication involves a vertical chew and a rotary action of the mandible in reference to the cranium. Vertical action is produced by simultaneous contraction of the mandibular elevators, while the rotary action, is produced by alternating contractions. Mastication is described fully in the chapter on deglutition.

Hyoid Bone

The *hyoid bone* is situated just below the mandible and is similarly "U" shaped. It serves as a framework for the root of the tongue and as a suspender for the larynx. It also provides structure for the anterior neck, supporting patency for the rostral opening of the laryngppharynx (Papadopoulos, Lykaki-Anastopoulou, and E Alvanidou, 1989). Morphologically, the hyoid has an anterior *corpus* or *body* that can be palpated just beneath the skin where the inferior, transverse plane under the mandible turns sharply downward into the anterior cervical area. This is approximately at the level of the C-4 vertebra.

Projecting from the corpus are four processes called *cornua*. Two *greater cornua* project posteriorly and two *lesser cornua* project superiorly. Within its arch, the epiglottis projects upward from its inferior attachment to the interior of the thyroid cartilage.

The hyoid does not articulate directly with any other bone or cartilage. A *stylohyoid ligament* connects its greater cornua to the styloid process of the temporal bone, and a complex of connective tissue ligaments and membranes bridges the gap between it and the thyroid cartilage. On the interior of the hyoid corpus, a *hyoepiglottic ligament* connects provides an attachment to the epiglottis.

The hyoid bone is an attachment for several muscles of the neck. These are all paired and categorized into a *suprahyoid* group and an *infrahyoid group*, depending upon the locations of their other ends.

Suprahyoid muscles included both bellies of the *digastric* muscles, the *geniohyoid, mylohyßoid,* and *stylohyoid* muscles. The infrahyoid muscles are the *sternohyoid, thyrohyoid, omohyoid,* and *sternothyroid* muscles. Origins and insertions of these muscles are described in Chapter 5. Note that the stylohyoid muscles are in the suprahyoid group but are not involved in opening the mandible against resistance, and the sternothyroid muscles do not attach directly to the hyoid bone, yet are conventionally grouped with the infrahyoid muscles.

MUSCLES OF MASTICATION

Muscles of Mandibular Elevation
  Temporalis
  Masseter
  Lateral Pterygoid
  Medial Pterygoid

Muscles that Depress the Mandible against Resistance
  From Suprahyoid Group
    Digastric (Both Bellies)
    Geniohyoid
    Mylohyoid

  Infrahyoid Group
    Sternohyoid
    Omohyoid
    Thyrohyoid
    Sternothyroid

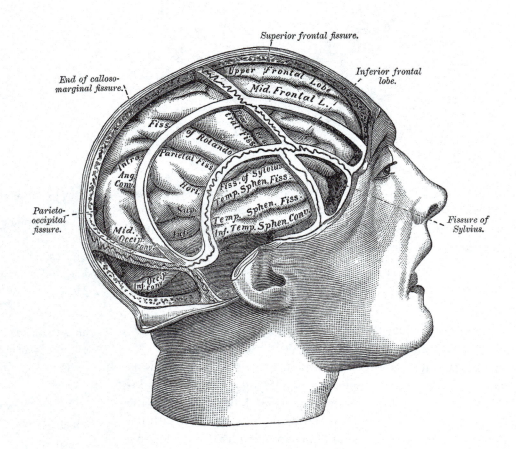

The Skull's Cavities in Detail

The cavities of the skull play important biological and communicative roles. By the virtues of its bony plates and resiliency, the skull protects the contents of these cavities.

Cranial Cavity

The cranial cavity is the largest cavity of the skull. It contains the brain and its supporting organs, including the tough outer brain membrane, the *dura mater*, with long, hollow conduits for the drainage of cranial blood, the *dural venous sinuses*. The dura mater is one of three *meningeal membranes* of the central nervous system. Numerous ridges, corresponding with the convoluted sulci and gyri of the cerebral surface, mark the inner surface of cranial cavity bones. Moreover, the inner surface is marked by grooves by which the bone supports several blood vessels. Superiorly and posteriorly, the grooves contain the superior sagittal and transverse dural sinuses. The superior sagittal groove also has relief for the arachnoid granulations. Along the lateral inner surface are grooves for the middle meningeal artery and vein.

The cranial cavity floor is formed by the articulation of the sphenoid, ethmoid temporal, and occipital bones. Major landmarks are three recessed areas, the anterior, middle, and posterior cranial fossae.

The anterior fossa contains the frontal lobes of the brain. It forms the major part of the orbital roof and is composed of the frontal bone, anteriorly, and the sphenoid bone, posteriorly. Between the orbits and forming the nasal cavity roof is the *cribriform plate (from Latin: "in the form of a sieve")* of the ethmoid bone, marked by numerous piercing, the *foramina of the cribriform* plate, through which the olfactory sensory end organs penetrate the nasal cavity and communicate with the olfactory nerves. A midline septal plate, the *crista galli (from Latin: "Crest of the cock")*, extends a short distance superiorly from the ethmoid bone and situates between the paired frontal lobes. The *falx cerebri*, a midline sagittal extension of the dura mater between the hemispheres, has its anterior attachment to the skull at this point. A shallow continuation of the crista galli, the *frontal crest*, separates the anterior fossa into right and left sections and separates in its middle to form a sulcus fur the *superior sagittal dural sinus*. Just anterior to the cribriform plate is a small foramen of the frontal bone, the *foramen cecum*, through which an emissary vein drains blood from the nasal cavity into the superior sagittal dural sinus. Since this vein communicates directly with the superior sagittal sinus, it is a likely route for transmission of infection from the nose to the cranial cavity. The posterior part of the anterior cranial fossa is bordered by the sphenoid bone's lesser wings.

The middle cranial fossa lies slightly lower than the anterior fossa, and contains the temporal lobes. It is formed laterally by the paired temporal bones. A central section is more elevated, and is formed by the body of the sphenoid bone. The lateral depressions in the middle fossae contain the temporal lobes. They are formed by the greater sphenoid wings and the squamous and petrous parts of the temporal bones.

A major feature of the middle cranial fossa is the *sella turcica*, or "Turkish saddle," formed by an eminence in the middle of the sphenoid bone. It is so named because of its resemblance to one of the large saddles used in the Middle East, having very large pommels and cantles. The *pituitary gland* is located in the "seat" of the saddle. It is also the location of the convergence of the twin *optic nerves* and of the *circle of Willis*, the conjunction of arteries that supply blood to the cerebrum.

The middle cranial fossa is riddled with canals and foramina. All are paired and allow nerves and blood vessels to pass from the cranial cavity to other parts of the skull through this middle part of the cranial cavity.

In the sphenoid bone, the *optic* canal allows the optic nerves to pass from the orbits to the brain. Inferior to the optic canals, the *superior orbital (also clinoid) fissure* allows the oculomotor, trochlear, and abducens nerves (cranial nerves III, IV, and VI), as well as a branch of the ophthalmic division of the trigeminal nerve (V), to pass to and from the orbits. The *foramen rotundum*, near the medial part of the lesser wing, allows passages for the maxillary division of the trigeminal nerve. The *foramen ovale* contains mandibular division of the trigeminal nerve, the accessory meningeal artery, the lesser superficial pertosal nerve and several emissary veins. Through the *foramen spinosum* pass the middle meningeal artery and meningeal nerve, which supply the dura mater. An inconstant foramen, the mastoid foramen, is sometimes present to transmit an emissary vein to the *lateral dural sinus* and sometimes an artery to the dura mater.

In the temporal bone is the *carotid canal*, a sigmoid aperture, through which runs the internal branch of the carotid artery, supplying fresh blood to the brain, and a plexus of sympathetic nerves. This canal courses through the petrous part of the temporal bone, near the inner ear and just anterior to the jugular foramen of the sphenoid bone. Slightly superior to the internal carotid canal opening is the *internal auditory meatus*, a tiny opening that allows the vestibulocochlear nerve to convey neural impulses generated by sound and by changing head position (hearing and balance impulses) to the brainstem. A *stylomastoid foramen* opens laterally, through the squamous part of the temporal bone. Through it the facial nerve emerges from the middle of the brain stem to supply the facial muscles and the stylomastoid artery enters to supply to the inner and middle ear cavities.

Several openings are created in the articulations between two or three bones of the middle fossa. The *foramen lacerum* is formed by a gap in the articulations of three bones: the greater wing of the sphenoid bone, the petrous portion of the temporal bone, and the basilar part of the occipital bone. It is blocked by cartilage during life. Regardless, the nerve and artery of the *pterygoid* (also called *vidian*) *canal* and some emissary veins emerge from the cranial cavity through the cartilage and toward the facial skeleton.

The *sphenopetrosal fissure* is formed by a gap in the articulation between the greater wing of the sphenoid bone and petrous part of the temporal bone. The lesser petrosal nerve, which provides sympathetic motor innervation to the *parotid salivary gland*, passes through this fissure in some cases.

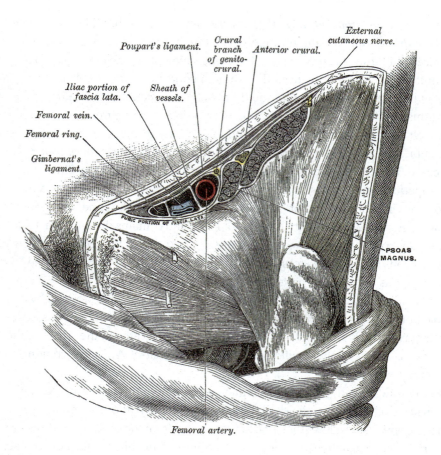

Gaps between the articulations of the temporal and occipital bones create the *jugular foramen*. The jugular foramen is the place at which the dural venous sinuses converge to pass their blood from the cranial cavity to the *internal jugular vein*. It also forms a passage through which the *glossopharyngeal, vagus,* and *spinal accessory nerves* exit the skull.

The posterior cranial fossa forms the lowest part of the cranial cavity. It extends from the dorsum sellae or "saddle's back," of the sphenoid bone to the posterior of the cavity.

The posterior fossa is divided by a horizontal sheet of dura mater called the *tentorium cerebelli*. It attaches anteriorly to the petrous portion of the temporal bone and extends posteriorly to the limits of the cavity, joining the falx cerebri and the outer dura. Below are the middle and lower parts of the brainstem and the cerebellum, so called *subtentorial* structures. Superior to the tentorium cerebelli are the parietal and occipital parts of the cerebral hemispheres.

The most prominent landmark of the posterior fossa is the large *foramen magnum,* an opening in the occipital bone at the base of the skull. It is situated approximately in its center, which represents the border delineating the borders of the brainstem and the spinal cord. Anterolateral to the foramen magnum, on either side, the hypoglossal canals form passages for the hypoglossal nerves on their way to supply motor innervation to the tongue. The anterior parts of the posterior fossa are bounded by the posterior walls of the petrous portion of the temporal bone. Piercing these walls are the *internal auditory meatuses,* via which the vestibulocochlear nerves connect the brain to the inner ears and vestibular mechanism. The facial nerves also pass through these openings, to provide motor control to the muscles of facial expression and taste sensation to the anterior two-thirds of the tongue. An internal auditory artery also passes through the internal auditory meatus.

The Pharyngeal Cavities

Anatomically, the pharyngeal cavities are not proper cavities of the skull. They exist only as soft tissue spaces at the rostral end of the respiratory and digestive tracts, with support offered by the basal skull and cervical vertebrae. Anteriorly and laterally, the hyoid bone and mandible offer an arched support.

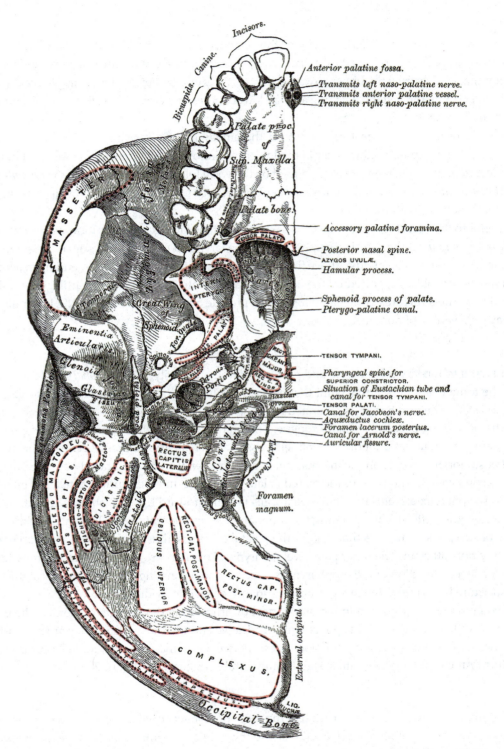

The importance of the pharyngeal cavities speech articulation cannot be overstated. They are the conduit connecting the respiratory system with the oral and nasal cavities and, creating important resonating bodies for the vocal tract. In addition, of course, their inferior division is the larynx, location of the mechanism for production of the voice and for glottal consonants.

Combined, three pharyngeal cavities form the *pharynx*. These are, from inferior to superior, the *laryngopharynx*, the *oropharynx* and the *nasopharynx*. Each cavity is usually considered separately, since each of the three has a certain structural and functional significance.

Laryngopharynx

The *laryngopharynx* is the most inferior cavity of the pharynx. It is described in detail in Chapter 5. The major structure of this part of the pharynx is the *larynx,* a complex muscular series of valves with an underlying cartilaginous skeleton that can obstruct the upper airway, either partially or completely. It is a complex cavity, beginning at the articulation between the first tracheal ring and the lower border of the cricoid cartilage and extending superiorly to the upper, free border of the epiglottis.

Completely closed, the laryngeal valves protect the airway from intrusive matter, including food or liquid. Alternatively, they can impound thoracic air in the lower respiratory system for certain biological functions.

Loose closure of the most inferior valve, formed by the vocal folds, allows creation of the *phonatory source* and three *glottal consonants,* two *glottal fricatives,* and a *glottal plosive* for speech, and all cavities rostral to the glottis contain air that resonates with these sources.

Just superior to the vocal folds is a small lateral widening of the laryngopharynx, forming the *laryngeal ventricle* or *Ventricle of Morgagni.* At the ventricle's upper extent, the laryngeal cavity narrows again, this time by the presence of a pair of loose ventricular folds. Above them is the *laryngeal vestibule* or *additus,* ending in a loose epithelial fold, the aryepiglottic folds, supported by the arytenoid, corniculate, and cuneiform cartilages and the epiglottis. Taken collectively, the structures of the laryngopharynx provide a means of separating the lower airway from the upper airway for a variety of non-speech functions.

Oropharynx

Superior or rostral to the laryngopharynx, and posterior to the oral cavity, is the *oropharynx.* Its major structure is the root of the tongue, and it is at this point that egressive pulmonary air flows through a passage common with the digestive system. The oropharynx is separated form the oral cavity by the posterior faucial arch, an elevated rim of mucosal tissue arching over the posterior tongue. This arch is also known as the *palatopharyngeal arch,* because the palatopharyngeus muscles underlie the epithelial tissue and create the eminence beneath the mucosal surface tissue. The posterior arch forms an incline that helps guide material to be swallowed past the laryngeal additus and into the esophagus. Contraction of the paired palatopharyngeus muscles during transfer of material from the oral cavity to the pharynx elevates the pharyngeal walls upward, toward the oral isthmus, and draws the left and right muscles together, effectively separating the oropharynx from the oral cavity. The palatopharyngeus muscles are also one of two pairs of muscles that oppose the palatal elevators, thus opening the velopharyngeal sphincter when gravity is not sufficient. The other paired velopharyngeal openers are the palatoglossus muscles.

The major speech function of the oropharynx is to serve as a resonating cavity for the vocal tract. In this role, its unique structure helps create the recognizable resonance of each individual's voice.

Pharyngeal consonants, produced in the oropharynx by articulating the tongue root and/or the epiglottis with the pharyngeal wall, take the form of approximants or fricatives (Maddieson, 2005). They are not part of Standard English phonology, but are important phonological components in some African, Middle Eastern, and indigenous northern North American languages, most significantly, Arabic (Elgendy and Pols, 2001).

Nasopharynx

The nasopharynx is the most superior or rostral of the three pharyngeal cavities, serving as a passageway between the nasal cavity and the upper respiratory tract. It is also the smallest, beginning at the imaginary horizontal plane of the palate and immediately curving upward and forward to end at the coronal plane formed by the edges of the choanae.

The major anatomical feature of the nasopharynx may be considered a part of the oropharynx as well. This is the *velum,* a muscular flap that forms part of the *velopharyngeal* sphincter. The velopharyngeal sphincter allows coupling or uncoupling of the nasopharynx from the rest of the vocal tract.

In anatomic position, gravity pulls the velum inferiorly, opening the passageway between the two pharyngeal cavities. This allows respiratory gasses to enter and exit the respiratory system through the nasal cavity. During

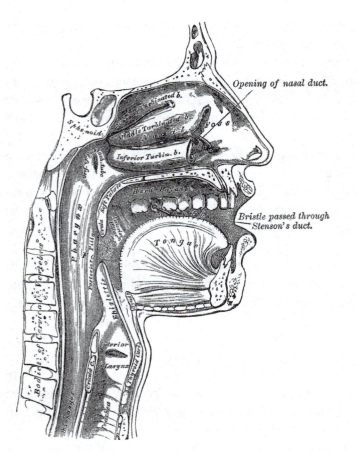

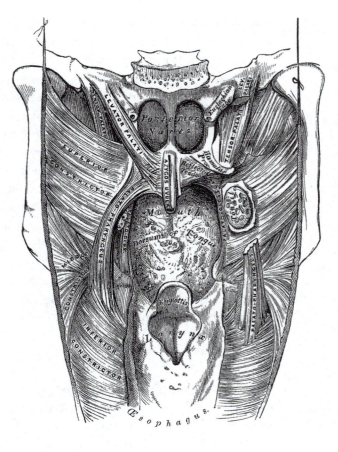

speech, the velum lowers, coupling the nasopharynx and nasal cavity to the vocal tract, and creating *nasal resonance*, essential to the production of *nasal phonemes* (in English, /m/; /n/; and/ŋ/).

A sling-like arrangement of muscular *palatal elevators* draws the velum upward and posteriorly to make contact with the lateral and posterior pharyngeal walls, separating the nasal cavity from the oropharynx when biological needs require variation in intraoral air pressure or when speech requirements are for an *oral resonance*. The palatal elevators are the paired *tensor velipalatini* and *levator palatini* muscles.

Two other important anatomic features of the nasopharynx are the openings (*ostia*) of the twin *Eustachian tubes (also called pharyngotympanic tubes or auditory tubes)*. Located just above and lateral to the velum, the Eustachian tubes are paired conduits between the tympanic cavities and the nasopharynx, whereby the former can maintain air pressure equilibrium with the ambient atmosphere. At the pharyngeal end, the tubes are formed of elastic cartilage and lined with ciliated pseudostratified columnar epithelial cells and goblet cells, which keep accumulated material away from the tympanic cavities. The cartilage of each tube creates a swelling or *torus tubarius* in the mucosa. Extending inferiorly from these are swellings of mucous membrane created by the salpingopharyngeus muscles. These muscles originate at the pharyngeal ends of the eustachian tubes and insert among the fibers of the palatopharyngeus muscles. Contractions of the paired

MANEUVERS TO CLEAR THE EUSTACHIAN TUBES

Valsalva Maneuver: With the nostrils pinched and the glottis closed, air pressure in the nasopharynx is elevated, causing a corresponding increase in tympanic air pressure and distention of the tympanic membranes. Excessive force during this action can rupture the tympanic membranes and cause blood pressure fluctuations.

Frenzel Maneuver: A low effort Valsalva maneuver accompanied by pharyngeal muscular contractions creates a more gentle action than the Valsalval maneuver alone, and helps avoid the negative consequences of the Valsalva.

Toynbee Maneuver: With the nostrils pinched shut, the patient swallows. This creates relatively negative air pressure in the nasopharynx, Eustachian tubes and tympanic cavities.

salpingopharyngeus muscles draw the lateral pharyngeal walls superiorly during swallowing, and may have the effect of opening the Eustachian tubes to facilitate tympanic ventilation.

A *pharyngeal* tonsil, or *adenoid*, is located on the posterior wall of the nasopharynx, immediately posterior to the nasal choanae. This mass of lymphoidal tissue is part of *Waldeyer's Ring*, along with the palatal and lingual tonsils, and protects the upper airway by reacting to bacteria and triggering immune responses. The pharyngeal tonsils are non-encrypted, and are usually most prominent during the first three years of life. Thereafter, they shrink and lose their effectiveness as immune system components.

Problems involving the pharyngeal tonsils are *adenoidal hypertrophy* or chronic infection. In extreme cases, a hypertrophic adenoid can completely occlude the nasopharynx, causing great difficulty or impossibility of passing air through the nasal cavities. The speech result of such a condition is *hyponasal* speech, or a loss of nasal resonance and the ability to produce the nasal phonemes. The voice under such a condition is sometimes said to be "Adenoidal." Adenoidal hypertrophy is rarely a problem in adulthood. Chronic infections can create respiratory and tympanic hazards if not managed medically.

If infections or hypertrophy of the pharyngeal tonsils are chronic, the glands are usually removed surgically, along with the palatal tonsils. Following tonsillectomy/adenoidectomy, the insufficiency of tissue in the velopharyngeal region may result in a transient *hypernasality* of speech. This condition is usually self-correcting, or is readily susceptible to speech therapy.

Oral Cavity

The oral cavity doesn't exist without the mandible to form its inferior aspect. This familiar cavity, commonly known as "the mouth," is inferior to the nasal cavity in the midline of the face. It is the entrance to the digestive tract, an alternate entrance to the airway, and the location of most of the speech articulators.

Anteriorly, the oral cavity is open to the environment, from which food, liquid, and air are admitted in the cavity's role as a respiratory and digestive tract organ. The open end, bordered by the highly moveable lips, and with the adjacent tongue is also convenient for the generation of phones or speech sounds.

The lips (*labia*) form the opening or the oral cavity. They are extremely sensitive and flexible, suiting them well for their alimentary function as well as for speech. The tissue of the lips is mainly of stratified squamous epithelium, superficial to a connective tissue *lamina propria* and fibers of the *orbicularis oris* muscle. The lips are addressed in detail among the facial articulators in this chapter.

Deep to the lips and cheeks at the anterior and lateral aspects of the oral cavity are the *superior* and *inferior dental arches*. Upper and lower *labial sulci* separate the anterior teeth from the lips, with *upper and lower labial frenulae* connecting the central inner part of each lip with the gums between the central incisors. The dental arches normally consist of up to thirty-two teeth contained in *alveolae*. The alveolae are spaces within the maxillary and mandibular *alveolar processes*. These potentially contain up to thirty-two teeth, the alimentary functions of which are to slice, grind, and mash food.

The soft tissue of both alveolar processes are the *gingiva* or "gums." This tissue is continuous with the rest of the oral epithelium being composed of stratified squamous tissue to resist wear. Most of the gingival tissue *alveolar* or attached *gingiva*, very tightly bound to the underlying bone of the alveolar processes. Around each tooth, however, a natural shallow pocket, or *gingival*, sulcus forms from the loose investment of *marginal* or *free gingiva*. Between pairs of teeth is the *interdental gingiva*. A gingival *lamina propria* of dense elastic connective tissue underlies the epithelial tissue. This connective tissue is continuous with that of the surrounding oral cavity.

The teeth erupt from alveolae in the upper and lower alveolar processes of the maxilla and the mandible. Outer surfaces of the dental arches are classified as *labial, buccal, lingual,* and *palatal,* according to which part of the oral cavity they face. Each tooth has a *crown* above the gum line and a *root*, below the gum line. An acellular, durable outer surface, the *enamel*, covers the crown. The root is covered with *cementum*, a less durable covering. A medial layer called the *dentine* comprises most of the tooth's bony mass. It is a calcium based complex formed mostly

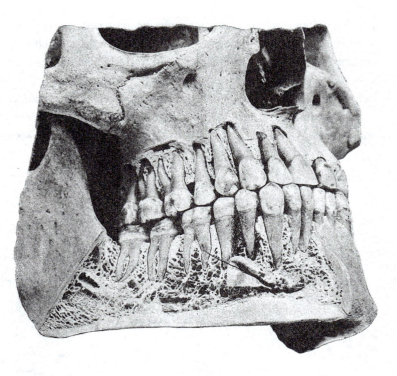

of many tubules, more numerous in the center than towards the outer surface. An inner core, the *pulp*, provides a structure through which nerve and circulatory supply is provided. The dentin-pulp complex provides the vital forces for the tooth's sustainability.

A *periodontal ligament* is situated between each tooth's root and the alveolar process, suspending it and protecting it in its bony socket. This ligament is a dynamic connective tissue structure, reacting to mechanical force vectors to self-repair and allowing repositioning of the tooth for more complete occlusion. Most of its *oxytalin* fibers are arranged radially around the tooth, attached at one end to the cementum and at the other to the alveolar bone. These fibers run in various directions, horizontally and obliquely up and down. Near the crown, a few periodontal ligament fibers connect the enamel to the free gingiva. It also contains tactile sensory end organs to provide sensory information for fine control of mastication, and a capillary network to nourish the adjacent alveolus and cementum.

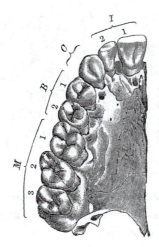

Each dental arch contains four *incisors*, narrow and sharpened, designed for slicing or stripping food. The two incisors on either side of the midline are *central* incisors, and the two situated lateral to the central incisors are *lateral* incisors. Following the dental arches around to the posterior teeth, the next pairs of teeth are *cuspids*, colloquially called *canines*, for their resemblance to the gripping teeth of carnivores. Two pairs of *bicuspids* or *premolars* are next, followed by three pairs of *tricuspids* or *molars*. The last pairs of molars are commonly called *wisdom teeth*, and usually do not erupt, if they erupt at all, until the late teens or early twenties. Molars are designed for grinding and mashing food, and for mixing it with saliva in preparation for swallowing.

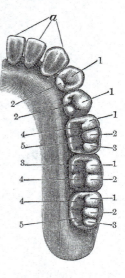

Human beings develop two complete sets of teeth during the lifetime. The first set is called *deciduous*, or *primary*, since it is shed before adulthood. Deciduous teeth erupt from their alveolae according to a schedule compatible with good nutrition during the rapid development of early life. First to erupt are the central incisors, usually at about 8 to 10 months, followed by the lateral incisors. Next, the cuspids and bicuspids, erupt. Tricuspids are not contained in the deciduous dental arches, leaving the total number of normally erupting deciduous teeth at 20. By the time the second upper premolars erupt, usually from 4 to 8 years of age, the incisors are shed. This allows the child to chew while the permanent incisors erupt. The permanent incisors are usually fully erupted by age 8, with the central incisors erupting first.

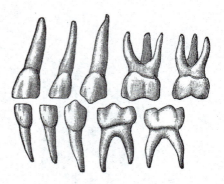

The speech function of teeth is to serve as fixed articulators, against which the tongue and lips are juxtaposed for *dental* and *labiodental* pulmonic consonants. These consonants are: /t/, /d/, /n/, /r/, /l/, /θ/, /ð/, /f/ and /v/. For speech purposes, only the upper incisors are needed.

The *dental* consonants are created by positioning the anterior tongue against the upper incisors. The *dental plosives, approximants,* and *nasals* are classified by the International Phonetic Association as among the groups of *dental-alveolar-postalveolar* consonants, since positioning of the tongue anywhere along that maxillary anatomic continuum has no contrastive (meaningful) effect on the syllable in which they are used. Among the fricatives, however, the differences are contrastive. Dental fricatives, /θ/ and /ð/, are contrastive in comparison with alveolar, /s/ and /z/, or postalveolar /ʃ/ and /ʒ/ fricatives. Both *labiodental* consonants used in English are fricatives, /f/ and /v/, differing mainly in terms of the voicing contrast.

The floor of the oral cavity is made up mostly of soft tissue, consisting of integument and musculature filling the interior of the U-shaped mandible. The mandible, itself, contains the dentition of the lower dental arch, which, although not used for speaking, bears against the upper dental arch for biting and chewing.

The largest visible structure in the oral cavity floor is the tongue. This important organ is highly flexible, mobile, and muscular, rooted to the interior of the mandible, to the hyoid bone, and to the pharyngeal walls. The main body of the tongue is formed by striated muscle fibers, intrinsic and extrinsic, supported by a loose areolar connective tissue *corium,* or *lamina propria,* which supports the superficial epithelial tissue. Internally, three connective tissue septa add structure and serve as attachments for musculature.

The most important and most often described septum is the median lingual septum. This single, midline, loose connective tissue structure is thicker at its bottom than it is near the tongue's dorsum, and thicker posteriorly, near the tongue's root, than it is near the apex. This gives it the most flexibility at its anterior or apical end. The median septum separates the paired bundles of genioglossus muscle fibers, and serves as the origin for the transverse lingual muscle.

Abd-El-Malik (1939) reported the presence of two other septa to be included in the connective tissue framework of the tongue. These were the paired *paramedian septa* and paired *lateral septa.* The *paramedian septa* extend horizontally on either side of the median septum, with the transverse and vertical intrinsic muscles above. They are broader posteriorly, narrowing to a point anteriorly, about two-thirds along the length of the median septum. This gives them the appearance of an arrow head in the inferior part of the tongue's substance, except that they curve inferiorly toward their posterior ends, where they attach to the hyoid bone. The lateral septa extend upward, then laterally, from the downward incline of the *paramedian septa* on either side, just below the intrinsic inferior longitudinal muscles.

The tongue's lateral and anterior margins are free, allowing these edges to reach almost any point inside the cavity and to protrude well beyond the rostral opening. The tongue's great mobility is suited to the alimentary functions of licking, chewing preparation of the food bolus and, importantly, transferring it to the oropharynx for swallowing. This high mobility is also an asset for speech.

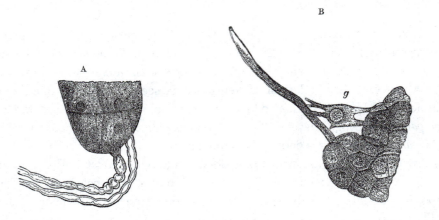

Extending inferiorly from the bottom of the tongue is an elevated tissue fold, the *lingual frenum* (Latin for "Little Bridle;" also *frenulum*), attaching the blade to the oral cavity floor. The frenum is manifested of various lengths, and a restrictively short example can limit tongue mobility. In such cases, surgical intervention may be required to restore tongue mobility and normalize speech articulation.

The tongue is covered with two types of sensory epithelial tissue to receive and convey the sensations of taste and touch to the central nervous system. Beneath it are two *sublingual folds,* through which open the ducts of the *sublingual* and *submandibular* salivary glands, embedded in soft epithelium. These digestive glands secrete saliva, a mixture of water, enzymes, electrolytes, mucous, several antibacterial compounds and opiorphin, a recently discovered pain inhibiter (Wisner, Dufour, Messaoudi, Nejdi, Marcel, Ungeheuer, and Rougeot, 2006).

The paired *sublingual salivary glands* create saliva and secrete it into the oral cavity through about a dozen openings called *Rivinus' ducts.* Several Rivinus' ducts unite at the anterior portion of the sublingual fold on either side of the frenum to form a pair of single larger salivary delivery tubes, *Bartholin's ducts,* which drain the anterior portions of the two sublingual glands. Two *submandibular salivary glands* are located posterior to the sublingual glands, near the hyoid corpus, and also provide saliva for the oral cavity. Their saliva is secreted through *Wharton's ducts* opening anterior to Bartholin's ducts within the sublingual fold.

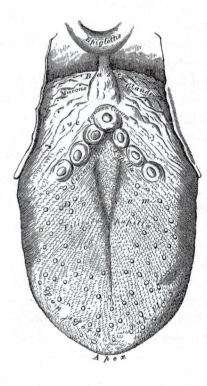

The *dorsum* or "back" of the tongue is that part most visible upon inspection of the open oral cavity and the part that contacts the oral roof directly when the mouth is closed. Superficial features of the dorsum include a tip or apex at its anterior extend, and a blade, or free edge extending laterally and posteriorly from the apex to the root.

On the dorsum's surface, clear histological differences can be seen between the anterior two-thirds and posterior third the surface. This difference is related to the different embryological origins of the tongue. The anterior two-thirds, or *oral part* of the tongue develops from the tissues of the mandibular arch, while the posterior one-third, or *pharyngeal part* of the tongue is derived from the third arch.

A *longitudinal sulcus* divides the tongue's convex dorsum into right and left halves from the apex back to the pharyngeal part of the tongue. The longitudinal sulcus ends abruptly at the shallow *sulcus terminalis,* a V-shaped groove with its apex in the center of the dorsum, pointing to the pharynx, at a point approximately 25 mm from the tongue's root. The legs of the sulcus terminalis extend anteriorly and laterally to the posterior end of the blade. The point of the sulcus terminalis is the *foramen cecum,* a vestigial depression remaining from the migration of the thyroid gland inferiorly to its ultimate location. A *thyroglossal duct* sometimes remains between the thyroid gland and the foramen cecum as evidence of this migration. The sulcus terminalis marks the separation of the anterior two-thirds and the posterior one-third of the tongue's dorsum.

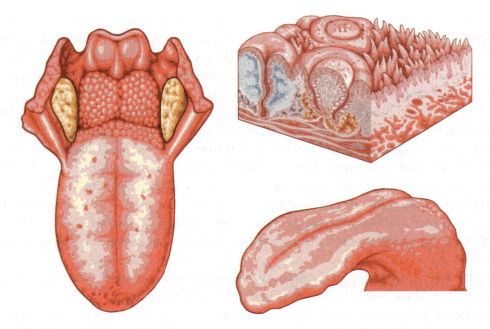

The superficial tissue of most of the tongue is a layer of stratified squamous epithelium, designed to resist the wear that is inevitable in the process of chewing and swallowing. This tissue is smooth on the tongue's inferior surface, but quite rough on the dorsum. On the dorsum, extending to the edges of the blade, the superficial tissue takes on the form of several types of finger-like projections, extending away form the surface. Underlying the superficial layer is a *lamina propria* of loose connective tissue, and beneath this are striated muscle fibers coursing transversely, vertically and longitudinally. Interspersed among the fibers are serous and mucosal glands, the later being most prominent at the tongue's root, as well as lymphatic and blood vessels and the tongue's nerve supply.

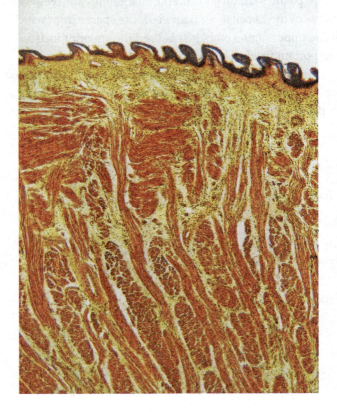

In addition to the squamous epithelium, the epithelium of the tongue's dorsum also takes on the form of variously shaped papillae or finger-like projections (from the Latin, *papulae*, or nipple). There are four types of papillae, three of which, the *fungiform, circumvallate,* and *foliate papillae,* are formed to provide foundations for taste organs. The fourth type, the *filliform papillae,* are rough structures, formed to provide mechanical support for mastication and swallowing. A more complete discussion of lingual papillae is contained in Chapter 7.

Deep to the epithelium of the tongue is its supportive connective tissue *lamina propria.* This layer provides circulatory support for the superficial epithelium and contains minor *Von Ebner's* salivary glands.

Deep to the lamina propria, and forming the bulk of the tongue, are several layers of intrinsic and extrinsic muscles. These are striated muscles, with fibers running in several directions. Intrinsic muscles have origins and insertions within the body of the tongue, whereas extrinsic muscles have origins outside the tongue's body.

The tongue's importance in the articulation of speech is vast. The chart of the International Phonetic Association indicates that forty-three of fifty-two pulmonic consonants require juxtaposition of the tongue with some other oral structure. To these might added an indeterminate number of vowels, depending upon phonetic context and the syllabic stress.

The oral floor is fortified by several layers of muscle, deep to the smooth epithelium and salivary glands, including those of the tongue. These play important roles in speech, mastication, and swallowing.

The roof of the oral cavity is also essential to speech articulation. It is bony for its anterior two-thirds, becoming soft at its posterior one-third. The foundation of the anterior two-thirds is the fused palatal processes of the maxillae, both fused posteriorly to the horizontal processes of the paired palatine bones. This part of the oral cavity is often referred to as the *hard palate,* and forms the floor of the nasal cavity just above. Superficial to the bones and mucoperiostuim of the hard palate are basement membrane and epithelium. An elevated ridge, the palatal suture, along the midline of the palate marks the fusion of its lateral halves. Several transverse folds in the epithelial tissue, so-called *palatal rugae,* are common and may be sufficiently individual in design to be a source of forensic identification (Robison, Summitt, Oesterle, Brannon, and Morlang, 1988). Palatal rugae are prominent in the embryo and steadily decrease in number after birth (Hauser, Daponte, and Roberts, 1989).

Around the arch of the hard palate's lateral and anterior margins are the teeth of the upper dentition. These are contained in the *alveolar process* of the maxillae, so named for the spaces from which the teeth arise. A prominent ridge at the anterior end of the upper dental arch, just behind the superior incisors, is familiar to phoneticians as the *alveolar ridge* and is an important fixed point for consonant articulation.

The posterior part of the oral cavity roof is soft tissue, formed by muscles covered with the continued epithelium of the anterior portion. This part of the upper oral cavity is the *velum,* popularly called the *soft palate* and is moveable to juxtapose with the posterior and lateral pharyngeal walls. The velum and pharyngeal walls form the

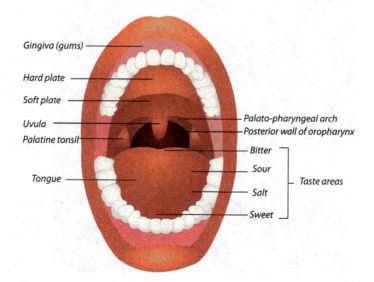

velopharyngeal sphincter, the action of which is to seal the nasopharynx from the rest of the distal respiratory tract, preventing ingress or egress through the nose. The velopharyngeal sphincter is quite active during speech, elevating to isolate the nasal cavity for formation of *oral phonemes* and lowering to couple the nasal cavity with the rest of the vocal tract, creating the distinctive resonance of *nasal phonemes* and their immediate temporal neighbors.

At the tip of the velum is an inferior projection of epithelium and muscle: the *uvula*. Its muscular infrastructure, the paired musculus uvulae, originates at the posterior nasal spine of the palatine bones and from the *palatine aponeurosis* and inserts into the uvular epithelial tissue. *Uvulae* contractions tighten the tissue of the uvula and velum, and may play a role in production of *uvular consonants*, none of which is standard in English.

The *oropharyngeal isthmus*, also called *isthmus of the fauces* (Fauces: Latin for "throat," or "narrow passage") in the posterior oral cavity marks the transition between the oral cavity and the oropharynx. It is bounded posteriorly by the *posterior faucial arch*, an arch of epithelium and the underlying *palatopharyngeus* muscle. At the anterior end of the oropharyngeal isthmus, the *palatoglossus* muscle gives shape to the *anterior faucial* arch.

The lateral walls of the oral cavity are formed by the soft epithelial tissues of the cheeks. This is stratified squamous epithelium with minor salivary glands interspersed. The openings of the *parotid ducts (Stenson's ducts)* are prominent features of the lateral oral epithelium, just lateral to the second upper molars on either side. The parotid ducts drain saliva produced by the *parotid gland*, the largest of the salivary glands. The parotid glands are wrapped around the mandibular rami bilaterally. Their ducts course anteriorly, through the buccal fat pads and superficial to the masseter muscles to pierce the oral epithelium. Between the dental arches and the lateral oral walls, upper and lower *buccal sulci* form pockets into which food or liquid can become lodged, the former often requiring mechanical assistance to remove. This is usually accomplished by the tongue.

A pair of *faucial arches* are prominent in the posterior oral cavity, created by the presence of musculature that draws the velum inferiorly, thus opening the velopharyngeal sphincter and coupling the nasal cavity with the vocal tract. The most posterior of these, the *palatopharyngeal arch* is the eminence formed by the palatopharyngeus muscle, originating at a medial tendon in the midline of the vellum and inserting among the fibers of the middle constrictor muscle. The anterior faucial arch marks the posterior limit of the oral cavity, and is created by the eminence in the epithelium created by the *palatoglossus* muscle. This muscle is considered by most as paired, although its fibers are continuous from the left to right side. They insert into the posterolateral tongue.

Roles of Oral Articulators in Speech

Oral articulators move in relationship with one another to perform two speech functions. First, they modify the resonating characteristics of the vocal tract for vowel, nasal, and approximant articulation. Second, they modify egressive pulmonary air to create plosive and fricative driving sources along an articulatory site continuum from the teeth to the velum. Between these two sites are the alveolar, postalveolar, and palatal places of consonant articulation.

The oral articulators provide the greatest degree of vocal tract flexibility of any of the speech articulators. This is mostly attributable to the great flexibility of the tongue. Tongue position is virtually infinite, allowing for continuous variability in oral cavity dimensions. This creates a similar variability in possibilities for vowel, approximant, and nasal articulation. As discussed previously, the vocal tract becomes a variable *double Hemholtz resonator*, with a large oral and pharyngeal resonating cavity, separated from a second, smaller oral resonating cavity by the constriction created by the tongue and the oral roof. As the tongue moves nearer and farther from the oral roof, the constriction becomes less or more open. The length of the lingual constriction varies, as well, creating more possibilities for speech sound variation. Anterior to the oral resonating cavity, another constriction created even more possibilities. This one is formed by the lips and is variable in area as well as length.

In addition to their flexibility in changing vocal tract resonance, the flexibility of the oral articulators allows them nearly or completely to occlude the oral cavity for fricative and plosive articulation at several locations. The most anterior oral site is the labiodental site, at which the lower lip articulates with the upper incisors. The most posterior oral articulation site is the velar site at which the back of the tongue elevates to meet the velum. Moving the position of the plosive and fricative driving sources within the elastic resonating system that is the vocal tract thus creates changes in resonating cavity dimensions in front of and behind the source.

Finally, the ability of the velum to separate or couple the oropharynx and nasopharynx combines with the tongue's ability to create variously dimensioned *cul de sac* resonating cavities in the oral cavity, allows for production of nasal phonemes.

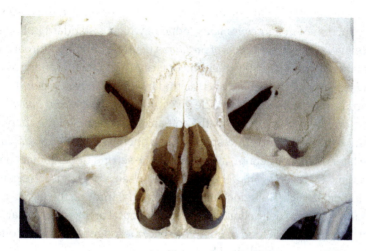

Nasal Cavity

The nasal cavity is located approximately in the middle of the facial skeleton. It is open to the outside anteriorly, and is the main entrance to the airway, seat of the organ of smell, and serves as an alternate vent and resonating cavity for speech. The nasal cavity opening is bounded by the twin nasal bones and the paired maxillae.

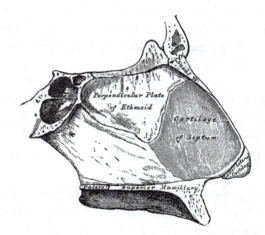

The bottom portion of its interior is formed by the same bones that form the superior oral cavity. These are the palatal processes of the maxillae, forming the anterior two-thirds, and the horizontal processes of the palatine bones, forming the posterior one-third. The processes of both bones are fused to their contralateral counterparts medially and to one another. Incomplete fusion of the palatal and horizontal processes results in an inability to separate the nasal cavity from the respiratory and digestive (vocal) tracts and has an abnormal effect on speaking and swallowing until treated.

The roof of the nasal cavity is formed anteriorly by the nasal bones and a small part of the frontal bone. Deeper into the cavity, the medial part of the roof is formed by the cribriform plate of the ethmoid bone. This plate is multiply pierced by the end organs for the olfactory sense that communicate directly to the olfactory nerve. Posteriorly, the nasal cavity roof slopes downward and is formed by the body of the sphenoid bone.

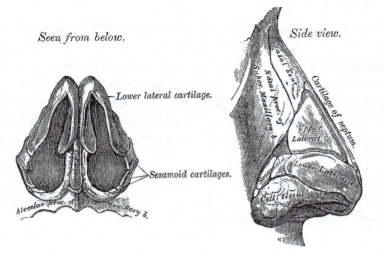

Seen from below.

Side view.

The nasal cavity is separated into two nasal passages by a *nasal septum*. The bony proximal portion of the septum is formed mainly by the perpendicular process of the ethmoid bone, descending down from the roof of the cavity and by the vomer, posteriorly and inferiorly. Small elevations created by the fusions of the palatal processes of the maxillae, horizontal processes of the palatine bones inferiorly, and by the fusion of the paired nasal bones superiorly, contribute to the septum, as well. The cartilaginous distal end of the nasal septum is formed by the septal cartilage and small projections of the greater alar cartilages, which form lateral margins of the nares. The anterior and lateral walls of the distal nasal cavity are formed by skin covered cartilage. The most prominent of these are the lateral and alar cartilages, which form the most distal part of the cavity. Two greater lateral cartilages attach superiorly to the nasal bones and inferiorly to the alar cartilages. The alar cartilages form the rounded nostrils or nares, and fold inward to contribute to the nasal septum. Several lesser lateral cartilages fill in the space between the alar cartilages and the maxilla. Fatty and fibrous connective tissue complete the form and characteristic shapes given to that end of the nose.

Around and behind the nasal cavity are four pairs of *paranasal sinuses*. These are spaces within the bony substances of the frontal, sphenoid, and ethmoid bones and the maxillae that communicate with the nasal cavity. They are present before birth, but grow with the facial skeleton and reach their full volumes at about age eighteen. The paranasal sinuses are lined with mucosal epithelium, the secretions of which flow into the nasal cavity through *ostia*. Interruption of this flow, commonly by swelling and inflammation of nasal cavity tissues, can cause discomfort and pain resulting from inflammation of the membranes lining the sinuses, or *sinusitis*.

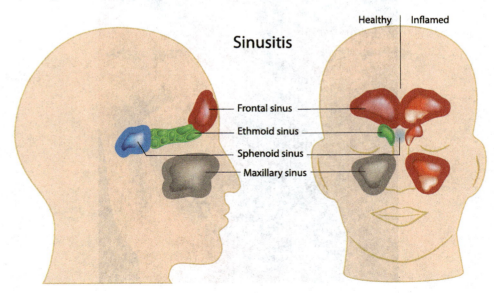

Role of the Nasal Cavity in Speech

A patent nasal cavity, coupled and uncoupled to the rest of the vocal tract by action of the velopharyngeal sphincter, is fundamental to normal speech. This is because the addition of an additional resonating cavity, creating so-called *nasal resonance* to the vocal tract system is perceived as a normal aspect of speech. Nasal resonance is normal not only for production of the *nasal phonemes* but for creating *nasality* to vowels or approximants in the contiguous phonetic environment of those nasal phonemes. If the nasal cavity is not patent, blocked, for example, by nasal secretions or by a hypertrophic pharyngeal tonsil, the resulting speech is said to be *hyponasal*. If the nasal cavity is coupled to the vocal tract excessively, the resulting speech is termed *hypernasal*.

Orbits

Lateral to the nasal cavity, on either side, are the paired orbits. They house the eyes, the extraocular muscles, lacrimal glands, and conjunctiva, the soft tissues that cushion the eye.

Bones of the Orbits

The orbits are formed by several bones, some of which contribute only a small portion of the inner walls. The bones of the orbits are the frontal and zygomatic bones and the maxillae, sphenoid, lacrimal, and ethmoid bones.

The outer margins of the orbits are formed by the maxillae and zygomatic bones inferiorly, the frontal and zygomatic bones laterally and superiorly, and the maxillae medially.

Contents of the Orbits

The interior of an orbit is roughly pyramidal in shape, with its apex deep inside and its base at the external margins. Through foramina in the apical end run important nerves and vessels. The optic nerve emerges from the eyeball through the *optic canal* to transmit action potentials associated with vision to the central nervous system. The ophthalmic artery, a branch of the internal carotid artery, uses the same canal to supply the eye and other orbital contents. A *superior orbital fissure* carries the oculomotor trochlear and abducens nerves, as well as the ophthalmic branch of the trigeminal nerve. The orbital veins also emerge from the orbits through the superior orbital fissures.

Deep in the orbits, surrounding the optic canals, two dense connective tissue rings that serve as the attachments for the origins of most of the six pairs of intrinsic eye muscles. In the superior and medial aspect of the orbit is a fossa at which attaches a connective tissue pulley for the tendon of the superior oblique muscle.

The orbits are lined with fascial connective tissues and a fatty cushion. A thin sheath of fascial tissue, *Tenon's capsule,* covers the eyeball as far as its cornea in front and extends to the exit of the optic nerve in the rear. The six paired extraocular muscles are each sheathed in fascia, continuous in some places with Tenon's capsule, and in others with *check ligaments* that limit eyeball movements. Orbital fat fills the rest of the lateral and posterior orbit, cushioning the organs within, and a varied layer of epithelial tissue, the *conjunctiva,* lines the interior of the eyelids continuing to cover the anterior parts of the eyeballs.

The roles of the eyes in communication are mainly as receptors of visual input, plus a relatively minor role in expression. Vision supports cognitive and linguistic communicative functions, ranging from the ability to orient visually to the environment or a communicative partner to the reception and ultimate association and integration of complex written material. Monitoring gestural or written communicative output is another fundamental role of the eyes. Finally, communicative expression can be enhanced by ocular gestures such as "rolling the eyes" or direction and diversion of eye contact in the communicative act.

The Facial Articulators

Facial articulators are those located at the extreme distal end of the vocal tract and on the anterior aspect of the skull. Operating on the bony foundations formed by the *frontal, nasal, maxillary, zygomatic,* and *mandibular* bones, these soft tissue structures create the unique attributes that are called "the face," distinguishing the individual both in their static characteristics and in their movements.

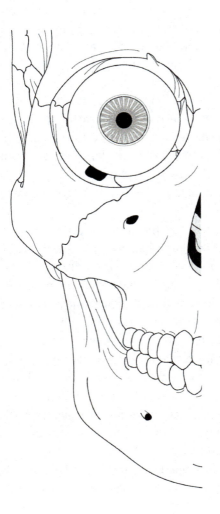

Muscles of Facial Expression

Groups of *muscles of facial expression* contract to create the movements of the soft tissues around the oral, nasal, and orbital cavities for speech and gestures. These muscles either constrict the openings of these cavities or oppose the constrictors to spread the openings. With a few exceptions, these muscles are very thin. Most originate at bony sites on the face, and all insert into fibers of other facial muscles or within the skin of the face.

The muscles of facial expression will be listed and described in detail with the other muscles of speech articulation.

Lips

The lips form a sensitive and flexible distal opening of the alimentary and vocal tracts, as well as an alternate opening to the respiratory tract. They are one of the most distinguishing facial features.

The lips begin at the base of the nose superiorly and at the mandibular border inferiorly, and extend laterally to the *oral commissures*, where the upper and lower lips meet. Extending diagonally from the lateral borders of the nose to the oral commissures are two *nasolabial folds*. These become more pronounced with age and represent the anatomic border between the lips and cheeks.

Immediately bordering the oral opening is *vermillion zone*. This is the smooth, pinkish labial fold, rich with tactile end organs. The outer, facial vermillion zone surface is covered by keratinized, stratified squamous epithelium to resist wear. This tissue contains no lubricating or moisture providing glands and is usually moistened by licking or by manual application of ointments.

On the inside, nearest the gums, the epithelial tissue remains squamous and stratified, but is non-keratinized. This tissue is continuous with that of the ventral tongue and oral floor, and contains small *labial salivary glands*, similar to those of the adjacent oral epithelium. Superior and inferior *labial frenula*, located in the areas between the central incisors of the upper and lower dental arches, provide attachment support between the inner lip tissue and the gingival tissue.

The deepest part the lips is a sphincteric skeletal muscle, the *orbicularis oris*. Between the orbicularis oris muscle and the superficial epithelium is a *lamina propria* of connective tissue. Capillaries from the *superior* and *inferior labial arteries* lie immediately deep to the vermillion zone surface and give it its characteristic color.

The lips are extremely sensitive to touch and are classified as erogenous zones. Labial sensitivity to touch is important in manipulation of materials to be ingested as well as in the appreciation of movements for speech and facial gestures. Sexual arousal causes the superficial capillaries to engorge, deepening the color and swelling the tissues.

Outside the vermillion zone, the tissue of the lips changes to tissue similar to that of the rest of the face, the *cutaneous zone*. Cutaneous zone tissue contains hair follicles, as well as sweat and sebaceous glands. Between the vermillion zone and the cutaneous zone is a narrow *transition zone*.

Twin elevations on either side of the tissue just inferior to the nasal septum create a midline depression called the *philtrum*. These elevations are formed by fusion of the *median nasal* and *maxillary processes* during facial development. Combined with the curvature of the upper border of the vermillion zone, the inferior curved projection of the filtrum's groove forms the *Cupid's bow*, which, in name which bears further traditional reference to the erotic role of the lips. The orbicularis oris muscle extends beyond the vermillion zone beneath the bordering skin.

Above the oral opening, the nasal cavity apertures extend beyond the plane of the face. Owing to its limited mobility, the nose is not particularly flexible as an articulator. Constriction of the nareal opening can be accomplished by contractions of certain fibers of the *nasalis muscle (compressor naris)*. Sufficient nasal cavity constriction can produce a *nareal fricative*, sometimes employed as an anterior fricative substitute in speakers who cannot create enough intraoral air pressure for the standard versions.

Movements of the eyelids, eyebrows, and eyeballs are widely used to produce facial gestures. The importance of these movements in communication is largely ignored until they become unavailable through illness or injury.

Roles of Facial Articulators in Speech and Gestural Communication

The speech role of the facial articulators is that of rounding or spreading the oral opening during vowel and sonorant consonant articulation and for approximating the lips either with one another or with the upper incisors for labial consonant articulation.

Rounding and spreading the lips has the effect of adding or removing length to the vocal tract. When the lips are rounded, the oral cavity opening narrows and extends, changing the dimensions of the front vocal tract cavity and the center frequencies of the first four formants of vowels or sonorant consonants (Stevens, 1998). Almost all vowels listed by the International Phonetic Association have rounded and unrounded counterparts, but lip-rounding is standard for articulation of only the back English vowels: /ɒ; ɔ; o; ʊ and u/, the oral opening narrowing progressively as tongue height increases from open to close configurations. Lip-rounding is also common during articulation of the dental-alveolar-postalveolar approximant, /r/ (U.S. symbol) or /ɹ/ (international symbol), and the postalveolar fricative cognates, /ʃ/ and /ʒ/.

The International Phonetic Association describes /w/ and /ʍ/, listed in the "Other" category, as *labial-velar*. This initially beguiling description is an attempt to account for their rounded labial configuration as well their close (back) lingual posture, probably related to their close relationship to the close back vowels (e.g., /u/). The distinct resonance of /w/ is accounted for by adding the term "approximant" to its terminology, thus, *labial-velar approximant*, but key to its articulation is dynamic change in the constriction, rather than its static posture. Taken to extremes, rounding also produces a continuous, aperiodic fricative source. This occurs as a natural result of narrowing the oral opening while air egresses. The occurrence of this additional driving source during articulation of /ʍ/ is covered by its description as a *"labial-velar fricative."*

An additional and essential contribution to communication is the creation of facial gestures. Smiles, frowns, winks, and other facial postures play important pragmatic roles in personal interactions. One only needs to consider the necessity of graphic "emoticons" used to enhance email messages. These typographic recreations of universal facial gestures are often employed to enhance or even clarify written communication in electronic media.

Muscles of Articulation

The muscles of speech articulation are grouped in to pharyngeal, oral, and facial categories. Most of the muscles that move the speech articulators also have digestive and respiratory biological functions, both of which will be included in their functional descriptions. Muscles of facial expression, considered in the facial group, also have gestural communicative functions.

Pharyngeal Muscles of Articulation

Muscles in the pharynx function in the production of glottal consonants and, of course, the creation of the phonatory source. The only English consonant phonemes produced at pharyngeal sites are gottal, including a *glottal plosive* /ʔ/ and two *glottal fricatives*, /h/ and /ɦ/. The muscles that create movements for glottal consonants are those that adduct and tense the vocal folds, located in the larynx. These are described in detail in Chapter 5.

Another speech function of the pharyngeal muscles of articulation is to change the dimensions of the vocal tract resonating cavities. Although this action does not create phonemic contrast, it does alter the acoustic characteristics of the vocal tract.

Biological or non-speech functions of the pharyngeal muscles are as muscles of deglutition and respiration. As muscles of digestion, they assist the bolus of food or volumes of liquid as they transfer between the oral cavity and the esophagus and protect the airway. Respiratory functions include protecting the airway from unwanted intrusion and blocking pulmonary air egress for airway maintenance and for inflating the thorax for non-speech functions requiring additional mechanical support, such as lifting and defecation. Digestive functions will be described in this section. Respiratory functions were described in detail in Chapter 5.

ZENKER'S DIVERTICULUM

Swallowing difficulties and halitosis can result from a *Zenker's Diverticulum*. A diverticulum is a *cul de sac* developing in a tubular organ, such as the pharynx or esophagus. A Zenker's diverticulum, described by Friedrich Albert Zenker in 1877, is technically a *pulsion diverticulum*, meaning it develops in the mucosal epithelium, rather than in the muscle tissue, between the muscles fibers of the cricopharyngeal and thyropharyngeal parts of the inferior constrictor muscle.

The condition is relatively rare, and most often reported in long-lived patients. Excellent results have been reported following surgical treatment (Bragg, 2006; Achkar, E. 1998).

The Pharyngeal Constrictors

The *pharyngeal constrictors* are the most superficial of the pharyngeal muscles. They form a sphincteric ring around the outer pharynx in three main bundles: the inferior pharyngeal constrictor, the middle pharyngeal constrictor, and the superior pharyngeal constructor. Hiatuses in the pharyngeal muscular wall appear at the lateral aspects between the three bundles.

The *inferior pharyngeal constrictor* originates on the lateral aspects of the thyroid and cricoid cartilages and the fibrous tissue between them. The fibers course posteriorly and meet in the midline at the *median pharyngeal raphe*, a long, narrow strip of connective tissue extending from its superior attachment at the occipital *pharyngeal tubercle* down the length pharynx to the esophagus. The median pharyngeal raphe is the insertion for the middle and superior pharyngeal constrictor muscles, as well. Contraction of inferior pharyngeal constrictor fibers facilitates swallowing by encircling the bolus and guiding it into the esophagus. The inferior pharyngeal constrictor muscle can be further divided into *cricopharyngeal* and *thyropharyngeal* parts. The most inferior portion, the cricopharyngeal portion, is also identified as the *cricopharyngeus muscle*. Thicker than the thyropharyngeal part, it arises on the lateral aspects of the cricoid cartilage, and serves as the sphincteric entrance to the esophagus. The cricopharyngeus muscle is under voluntary control and is used by alaryngeal speakers to produce the *esophageal voice* source for speech.

The *middle pharyngeal constrictor* originates on the horns of the hyoid bone and the inferior part of the stylohyoid ligament. Fibers insert at the median pharyngeal raphe. Contraction of middle constrictor fibers facilitates movement of the bolus during swallowing.

The *superior pharyngeal constrictor* takes its origins from multiple sites in the oropharynx and nasopharynx. From inferior to superior, these include the posterolateral aspects of the tongue and the mandible, just posterior to the last molar; the connective tissue *pterygomandibular raphe*, and, at its most superior extent, the medial pterygoid plate of the sphenoid bone. Laterally and inferiorly, the fibers of the superior pharyngeal constrictor become continuous with fibers of the *palatopharyngeus* muscle, passing anteriorly and superiorly to the palatine aponeurosis to form the *posterior faucial (palatopharygeal) arch*, at the border of the oropharynx and oral cavity. Other superior pharyngeal constrictor muscle fibers, at least in some specimens (Saigusa, Yamashita, Tanuma, Saigusa, and Niimi, 2004), have been reported to be continuous with those of the transverse lingual muscle, forming a sphincter at the base of the tongue and acting in opposition to the genioglossus muscle. Remaining superior constrictor fibers insert posteriorly at the median pharyngeal raphe and, at its most superior extent, the pharyngeal tubercle of the occipital bone.

Simultaneous contractions of the superior constrictor and the palatopharyngeus muscle help propel the food bolus inferiorly during swallowing, and give rise to their being combined as muscles of the *palatopharyngeal sphincter*. The palatopharyngeal sphincter is distinguished from the velopharyngeal sphincter in that its function is to narrow the oropharyngeal lumen, while the velopharyngeal sphincter, located just above, narrows and closes the opening between the oropharynx and the nasopharynx.

The *stylohyoid* muscles have their origins at the base of the styloid process of the temporal bone, just medial to the external auditory meatuses on either side. Their fibers course inferiorly and anteriorly to insert on the hyoid bone, where the greater cornua join the corpus. Contraction of the stylohyoid muscles draws the hyoid bone upward and backward, taking with it the larynx and facilitating the approximation of the epiglottis and aryepiglottic folds during swallowing.

The *stylopharyngeus* muscles are another pair of muscles that facilitate swallowing by elevating the pharyngeal walls. They originate, as their name implies, at the styloid processes, slightly posterior and medial to the origin of

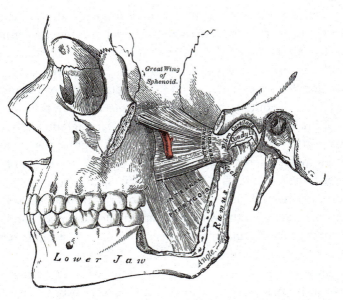

the stylohyoid muscle. Some of their fibers descend to intermingle with those of the superior pharyngeal constrictors, and others insert onto the posterior edges of the thyroid cartilage laminae.

The *lateral pterygoid* and *medial pterygoid* muscles originate on the undersurface of the sphenoid bone near the pterygoid plates. Both move the mandible in relationship to the sphenoid bone.

The *lateral pterygoid* muscles are the most superficial of the two pterygoid muscle pairs. They have two heads, one upper (superior) and one lower (inferior). The superior heads originate on the infratemporal crests of the right and left greater sphenoid wings. The inferior heads originate on the lateral surfaces of the lateral pterygoid plates. Both heads insert on the condylar process of the mandible. The lateral pterygoid muscles are distinguished in that they are the only muscles of mastication to contribute to *depressing the mandible*. They also contribute to mandibular lateralization by alternating contractions during so-called, "rotary chewing" and can move the mandible anteriorly through paired contractions.

The paired *medial pterygoid* muscles also have two heads, one deep and one superficial. The deep heads form the bulk of the muscles. They originate on the medial surfaces of the lateral pterygoid plates of the sphenoid bone. The small superficial heads originate partly on the lower parts (pyramidal process) of the palatine bones and partly on the maxillary tuberosities, just posterior to the last molars. Both heads insert in long swaths along the medial surfaces of the mandibular rami. Contractions of one or the other medial pterygoid muscle rock the mandible from left to right and upward, while simultaneous contractions elevate the mandible.

Muscles of the Velopharyngeal Sphincter

The muscles of the velopharyngeal sphincter system open or close the velopharynx, coupling or separating the nasopharynx with the rest of the vocal tract. The effect of coupling the nasopharynx with the vocal tract is that of creating an additional resonating cavity and thus, a distinctive spectrum for nasal phonemes.

Velopharyngeal musculature can be further subdivided into those that separate the oropharynx and nasopharynx, and those that open the sphincter to create nasality and allow air to flow through the nose. Muscles that close the velopharynx and separate the nasopharynx from the oropharynx all originate above the plane of the palate, thus lifting and pulling the soft tissues rearward. These are the *levator velipalatini* and *tensor velipalatini*. In some cases, particularly where the velum is short, the superior pharyngeal constrictor moves the pharyngeal walls medially to assist.

The paired *levator veli palatini* muscles elevate the velum and draw it toward the posterior and lateral pharyngeal walls. They originate on the undersides of the temporal bones and on the medial surfaces of the cartilaginous portions of the pharyngotympanic tubes, their fibers course down and forward to meet and interdigitate with one another in the midline of the velum.

The *tensor velipalatini* muscles form a sling that, when contracted, draws the sides of the velum laterally, toward the medial pterygoid plates. This not only tightens the velum, creating a firmer closure, but also draws it slightly upward. In its contacted tonic state, the tensor velipalatini also provide a stiffer platform against which to press the tongue back during articulation of velar phonemes. Tensor velipalatini muscles take and interesting course, and understanding them is essential to proper understanding of velopharyngeal sphincter function. The paired muscles have fairly wide origins in the *scaphoid fossae*, between the pterygoid plates of the sphenoid bones, the nearby *spines* of the sphenoid bone, and also on the cartilaginous (pharyngeal) ends the Eustachian tubes. These sites are superior to the plane of the hard palate and slightly posterior to its pharyngeal edge.

Tensor velipalatini fibers course inferiorly and anteriorly to the inferior edges of the medial pterygoid plates, where tendons form to create their insertions. These tendons wrap around the hammular processes and inferior edges of the medial pterygoid plates and fuse in the midline to form the *palatine aponeurosis*.

The palatine aponeurosis is central to the velar function, for all the superior and inferior muscles of the velopharyngeal sphincter, except the superior pharyngeal constrictor, attach to it, either partially or completely. It forms the dense, fibrous connective tissue foundation of the velum.

In anatomic position, gravity opens the velopharyngeal sphincter and muscular action is required to close it. However, in other positions, such as supine, a little muscular assistance may be required to open the passage between the nasopharynx and the oropharynx. Active contraction may also be required to ensure nasopharyngeal coupling during velar nasal (/ŋ/) articulation, as the tongue pushes against the lowered velum. This is provided by the *palatopharyngeus* and *palatoglossus* muscles, assuming the levator velipalatini and tensor velipalatini muscles are relaxed.

The *palatopharyngeus* muscles have already been described as pharyngeal elevators and as muscles that narrow the *palatopharyngeal sphincter*, the muscular portal through which food boli transit on the way from the oral cavity to the pharynx and espohagus. Palatopharyngeus muscles originate at the palatine aponeurosis and the horizontal plates of the palatine bones and pass inferiorly inside the pharyngeal constrictors. Some fibers pass inferiorly and posteriorly to intermingle with those of the superior pharyngeal constrictor. Other palatopharyngeus fibers continue downward to insert on the posterior edges of the thyroid cartilage laminae. Contractions of the palatopharyngeus muscles can draw the velum downward, toward the thyroid cartilage, narrowing or closing the oropharyngeal isthmus.

The *palatoglossus* muscles originate at the same palatine aponeurosis as does the palatopharyngeus muscle, only more anteriorly, with some of one side's fibers being continuous with those of the contralateral *palatoglossus*. They insert into the muscular body of the tongue, where some of its fibers are continuous with those of the transverse lingual muscle (Ekberg, Aksglaede, and Baert, 2004). Their actions can either pull the velum inferiorly, or draw the back of the tongue superiorly.

The paired *salpingopharyngeus* muscles have dual functions, both coincident with swallowing. Originating in the cartilaginous portion of the Eustachian tubes, where they open into the nasopharynx, these paired muscles fan out with fibers that insert among those of the palatopharyngeus, which itself, inserts into the lateral fibers of the inferior constrictor muscle and the posterior edges of the thyroid laminae. Contractions of salpingopharyngeus muscles during swallowing can draw the pharynx upward and pull down on the pharyngeal openings of the Eustachian tubes. This both assists transfer of swallowed material into the esophagus and provides an active means of equalizing the air pressure between the tympanic cavity and the nasopharynx by drawing open the lumina of the Eustachian tubes. Many readers are familiar with the process of dry swallowing to relieve the middle ear discomfort associated with rapid elevation or altitude changes, and it is the salpingopharyngeus muscles that facilitate this relief.

Oral Muscles of Articulation

Oral musculature includes mainly the intrinsic and extrinsic tongue muscles. Actions of these muscles give the tongue its great flexibility, allowing the almost limitless variation in shape and position within the oral cavity to modify resonance and create plosive and fricative sound sources at the various places of articulation. In addition to the tongue musculature, two muscles in the posterior oral cavity are parts of the velopharyngeal sphincter.

Tongue muscles can be categorized as *intrinsic* or *extrinsic*, with intrinsic muscles having both origins and insertions within the body of the tongue. Extrinsic tongue muscles originate outside the tongue.

Intrinsic Tongue Muscles

Intrinsic tongue muscles alter the shape of the tongue by shortening the distances between its longitudinal, transverse, and vertical dimensions. Contractions of the intrinsic muscles bring about elevations of the tongue tip, longitudinal or lateral curling of the blade, and thickening or flattening the body. Such movements are essential to dental, alveolar, and postalveolar plosive and fricative articulation, as well as to palatal approximant articulation and to vowel articulation. In addition to speech-related movements, intrinsic muscles allow licking movements for food management within and outside the oral cavity.

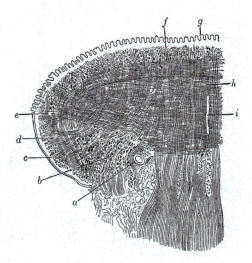

Intrinsic tongue muscles are arranged in four pairs, one member of which is located on either side of the tongue's midline fibrous septum and the fibers of the paired, extrinsic *genioglossus* muscles. There are two longitudinal muscle pairs, one superior and one inferior, a pair of vertical muscles, and a pair of transverse muscles (Abd-El-Malek, 1939).

The longitudinal muscle fibers run the length of the tongue's body, just beneath the epithelium on the superior surface (dorsum) and inferior surface, from root to tip. Contractions of the *superior longitudinal* muscle curl the tongue's tip superiorly, bringing it into contact with the oral roof: the upper teeth, superior alveolar ridge, and palate. Contractions of the *inferior longitudinal* muscle curl the tip inferiorly, opposing the action of the superior longitudinal muscle.

The *vertical muscles,* as their name suggests, span the superior-inferior dimension of the tongue. They originate and insert in the mucosa of the dorsum and inferior surface, and their fibers pass through those of the superior and inferior longitudinal muscles (Saito and Ito, 2003). Vertical muscle contractions flatten the blade, opposing those of the transverse muscles.

The *transverse muscles* originate at the fibrous septum and insert in the lateral mucosa, along the length of the tongue blade. Like the vertical muscles, the transverse muscle fibers penetrate the finer bundles of the longitudinal muscles as they cross through (Saito and Ito, 22003). Transverse muscle contractions pull the lateral aspects of the tongue towards each other, shortening the distance between the sides. The intrinsic muscular morphology of the tongue appears to be of thin layers, or lamellae (Abd-El Malek, 1939; Takemoto, 2001). If the full compliment of transverse fibers contract, they will increase the vertical dimension of the tongue's body, drawing the lateral aspects of the lingual body medially, toward the median septum, and pushing the tongue's dorsum closer to the palatal vault. This posture is appropriate for the palatal approximant, /j/. If contractions are limited to the upper fibers, the tongue's sides curl inward, in coordination with contractions of the superior longitudinal muscles, the blade and apex will be brought into contact with the lateral and anterior teeth. This lingual posture produces the proper oral cavity shape for dental-alveolar-postalveolar approximant, /r/ (/ɹ/).

Extrinsic Tongue Muscles

Extrinsic tongue muscles originate outside the body of the tongue. Their function is to move various parts of the tongue in anterior/posterior or superior/inferior directions. The extrinsic tongue muscles are the *genioglossus, hyoglossus, palatoglossus,* and *styloglossus.* Genioglossus muscles draw the tongue anteriorly, whereas the other muscles draw it posteriorly.

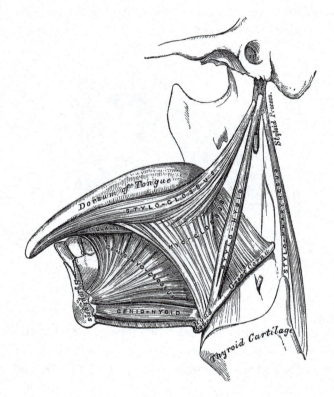

The paired *genioglossus muscles* originate at the superior mental spine on the anterior, inner surface of the mandible. This location is inside the chin, the Latin word for which is *geneion.* The two muscles are separated by the fibrous median septum. Genioglossus fibers form the main bulk of the tongue's core body, coursing posteriorly from their origins and radiating upward to fan out all the way to the apex. Some inferior fibers insert on the hyoid corpus, whereas the great majority insert into the tongue mucosa.

The genioglossus muscles draw the tongue forward, protruding the tip outside the oral cavity when simultaneously contracted. Unilateral contractions draw the tongue forward and laterally, toward the side of the contracting muscle. Thus, if the right genioglossus muscle contracts, the tongue will deviate to the right side. Tongue lateralization within the oral cavity may also be assisted by intrinsic musculature (Riggs, 1984).

In addition to its function as a lingual protractor, the genioglossus muscles are part of the *pharyngeal dilator* muscle group. These muscles are active during inspiration and when increased upper airway patency is required.

The *hyoglossus* muscles take their origins along the upper margins of the right and left sides of the corpus and of the greater cornua of the hyoid bone. The muscle belly has a flat and roughly square shape. Its fibers ascend to insert into the right and left sides of the tongue's inferior body, interdigitating with fibers of the styloglossus muscles and the inferior longitudinal muscles.

Contractions of the hyoglossus muscles draw the hyoid bone and the tongue closer to each other. The effect of this is to retract and depress the tongue.

The two *palatoglossus* muscles, also referred to as *glossopalatine* muscles, originate at the palatine aponeurosis and insert into the sides of the posterior. Some of their fibers are continuous with those of the intrinsic transverse lingual muscle. This pair of muscles is also properly grouped as a muscle of the velopharyngeal sphincter as well as a pharyngeal dilator, since their contractions can elevate and retract the tongue's back as well as depress the velum and open the nasopharynx to the rest of the upper airway.

Contractions of the palatoglossus muscles juxtapose the tongue and the velum, creating the necessary lingual postures for the velar consonants, /k/;/g/ and /ŋ/, and the close back vowels, such as /u/ and /ʊ/.

The *styloglossus* muscles originate, as their name suggests, on the anterior surface of the styloid processes of the temporal bones and from the stylomandibular ligament. They divide into two muscular bundles, the insert into the lateral and inferior aspects of the posterior tongue. Styloglossus contractions retract the tongue for posterior bolus transfer in deglutition and for articulation of back vowels and velar consonants in speech.

Facial Muscles of Articulation

The facial muscles of speech articulation are parts of the general group of facial muscles called the *muscles of facial expression*. For speech purposes, they move the lips for modification of the breath stream and for changing distal vocal tract resonance. Such movements spread and round the lips for vowel articulation and occasionally for articulation of certain consonants such as /r/ or /ʃ/ and create the *labial consonants*.

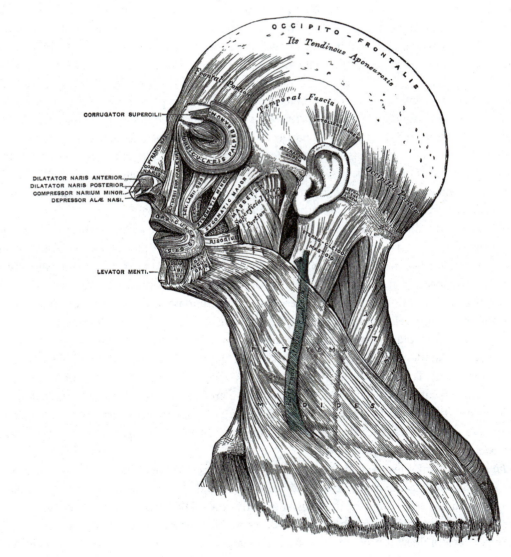

The muscles of facial expression are very thin and function to narrow or open the oral cavity, nasal cavity, and orbital openings or to elevate or lower the eye brows. They are sometimes considered as parts of a complex system of muscles and connective tissue, the *superficial muscular aponeurotic system, "SMAS,"* (Mitz and Peyronie, 1976) that extends from the lateral skull to the face. In many cases the fibers of one muscle are continuous with those of another. Certain members of this muscle group are the subjects of considerable confusion in the anatomical literature, and where it is agreed that a specific muscle exists, the confusion extends to which name to call it. Further, some members of the facial muscle group are not present in all individuals, leading to further confusion about their anatomy and not at least, to the great individual variation in facial characteristics.

The muscles that move the lips are most pertinent to the focus of this work and will be considered first. They are essential for articulation of rounded vowels and labial consonants. In addition, they create numerous facial gestures, the most recognizable of which is the smile. Their biological functions are to open or close the oral cavity for the admission and retention of food and liquid and to form a tight sphincter around a nipple or drinking straw for sucking.

These muscles may be distinguished according to whether they tighten or open the labial sphincter. Of these, *orbicularis oris* and the paired *incisivus superior* and *incisivus inferior* tighten the sphincter.

Orbicularis oris is a complex sphincteric muscle, the fibers of which surround the oral opening. Easily recognized by its broad band of fibers surrounding the mouth, its contractions narrow or close the opening. Orbicularis oris fibers are continuous with those of many other facial muscles, including those which act in opposition.

For English speech purposes, approximating the lips and creating a progressively tighter rounded opening is part of *lip rounding*. Lip rounding is required for English back vowels as an accessory modification of the anterior oral cavity, complimented by changes in the shape of the posterior oral cavity produced by progressively moving the back of the tongue toward the oral roof. The degree of rounding increases for the various vowel postures through further orbicularis contraction, as the tongue's back, being drawn primarily by the palatoglossus muscles, approaches the velum. Lip rounding not only narrows the oral sphincter, but slightly lengthens the constriction anterior to the tongue, creating a distinctive resonance, and in extreme cases, a fricative source.

Contractions of orbicularis oris are also necessary for *labial consonant* production. Labial consonants are either *bilabial*, requiring approximation of both lips, or *labiodental*, for which the lower lip is approximated with the incisor teeth. Bilabial English phonemes are /p/, b/, /m/ .w. and /ʍ/. Two additional thin muscular slips, the *incisivus superior* and *incisivus inferior* muscles, assist orbicularis oris in producing labial consonants (Joseph, 1976) by pulling the lips closer to the maxillary and mandibular processes. It is this inward tension, combined tonus of the orbicularis oris muscle, which resists air pressure created during bilabial plosive and labiodental fricative production.

The *incisivus superior* muscles arise on the maxillary alveolar processes, at the eminences created by the cuspid teeth (*canine jugae*) and insert from behind into the fibers of the orbicularis oris muscle, at both sides of the upper lip, near the angles of the mouth. The *incisivus inferior* muscles originate at the eminences created by the lower lateral incisors and insert into obicularis oris fibers at the oral angles (Bardeen, 1907; Zemlin, 1998).

Muscles that oppose orbicularis oris do so by contractions along lines radiating from the oral opening at various angles. These muscles take their origins at sites on the maxilla, zygomatic bones, and mandible, and insert among the fibers of orbicularis oris. They pull the lips upward and downward, in both vertical and oblique directions, and pull the oral commissures laterally. Combined, they comprise a group called the *lower muscles of facial expression*.

The labial *levators* (from Latin *levare*, to lift) oppose the orbicularis oris by drawing the upper lip superiorly, toward the nose and cheeks. They include *levator labii superioris, levator labii superioris alaeque nasi, zygomaticus minor, zygomaticus major*, and *levator anguli oris*.

Levator labii superioris muscles are wide and flat muscle with three distinct groups of fibers, each of which has its own name: *labii superioris alaeque nasi, levator labii superioris*, and *zygomaticus minor*. These bundles are most often considered separately, an all draw the upper lip superiorly, toward the eyes. The most medial muscular bundle, *labii superioris alaeque nasi*, originates on the frontal process of the maxilla, at the root of the nose, near the orbit. The middle group of levator fibers is properly named *levator labii superioris*. It originates on the maxilla, between the orbit, above, and the infraorbital foramen, below. The most lateral group is named *zygomaticus minor*. These muscles originate on the zygomatic bones, near their sutures with the maxillae. All fibers insert into the soft tissue of the lip and muscle of the orbicularis oris, with some fibers becoming indistinguishable from those of the oral sphincter.

Zygomaticus major fibers originate on the zygomatic bone, lateral and deep to those of zygomaticus minor, and insert into the muscle and epithelium of the upper lip. They draw the oral angles superiorly.

The *levator anguli oris* muscles are the deepest of the labial levators. They originate on the maxilla, near the point where the zygomatic process turns outward, and just below the infraorbital foramen. These paired muscles are sometimes referred to as *caninus*, since their origins are in the canine fossae of the maxillae. As is the case with the other labial levators, fibers merge with those of the orbicularis oris, except that these become continuous with those of the lower lip.

The paired labial *depressors* pull the lower lip down. These muscles include the *depressor labii inferioris, depressor anguli oris*, and the *platysma*.

The *depressor anguli oris* muscles originate on the mandible, in an area along the edge of the mandibular body, inferior to the mental foramen and the second bicuspid and first molar teeth. The fibers begin in a fan shape, then converge superiorly and create a shallow arc, becoming a narrow band and inserting among fibers of orbicularis oris. Contraction of these fibers draws the oral commissures inferiorly.

Deep and medial to the origins of the depressor anguli oris muscles, the fibers of the *depressor labii inferioris* originate. They take a superior and medial course toward the middle of the lower lip, where they insert. Contractions of depressor labii inferioris fibers draw the lower lip toward the chin.

The broad, flat *platysma* muscle has multiple functions, one of which is to draw the lower lip inferiorly. Originating in a the deep connective fascial tissue inferior to the clavicles and superficial to the pectoralis major and deltoid muscles, its wide but shallow swath of fibers course superiorly and insert at a broad area on the anterior skull. From medial to lateral, this broad area includes the fibers of the contralateral platysma muscle, the mandibular corpus from mental symphysis to the area along the oblique line, and fibers of the ipsilateral lower facial muscles. Gray (1977) reported that, in some instances, insertion points included the zygomatic arches and fibers of the ipsilateral orbicularis oculi muscles.

Several muscles of facial expression control the opening of the nasal cavity. These muscles either dilate or constrict the nares, but none completely closes them. Changes in the nareal opening are usually involuntary functions, mediated by the autonomic nervous system, increasing or decreasing impedance to air flow according to respiratory demands (Mann, Sasaki, Fukuda H, Mann DG, Suzuki M, and Hernandez JR., 1977). Such muscular contractions can also be voluntary or involuntary contributors to facial expressions.

Treatment of the anatomy of the nasal musculature, particularly that of the tip, has been subject to great confusion over the past hundred years. Gray (1977) identified several small muscles that control the nareal openings. Gray's taxonomy is used by most for general application, but other sources (Sinclair, 1972 and others) describe alternate nomenclature and attachments. Figallo, Eleazar, and Acosta (2001) reviewed the literature of over a century and reconciling that research with their dissections of twenty-four cadavers, attempted to present a consolidated reference that is used in the present work.

The muscles of the nose's tip are very small and difficult to dissect, and their fibers intermingle with those of other small muscles. They are generally not considered part of the superficial muscular aponeurotic system, but Figallo, Eleazar, and Acosta (2001) felt they had a muscular aponeurotica system of their own. Their very small size and intricacy renders them especially susceptible to being compromised as a result of injury or as an unintended result of cosmetic surgery.

The bilateral *nasalis* muscles have two sections, an *alar portion* (also called *dilator naris posterior*) and the *transverse portion*, (also called *dilator naris anterior*). The alar portion is the largest part, originating on the maxillae, beneath the fibers levator labii superioris, and inserting into the skin at the distal end of the nose, interdigitating with fibers of the myrtiform, levator labii superioris alaquae nasi and procerus along the way. The transverse portion originates on the alar cartilage at the distal end of the nose and inserts into the skin in the same area. Medial fibers of the two *levator labii superioris alaquae nasi* muscles insert in the alar cartilages and can dilate the nasal openings when contracted. The nasalis muscles can also compress the nose across the transverse portion.

The *myrtiformis* (from Latin: *"shaped like as myrtle leaf."*) muscle is a complex muscle, located at the base of the nose, which acts in opposition to the muscles that lift the nasal tip: procerus, anomalous, and levator labii superioris alaquae nasi. The name *myrtiformis* is omitted in many texts, including several editions of Gray's *Anatomy, Descriptive and Surgical* (Gray, 1977). However, it reappeared in the 1992 edition (Figallo, Eleazar, and Acosta, 2001). This myrtiformis is subject to considerable anatomical variation (Rohrich, Huynh, Muzaffar, Adams, and Robinson, 2000), leading it to be variously described in the literature as the *depressor septi* a composite of fibers named by variously as *depressor alae nasi, compressor nasi* and *compressor narium minor*. It originates on the maxillae, just above the central incisors. Sinclair (1976) illustrated this origin, naming it *depressor septi,* and its fibers separated into inner and outer divisions. Inner fibers mix with those of the nasalis, levator labii superioris alaquae nasi, and orbicularis oris to depress the nasal tip.

The *procerus* (from Latin, *"stretched"*) muscles cover the dorsum of the nose. Combined, they are roughly rectangular in shape or, considered three dimensionally, pyramidal in shape. In consideration of the latter shape, they are sometimes called *pyramidalis nasi*. Procerus can be considered in two sections. The *glabellar portion* is situated in the area between the eye brows. Its fibers originate in the tin connective tissue covering the distal portion of the nose and insert in the skin of the glabella. Some fibers blend with those of *occipitofrontalis*. Contractions of the glabellar fibers produce a wrinkling of the skin on the upper part of the nose. The *lateral portions* of the procerus originate at the lower edges of the nasal bones and lateral nasal cartilages and send fibers downward to blend with those of the transverse portion of the nasalis muscles and the lavator labii superioris muscles. Their actions lift and dilate the nareal openings.

The *anomalus* (from Greek, *"Irregular"*) muscles are thin strips of muscle tissue originating at the frontal processs of the maxillae, just medial to the orbits, and extending downward to merge with the procerus, nasalis, and levator labii muscles. It existence is disputed (Figallo, Eleazar, and Acosta, 2001), leading to the suspicion that it only present in certain individuals. Its action would be to bring the nose tip up toward the orbits, dilating the nostrils, shortening the length of the nose and wrinkling the glabellar skin as it does so.

The upper muscles of facial expression are those that serve the skin around and between the orbits and the forehead. They include the *orbicularis oculi, levator palpebrae superioris, corrugator supercilli,* and *frontalis* muscles.

The paired *orbicularis oculi* muscles form sphincters round the external orbital openings. Each orbicularis oculi has three parts: an *orbital part,* which forms a wide but shallow muscular ring around the outside of each orbit, a *palpebral* part, which gives muscular substance to each eyelid, and a lacrimal part, which draws the medial ends of the eyelids inward. Working in unison, all three parts allow tight closure of the orbits to the external environment. Paralysis of the orbicularis oculi muscles causes inability to close the eyes, often requiring mechanical assistance to avoid eye drying or further eye injury.

The *orbital parts* are the biggest portions of orbicularis oculi. They originate at the frontal processes of the maxillae, at the medial center of each orbit. Orbital portion fibers course completely around the orbit to meet again at the opposite side of the same ligament.

Orbicularis oculi-palpebral parts fibers originate at the *median palpebral ligaments*. These ligaments form the connective tissue edges of the upper and lower eyelids, just beneath the skin. These ligaments form the edges of the *tarsal plates,* dense connective tissue structures within the eyelids. The ligaments attach medially to the frontal process of the maxillae and course laterally, dividing almost immediately. *Palpebral part* fibers divide into superior and inferior bundles for the upper and lower eyelids. Fibers of the bundles meet again at the *lateral palpebral raphes,* on either side.

The *lacrimal portion* is the smallest part of the orbicularis oculi. Fibers originate on the lacrimal crest and adjacent fascia and insert at the medial edges of the tarsal plates.

Opposing the orbicularis oculi function of closing the eyelids are the *levator palpebrae superioris* muscles. These originate deep in the orbital interior, on the under surface of the lesser sphenoid wing. Fibers spread out from their narrow origins and insert on the upper edge of the superior tarsal plates. Contraction of levator palpebrae superioris draws the upper eyelids superiorly.

Between the eyebrows are the paired *corrugator supercillii* muscles. Originating on the frontal bone, at the medial end of the supercilliary arch, their fibers spread superior-laterally to insert into the skin of the eyebrow region, in the area of the supraorbital notch. Contraction of *corrugator supercillii* draws the eyebrows toward the root of the nose, in a frowning gesture.

Depressor supercilii muscles also draw the eyebrows inferiorly. Historically confused with the medial head of the orbital part of orbicularis oculi or, alternately, with the depressor supercilii. Cooke, Lucarelli, and Lemke (2001) addressed the confusion and concluded from their study of eighteen orbits (on both sides of nine cadavers) that it was, indeed, a separate muscle. Situated deep in the soft tissue of the forehead, depressor supercilii muscles originate just above the medial palpebral ligaments, near the nasal part of each orbit. Fibers course vertically about 13 to 14 mm and insert in the dermis.

The frontalis muscles are the anterior parts of the *occipito-frontalis muscles*, each pair being located at the anterior-posterior extents of the *galea aponeurotica*, the broad connective tissue overlying the calvaria. The frontalis muscle fibers originate on the galea aponeurotica, and insert in several locations in the skin and among the fibers of the upper facial muscles. Medially, their fibers are continuous with those of the procerus muscles, extending upward from the bridge of the nose. Other medial fibers join the corrugator supercilii and orbicularis oculi muscles. In the glabellar region, fibers of these paired muscles are joined medially. Further upward, they diverge, and are separated by the galea aponeurotica. Frontalis fibers end in the galea aponeurotica approximately in the area of the coronal suture. Laterally, frontalis fibers are continuous with the orbital parts of the orbicularis oculi muscles. Their actions draw the eyebrows and skin of the forehead superiorly, "raising the eyebrows" to withdraw them from the visual field when looking upward or simply as a facial gesture.

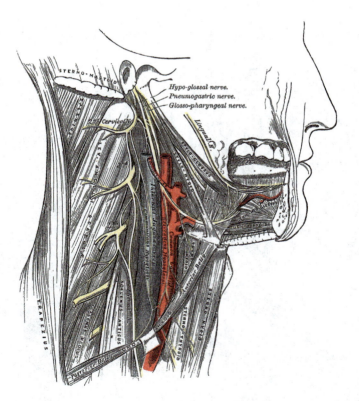

Auricular Muscles

Three muscles surround the opening of the external auditory meatuses, providing limited mobility for the pinnae or auricles. These thin slips of muscle are the *anterior, posterior,* and *superior auricular* muscles. The anterior and posterior auricular muscles pull the auricles anteriorly and posteriorly, respectively, whereas contractions of the superior auricular muscle draw them superiorly. Functionally, however, only superior and posterior movements have been observed empirically (Fortinguerra, 1993) in conjunction with yawning and smiling, and their activation during these actions is the reason for their being included in among the facial muscles. Voluntary control of these muscles is rare in human beings.

The anterior and superior auricular muscles originate at the lateral aspects of the galea aponeurotica and inserts into the cartilage of the auricle. The anterior auricular muscles insert into the auricles near their helices, whereas the superior auricular muscles insert at the upper or cranial parts of the conchae. The posterior auricular muscles take their origins on the lateral surfaces of the mastoid processes, bilaterally, and insert on the auricular cartilages at the posterior areas of their conchae.

Neurology of the Articulatory Mechanism

The articulatory mechanism's neural connections to the central nervous system are via the cranial nerves and the cervical branches of the sympathetic trunk. These peripheral nerves convey afferent and efferent action potentials between the central nervous system and the pharynx, the oral and nasal cavities, and the face, including the orbits.

Peripheral nerve distribution of the articulatory mechanism may be easier to appreciate if it will be remembered that a pair of cranial nerves is associated with each of the five branchial arches in the developing embryo. Postnatal structures deriving from these arches will generally be served by their associated cranial nerves with some overlap.

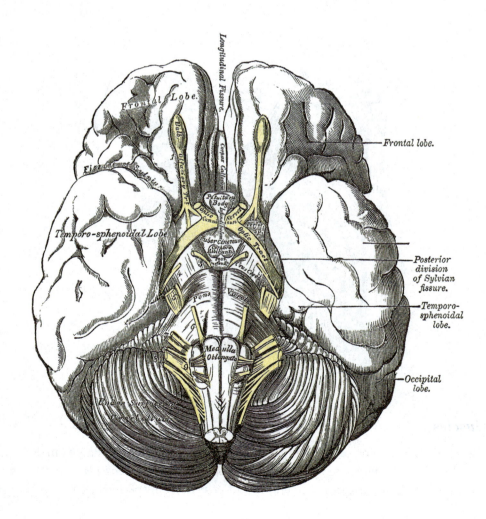

The nerve of the first arch is the trigeminal nerve (V), whereas the nerve of the second arch is the facial nerve (VII). The nerve associated with the third arch is the glssopharyngeal nerve (IX) and the fourth and sixth branchial arches share innervation from the vagus and accessory nerves (X-XI), with the fourth arch receiving the superior branch of the laryngeal nerve and the sixth arch receiving the recurrent laryngeal branch.

Nuclei of IX, X, and XI can be grouped together inasmuch as they all contribute to innervation of later branchial arch derivatives, especially involving respiratory, cardiovascular, and visceromotor functions.

Afferent Sensations

Afferent sensations from the head and neck are diverse, since the special senses of vision, taste, smell, and hearing have their end organs in the cranial region. Special senses of taste and smell are generated by end organs in the oral and nasal cavities, respectively, while the orbits contain the end organs for the visual sense.

This region is also rich in end organs from the various senses of touch. All structures, from the pharynx to the vertex of the skull and all the cavities in between have somesthetic receptors that convey action potentials associated with fine and gross touch, including movement, position senses where structures are highly moveable, the presence of objects in the pharynx, pain, and temperature.

Articulatory Afferent Neurology by Region

Pharynx

Action potentials associated with the somesthetic senses (touch, kinesthesia, and proprioception) are conveyed from the pharynx to the CNS via the *pharyngeal plexus*. The pharyngeal plexus is a complex of nerves created by branches of the *glossopharyngeal* and *vagus* cranial nerves (IX and X) along with the *cervical sympathetic ganglion*. Even though multiple nervous contributors to the plexus, it is customary, if not somewhat confusing, to specify pharyngeal innervation in terms of branches of the vagus nerve.

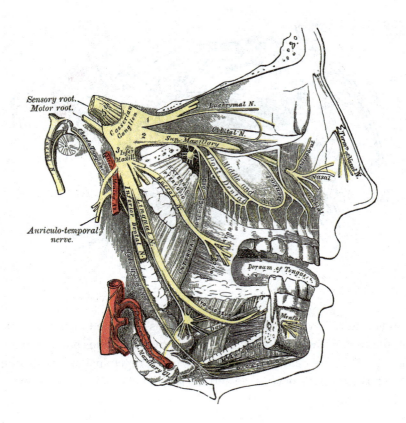

The recurrent branches of the vagus nerves contribute to the pharyngeal plexus below the level of the vocal folds, while tissue above the vocal folds, all the way to the posterior part of the tongue's dorsum, receives its sensory innervation with fibers from the internal branch of the superior laryngeal nerve.

Oral Cavity

Afferent sensations from the oral cavity include somesthesis and taste. Oral cavity somesthesis is of particular importance to practitioners in speech-language pathology, for it is by this sense that the speaker becomes aware of the dynamics and positioning of the oral and labial articulators. While taste or gustatory sensation's contribution to the speech, language and hearing clinician's practice may not be as immediately obvious, its role in swallowing should not be overlooked.

From the posterior one-third of the oral cavity, that is, the portion of the oral cavity posterior to the *sulcus terminalis*, somesthetic stimuli are conveyed to the central nervous system via the glossopharyngeal nerve. Touch sensations form the anterior two-thirds, including the tongue, oral floor, "hard palate," teeth, gums, and inner cheeks are conveyed by the *maxillary* and *mandibular* branches of the *trigeminal* nerve (V). These branches are also known by their own names, that is, the *maxillary nerves* and *mandibular nerves*.

The tongue tip and blade areas appear to be the most tactilely sensitive areas of the oral cavity, followed by the tongue dorsum, whereas the least sensitivity is associated with the floor of the oral cavity, beneath the tongue and at the velum (Sinha, Rhee, Alcaraz, and Urken, 2003; Zur, Genden, and Urken, 2004).

Gustatory sensation from the posterior one-third of the tongue's surface, behind the sulcus terminalis, as well as those from the velum and fauces are conveyed to the central nervous system by fibers of the glossopharyngeal nerve. Taste sensation form the anterior two thirds of the tongue's dorsum are conveyed by afferent fibers of the facial nerve (VII).

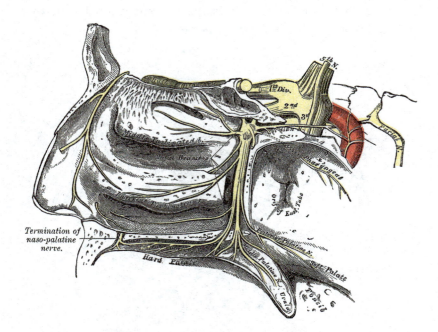

Nasal Cavity

Olfactory information is received by special sensory end organs contained in *olfactory epithelium* on either side of the septum and in the roof of the nasal cavity. It is then conveyed to the central nervous system via the olfactory nerve (I). Tactile sensations are delivered via branches of the trigeminal nerve, the *nasopalatine branch* of the maxillary nerve for the posterior parts of conchae and the *anterior ethmoidal nerve*, a branch of ophthalmic nerve for anterior parts of the conchae.

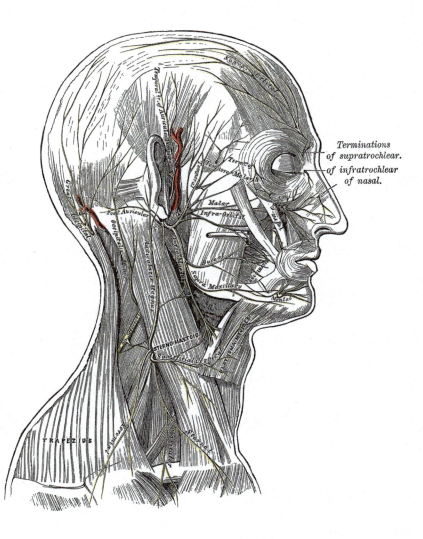

Terminations
of supratrochlear.
of infratrochlear.
of nasal.

Face

Tactile sensations of movement, position sense and pain, temperature, and gross touch are received by sensory receptors arranged over the facial surface in regional *dermatomes*. These dermatomes are delineated and named according to the regions served by the three branches of the trigeminal nerve (V): the *ophthalmic, maxillary,* and *mandibular* branches. Smaller branches ramify from each of these nerves to supply various facial structures, both external and internal.

The ophthalmic branch, sometimes referred to as "V-1," begins at the top of the skull, approximately over the external ears, and extends forward to the lower borders of the orbits, including their contents. Action potentials originating in the retinas are conveyed to the central nervous system via the optic nerve (II). A narrow extension of the ophthalmic dermatome covers the dorsum of the nose.

The maxillary dermatome, or "V-2," covers the maxillary region, beginning laterally behind and below the orbits and including the facial area of the upper lip, extending laterally to the zygomatic region. This dermatome also covers of the anterior two-thirds of the oral cavity, including the upper teeth and maxillary part of the palate.

The mandibular dermatome, "V-3," includes the temples and lower face. It begins posterior to the maxillary dermatome and inferior to the lateral extent of the ophthalmic dermatome and extending downward to cover the skin approximately in the region that would be covered by a male's beard. This dermatome also serves the anterior, inferior oral cavity, including the tongue.

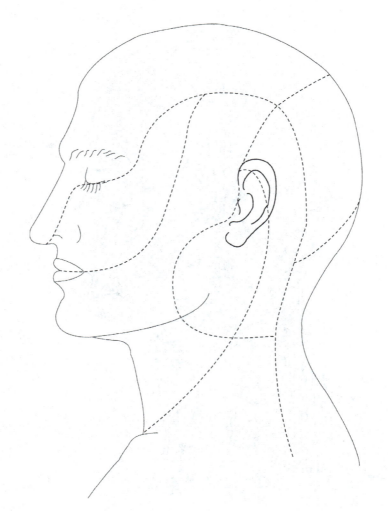

Articulatory Motor Effectors

Muscles involved in speech articulation communication are located in the pharynx, oral cavity, and face. These muscles alter the dimensions of the vocal tract to modify its resonating characteristics and create sound sources by tapping the power of pulmonic air pressure.

Motor functions are conveyed from the central nervous system to the head, neck, and face by branches of cranial nerves III through VII, and IX through XII. These lower motor neurons are the "final common pathways" for voluntary and involuntary efferent impulses to the musculature of the head and neck.

Extrinsic eye muscles are innervated by branches of the oculomotor, trochlear and abducens (III, IV, and VI) nerves. Control of the eyes is essential to orientation and to focusing on communicative targets, and ocular movement is adjusted both voluntarily and involuntarily.

Those muscles of mastication that elevate the mandible and move it laterally receive their motor innervation from the trigeminal nerve (V). These muscles are the *temporalis, masseter,* and the two *pterygoid* muscles, lateral and medial. Ability to elevate the mandible to various degrees is essential to normal articulation of most consonants and vowels. It is also important for vertical and rotary chewing. The trigeminal nerve also provides motor fibers for the *tensor veli palatini,* a muscle of the velopharyngeal sphincter, *mylohyoid* and *digastric (anterior belly)* muscles, included in the suprahyoid group, and *tensor tympani* muscle, a muscle of the middle ear.

Motor fibers of the facial nerve (VII), as its name suggests, are of most importance to speech by virtue of their supply of motor fibers to the group of facial muscles most commonly called *muscles of facial expression.* Motor fibers are distributed to the facial expression muscles through six paired main terminal branches: *temporal, zygomatic, maxillary buccal, mandibular*, and *cervical*. Facial muscles play important communicative roles in the articulation of labial consonants, rounded vowels in addition to their roles in creating the gestures of facial expression. They also provide the ability to close the orbits, constrict the nares, and close the oral cavity. Note that the eye blink response to tactile stimulation of the cornea or of the palpebral region has its afferent limb through the trigeminal nerves and its efferent limb via the facial nerves.

Additional branches of VII give motor innervation to the posterior belly of the *digastric* muscle and the *posterior auricular* muscle, located behind the pinnae and not usually grouped among muscles of speech articulation.

The glossopharyngeal (IX) nerves supply the motor functions of the *stylopharyngeus* muscles. While not essential to production of a particular group of phonemes, this muscle elevates the lateral pharyngeal walls and can by doing so, alter vocal tract resonance to some small extent. Its main importance comes from its role in swallowing, wherein it assists transfer of the bolus through the pharynx.

Among the widely diverse functions of the vagus nerves (X) are as the final common motor pathways for most pharyngeal muscles. Such functions are visceral as well as somatic, voluntary as well as involuntary.

The pharynx includes the larynx, covered in more detail in Chapter 5 as the location of phonatory source generation, but also the site at which glottal consonants are produced. Muscular contractions in the pharynx alter resonance in the lower vocal tract and create the phonatory, plosive and fricative sound sources. The recurrent laryngeal nerve, which branches from combination of fibers from the vagus nerve and the cranial division of the accessory nerve (XI) provides motor innervation for most of the intrinsic laryngeal muscles, including those that control the glottal aperture. The external branch of the superior laryngeal nerve, another ramification of X-XI complex, provides final motor innervation for the cricothyroid muscles, which help control the tension and length of the vocal folds.

It is important to clarify the combination of vagus and accessory nerves in motor functions is desirable. For voluntary control of these muscles in speech and the initial pharyngeal stages of swallowing, vagus nerves bring with them fibers of the accessory nerves (XI), which joins them where they leaves the skull through the jugular foramena. Thus, the accessory nerves provide voluntary control of phonation and most of velopharyngeal action, leading to the grouping of the two nerves together as "X-XI Complex."

Smooth muscle contractions are necessary for vasomotor control, secretory functions, and involuntary contraction of pharyngeal muscles common in swallowing and other airway protection actions. The vagus nerves supply special visceral efferent fibers for these purposes by bringing with it postganglionic autonomic fibers.

Final common motor innervation of all of the intrinsic and extrinsic tongue muscles, except one pair, the palatoglossus muscles, for speech articulation and managing food and liquid intake is delivered by branches of the hypoglossal nerves (XII). These nerves supply the intrinsic and extrinsic tongue muscles. The paired palatoglossus muscles receive their lower motor neuron innervation through X-XI complex. As is the case within the larynx, branches of the accessory nerve (XI) "hitch hike" with branches of the vagus nerves for motor innervation of the palatoglossus muscles.

Developmental Anatomy of the Articulatory Mechanism

During the embryonic period, starting at the fourth week of gestation, all major external and internal respiratory and digestive tract structures begin to develop. The central nervous and cardiovascular systems had begun to differentiate during the previous week.

Up to this point, the embryo had been formed in disc and was relatively flat. Now, it changes from that of a three-layered circle to that of a three-layered pear, being larger at one end. The larger *cephalic* end will become the head.

Rapid growth of the embryo results in folding at each end, since the cells sides do not multiply as fast as those in the long axis. The entire embryo curls, with its larger cranial region bending ventrally along. The tail or caudal region also folds ventrally. This posture will persist until birth.

Cephalic folding is a critical development, for with it, the cranium becomes positioned above the cardiovascular and respiratory systems. Folding in the region also creates an enclosure for the primitive distal alimentary tract when part of the yolk sack is separated and incorporated into the inner embryo. This occurs as the primitive

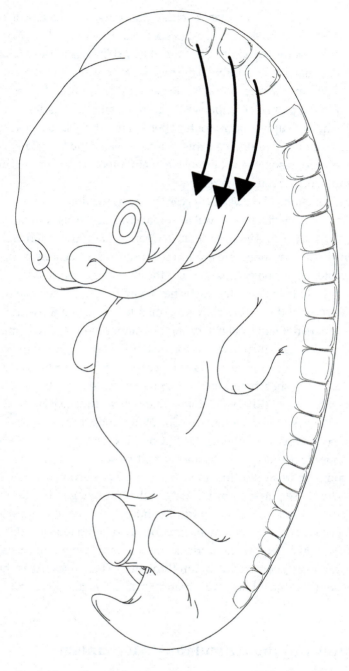

prochordal plate, the rostral meeting of endoderm and ectoderm, is folded under the head. It now becomes the *buccopharyngeal* (also called *oropharyngeal*) membrane, and is composed of the two original endodermal and ectodermal layers, separating the frontal end of the cranium from the first branchial arch. Soon, the respiratory tract will split off from the alimentary tract and follow its own developmental course.

The distal alimentary tract develops as the *foregut*, formed of endoderm. It begins at the rostral end of the embryo and develops in a caudal direction. It will develop to become the respiratory system and the rostral part of the alimentary system, ultimately ending at the duodenum, but separated from the future oral cavity by the buccopharyngeal membrane.

Meanwhile, the primitive mouth appears as a slight depression on the surface ectoderm, called the *stomodeum*. At first the stomodeum is separated from the primitive pharynx by the buccopharyngeal membrane. This membrane ruptures on about the 24th day, bringing the primitive gut into communication with the amniotic cavity.

The vocal tract being the distal part of the respiratory and alimentary tracts, its structures are derived from branchial (pharyngeal or gill) arches, with their associated grooves and pouches. These, it will be remembered, appear in the embryo beginning about day 22 of gestation, with the first appearing at the cranial end of the primitive foregut and the last five developing cranio-caudally.

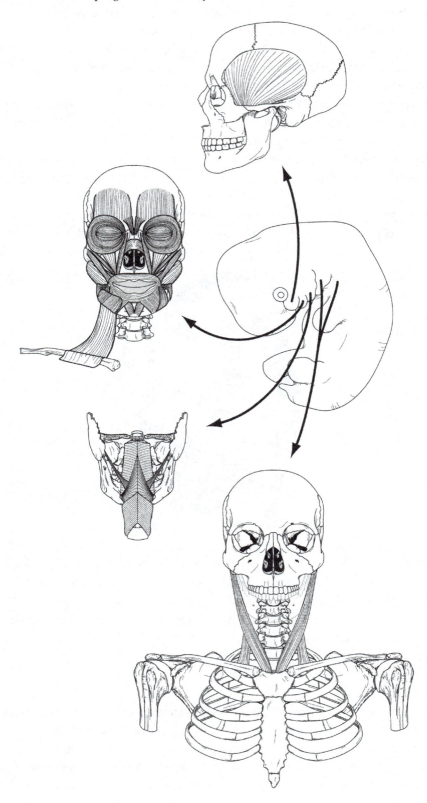

Both upper and lower jaws, along with several other skull bones, will be formed from divisions of the first branchial (Meckel's) arch. As the first arch develops, two distinct cartilaginous divisions appear. From the more caudal of the two, *Meckel's cartilage,* will come the maxilla and the mandible, as well as the zygomatic bones malleus and squamous portions of the temporal bones.

Most of the muscles of mastication, the temporalis, masseter, and pterygoid muscles evolve from the first arch. Other distal vocal tract muscles, including the tensor veli palatini, anterior belly of the digastric and the mylohyoid have their sources in the embryonic tissues of this arch, as does the tensor tympani of the middle ear, mentioned here to emphasize the close relationship between vocal and auditory anatomy. The trigeminal nerve, associated with the first brancial arch, provides motor innervation for these muscles.

The second branchial arch, also called the hyoid arch, gives rise to the lesser cornua of the hyoid bone and that part of its corpus locate between them. The stylohyoid ligament, a strong filament of connective tissue connecting the lesser cornua of the hyoid bone and the styloid process of the temporal bone, also develops from the cartilage of the hyoid arch, as does the styloid process, itself, and the tiny stapes, the most proximal bone of the ossicular chain.

Muscles of the second branchial arch are those of facial expression, buccinator, platysma, posterior belly of the digastric, and the stylohyoid. The stapedius, a muscle of the middle ear, also has its origins in the second branchial arch. The cranial nerve associated with this arch is the facial nerve, and naturally, it provides the common final pathway for motor control from the central nervous system.

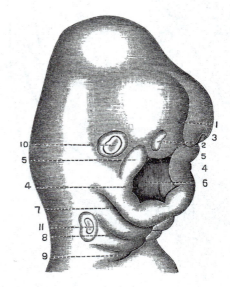

The third branchial arch's cartilage is the source of the remainder of the hyoid bone, including the lower corpus and the greater cornua. The glossopharyngeal nerve (IX) provides motor innervation for the only muscle derived from the third arch, the stylopharyngeus.

The remaining branchial arches, numbered four through six (and omitting five), are the embryonic sources for the lower oropharynx, including the epiglottis, and the larygopharynx. These are discussed fully in Chapters 3 and 5.

Formation of the Face

From the fourth to eighth gestational weeks, the face begins development. From a simple opening around the stomodeum, the eyes, nose, and lips will form the features that distinguish the individual and allow respiraton and the intake of food and drink. Tissue around the stomodeum begins to change with the appearance of five small tissue buds called *facial primordia.* These consist of a single median *frontonasal prominence* and paired *maxillary* and *mandibular prominences.* They will grow and move, eventually fuse, and the result will be the face.

The frontonasal, prominence formed by the proliferation of neural crest mesenchyme ventral to the forebrain, constitutes the superior or cranial boundary of the stomodeum. After further differentiation and the joining of more primordia, it will enlarge and grow downward, becoming the forehead and the nose, including the nasal cavity. The frontonasal prominence appears first as a slight elevation in the middle of the developing forehead and on the ventral aspect of the embryo.

Flanking the opening of the stomodeum and flanking it two maxillary prominences appear, and below them appear the two mandibular prominences. These are developments of the mandibular arch, and, with the frontonasal process, they grow to surround and form the boundaries of the stomodial opening.

The mandible is the first part of the face to form, resulting from merging of the medial (ventral) ends of the two mandibular prominences during the fourth week. Thus, the chin, or mental eminence, is the earliest recognizable feature of the human face.

By the end of the fourth week, the inferior part of the frontonasal process has extended downward, forming a small projection in the stomodeal opening. At the same time, bilateral oval thickenings, *nasal placodes*, have developed on each side of this inferior part of the frontonasal prominence. Mesenchyme proliferates at the margins of these placodes, producing horseshoe-shaped elevations, the sides of which are called the *medial* and *lateral nasal prominences*. These will become the dorsum and sides of the nose. By proliferation and elevation of tissue at their circumferences, the centers of the nasal placodes now become *nasal pits*, ultimately to be halves of the inner nasal cavity.

The maxillary prominences also enlarge and grow medially toward each other and the medial nasal prominences. Medial migration of the maxillary prominences causes the medial nasal prominences also to move toward the median plane and toward each other. Each lateral nasal prominence is separated from the maxillary prominence by a cleft called the *nasolacrimal groove*. An *intermaxillary segment* appears between the medial nasal processes, destined to become the nasal septum.

By the end of the fifth week, each maxillary prominence has merged with the lateral nasal prominence the same side, along the line of the nasolacrimal groove. This establishes continuity between the side of the nose, formed by the lateral nasal prominence, and the cheek region, formed by the maxillary prominence.

Development of the Palate

Separation of the oral and nasal cavities occurs when the palate forms, creating a boundary that is the roof of the mouth and the floor of the nasal cavity. The palate develops from two primordia: the *primary palate* and the *secondary palate*. This development, referred to as *palatogenesis* begins toward the end of the fifth week of gestation. Development is not complete until about the twelfth week.

The Primary Palate

The primary palate develops at the end of the fifth week from the deep intermaxillary segment of the maxilla. This segment is formed by the merging of the medial nasal prominences, brought about by the medial growth of the two maxillary prominences. The intermaxillary segment fuses with the frontonasal prominence. The primary palate will develop into the upper lip and philtrum, the incisor teeth and a wedge-shaped anterior segment of the oral roof, anterior to the incisive foramen.

The Secondary Palate

The *secondary palate* is the hard and soft parts of the palate, extending posteriorly from the incisive foramen. It develops from two horizontal mesenchymal primordial projections that extend from the internal aspects of the maxillary prominences, called lateral palatine processes. These project inferiorly on each side of the tongue.

Growth of the oral cavity and cervical region create room for interior development. As the jaws and neck develop, and the oral cavity becomes larger and deeper, the tongue becomes relatively smaller and has room to move inferiorly. Where it previously occupied the space between the maxillary prominences, the tongue's downward migration allows space for the lateral palatine processes to fuse. The lateral palatine processes elongate and move to a more horizontal, domed configuration superior to the tongue during the seventh week.

The processes approach each other and fuse in the midline. They also fuse with the primary palate and the nasal septum.

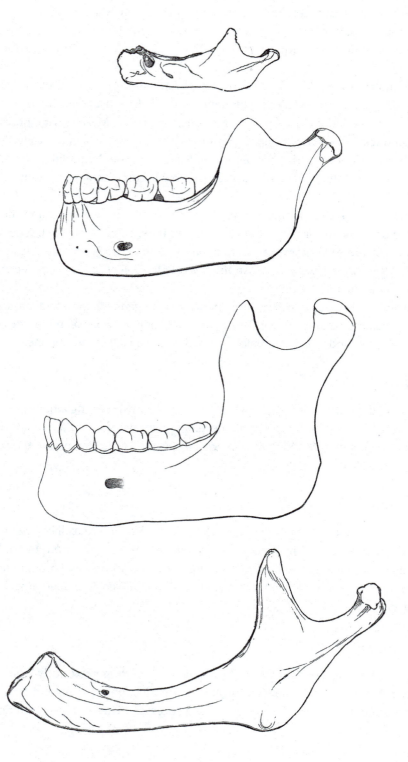

The nasal septum develops as a deep downward growth from the merged medial nasal prominences, themselves created by development of the nasal placodes on the lateral aspects of the frontonasal prominence. The fusion between the nasal septum and the palatine processes begins anteriorly during the ninth week and is completed posteriorly in the region of the uvula by the twelfth week.

Ontogeny of the Speech Articulatory Mechanism

Speech Articulation Changes in Childhood

During childhood, rapid growth in nervous system function, coupled with physical changes in the head and neck, underlie transformation of speech from incomprehensible, undifferentiated phones to mastery of the syllabic patterns of adult speech. Speech articulation development generally proceeds from open vowel to closed consonant vocal tract configurations, beginning with meaningless vocal play, and continuing to gradual assignment of meaning to sounds and combinations of sounds. Use of consonants to form syllable boundaries usually proceeds from front to back, and from phonemes requiring fewer muscle groups and less precision to those requiring more complex combinations. Thus, labial nasals and plosive phonemes appear first, followed by alveolar and velar ones.

The most rapid development of speech articulatory abilities occurs from birth to age seven or eight, but further refinement appears to continue on into adulthood (Sadagopan and Smith, 2008; Wohlert and Smith, 2002). Refinement of muscular contractions and auditory discriminatory abilities brings on the appearance of more complex articulatory postures and syllabic combinations. Approximants become distinguished from vowels, fricative and affricate consonants become meaningful parts of syllables, and diphthongs, triphthongs, and consonant clusters are employed (Tanner, Culbertson, and Secord, 1997).

Physical development of the head and neck accompanies the rapid development of speech. Perhaps the most dramatic changes come about in childhood as the primary dentition erupts and sheds to be replaced by the permanent dentition.

Primary anterior teeth erupt first, usually beginning with the central incisors, during the first year. Eruption of the upper incisors presents the possibility of articulating of dental sounds, /f/; /v/; /θ/ and /ð/, and may also help the young speaker distinguish /s/ from /θ/ and /z/ from /ð/. Primary teeth are shed with the appearance of the molars, at about age six, leaving the juvenile speaker at a loss for those articulatory surfaces until about age seven or eight.

Dental changes also bring about changes in the dimensions of the mandible. The angular relationship of the ramus and corpus, called the *angle of the mandible*, becomes smaller with the development of posterior teeth. The first and second primary molars appear during the second year, to be replaced after the permanent incisors appear, around age six. The permanent first molars are in place by about age six, and the third molars, if they appear at all, are in place by adulthood.

Speech Articulation Changes in Puberty

Changes in the physical dimensions of the head, neck, and thorax bring about acoustic changes related to vocal tract length (Petersen and Barney, 1952; Varparian and Kent, 2007) and enhanced respiratory capacities. Facial development includes the anterior and inferior growth of the facial skeleton, whereas cervical development includes the inferior migration of the larynx, resulting in a longer vocal tract and chances in the resonating characteristics of the skull. This, coupled with changes in the phonatory source (see Chapter 5), results in perceptible differences in the speech of the maturing individual. In addition, thoracic growth creates increased lung capacity for longer sustained utterances.

Maturation of the nervous system brings with increasing facility with contraction and coordination of muscle groups. This progression continues It also brings about improved auditory discriminatory ability, and hence, increases phonological inventory and diadochokinetic agility (Sadagopan and Smith, 2008; Walsh and Smith, 2002).

Speech Articulation Changes in Old Age

The same physical and nervous system changes that brought about vocal tract resonance changes and enhanced speech articulation facility as they advance bring about degeneration of speech articulation skills as they retreat in the long-lived person. Tissue changes, loss of dentition, and decreases in cognitive and neuromuscular control reverse the advances of early life.

The tissues of the vocal tract become thinner and less flexible over time and with exposure to disease and/or environmental effects. Perhaps the major source of articulatory dysfunction associated with anatomical tissue changes of later life are loss of dentition through decay or disease. With the loss of posterior teeth, the mandible begins to take on, once again, the greater angle seen during infancy, changing with it oral cavity and facial dimensions and vocal tract resonating characteristics. The loss of superior incisors removes that articulatory surface against which to juxtapose the lip or tongue for articulation of labiodental or dental fricatives. Resulting substituent phonemes may be bilabial or alveolar fricatives. The presence of dentures restores the incisive articulation surface, but brings with it possible additional oral sources if the dentures are ill-fitted.

Reduced speaking rate, longer vowel and consonant duration, and possibly reduced fluency have been reported to characterized the speech of long-lived individuals. Ramig (1983), for example, reported significantly slower speaking and reading rates in older subjects, as did Durchin and Mysak (1987) and Benjamin (1982).

Diseases of the central or peripheral nervous systems bring with them the constellation of speech articulation disorders known as *dysarthria*. Dysarthric speech articulation can be characterized by varying degrees of slowed or increase speaking rate, and by inaccuracy in articulatory targeting and ballistics, depending upon which parts of the nervous system are affected.

CHAPTER SUMMARY

This chapter began with a treatment of the dynamics and acoustics of speech to lay a foundation for understanding the relevance of subsequent details of the skeleton, soft tissues, and nervous support of the articulatory mechanism. The gross anatomy of the vocal tract was then presented, beginning with the basic dimensions of the vocal tract, progressing through its bony infrastructure, cavities, histology, musculature and neurology. The chapter concluded with an overview of the development and ontogeny of head and neck structures employed for creating the modality of language expression we call speech.

SUGGESTED READING AND WEBSITES

Abd El Malik, S., (1939). Observations on the morphology of the human tongue. *Journal of Anatomy, 73*, 201–210.

Achkar, E. 1998. Zennker's diverticulum. *Digestive Diseases, 16* (3), 144–151.

Arvidson, K., and Friberg, U. (1980). Human taste: response and taste bud number in fungiform papilla. *Science, 15*, 807–808.

Bardeen, A.B. (1907). The musculature. In Morris, H.M. and Lond., M.B. (Eds.) Morris's *Human Anatomy* (4th ed., Pt. 2, p. 326).

Benjamin, B.J. (1982). Phonological performance in gerontological speech. *Journal of Psycholinguistic Research, 11*, 159–167.

Bragg, J. (2006). Espophageal diverticula. *eMedicine specialties*. Retrieved June 8, 2011, from http://www.emedicine.com/med/topic736.htm.

Cook, B. E., Lucarelli, M. J., and Lemke, B. N. (2001). Depressor supercilii muscle: anatomy, histology, and cosmetic implications. *Ophthalmic Plastic and Reconstructive Surgery, 17,* 404–411.

Sandra W. Duchin, Edward D. Mysak (1987). Disfluency and rate characteristics of young adult, middle-aged, and older males. *Journal of Communication Disorders, 20,* 245–257.

Edwards, H. (1998). *Phonetics: The Sounds of American English.* San Diego, CA.: Singular.

Ekberg, O., Aksglaede, K., and Baert, A.L. (2004). *Radiology of the Pharynx and the Esophagus.* New York: Springer.

Elgendy, A.M., and Pols, L.C. (2001). Mechanical versus perceptual constraints as determinants of articulatory strategy. *Proceedings of the Institute of Phonetic Sciences, Amsterdam, 24,* 62–74. Retrieved June 8, 2011, from http://www.fon.hum.uva.nl/Proceedings/Proceedings24/Proc24_artikelElgendy.html

English, W.R., Robison, S.F., Summitt, J.B., Oesterle, L.J., Brannon, R.B., Morlang, W.M. (1988). Individuality of human palatal rugae. *Journal of Forensic Sciences, 33,* 718–726.

Fant, G. (1970): Acoustical Theory of Speech Production. The Hague: Mouton & Co.

Figallo, Eleazar E. M.D.; Acosta, Jaime A. M.D. (2001). Nose muscular dynamics: The tip trigonum. *Plastic and Reconstructive Surgery, 108,* 1118–1126. Retrieved July 9, 2011, from http://ovidsp.tx.ovid.com/spb/ovidweb.cgi

Fortinguerra, B. F. (1993). EMG study of the anterior, superior and posterior auricular muscles in man. *Annals of Anatomy, 175,* 195–197.

Gray, H. (1977). *Anatomy, Descriptive and Surgical,* (Revised American Version of the 15th ed.) New York: Bounty Books.

Hauser, G., Daponte, A., and Roberts, M. J. (1989). Palatal rugae. *Journal of Anatomy, 165,* 237–249.

International Phonetic Association (2011). Retrieved June 8, 2011, from http://www.arts.gla.ac.uk/IPA/ipa.html

Jahan-Parwar, B., and Blackwell, K. (2007). Lips and Perioral Region Anatomy. *eMedicine Specialities.* Retrieved June 8, 2011, from http://www.emedicine.com/Ent/topic7.htm

Jakobson, R. (1940–42/1972) *Child Language Aphasia and Phonological Universals.* The Hague & Paris: Mouton. Originally published in German (1940–42).

Jopseph, J. (1976). Locomotor system. In Hamilton, W. *Textbook of Human Anatomy* (2nd ed., pp. 20–200). St. Louis, MO: C.V. Mosby.

Lisker, L.& Abramson, A.S. (1964). A cross-language study of voicing in initial stops: acoustical measurements. *Word, 20,* 384–422.

Maddieson, I. (2005). Presence of uncommon consonants. In Haspelmath, M., Dryer, M.S., Gil, D., and Comrie, B. (Eds.). *The World Atlas of Language* (pp. 82–85). Oxford: Oxford University Press.

Mann, D.G., Sasaki, C.T., Fukuda, H., Mann, D.G., Suzuki, M., Hernandez, J.R. (1977). Dilator naris muscle. *Annals of Otology, Rhinology and Laryngology 86,* 362–370.

Minifie, F. (1973) Speech Acoustics. In Minifie, F., Hixon, T., and Williams, F. (1973). *Normal Aspects of Speech Hearing and Language.* Englewood Cliffs, N.J.: Prentice-Hall.

Mitz, V. and Peyronie, M. (1976). The superficial musculo-aponeurotic system (SMAS) in the parotid and cheek area. *Plastic and Reconstructive Surgery, 58,* 80–88.

Papadopoulos, N., Lykaki-Anastopoulou, G., and Alvanidou, E. (1989). The shape and size of the human hyoid bone and a proposal for an alternative classification. *Journal of Anatomy, 163,* 249–260.

Peterson, G. E., and Barney, H.L. (1952). Control methods used in a study of the vowels. *Journal of the Acoustical Society of America, 24,* 175–184.

Riggs, J.E. (1984). Distinguishing between intrinsic and extrinsic tongue muscle weakness in unilateral hypoglossal palsy. *Neurology, 34,* 1367.

Ramig L. A.(1983). Effects of physiological aging on speaking and reading rates. *Journal of Communication Disorders, 16*, 217–226.

Rohrich RJ, Huynh B, Muzaffar AR, Adams WP Jr, Robinson JB. (2000). Importance of the depressor septi nasi muscle in rhinoplasty: anatomic study and clinical application. *Plastic and Reconstructive Surgery, 105*, 376–83. Retrieved June 6, 2011 from http://ovidsp.tx.ovid.com/spb/ovidweb.cgi

Saigusa, H., Yamashita, K., Tanuma, K., Saigusa, M., and Niimi, S. (2004). Morphological studies for the retrusive movement of the human adult tongue. *Clinical Anatomy, 17*, 93–98.

Sadagopan N, Smith A. (2008). Developmental Changes in the Effects of Utterance Length and Complexity on Speech Movement Variability. *Journal of Speech Language and Hearing Research, 51.* Retrieved June 8, 2011, from http://libproxy.nau.edu:2081/pubmed/18664705?ordinalpos=1&itool=EntrezSystem2.PEntrez.Pubmed. Pubmed_ResultsPanel.Pubmed_RVDocSum

Saito, H., and Ito, I. (2003). Three dimensional architecture of the intrinsic tongue muscles, particularly the longitudinal muscle, by the chemical maceration method. *Anatomical Science International, 78* (3), 168–176.

Jeffrey P. Searl, Rodney M. Gabel, J. Steven Fulks (2002). Speech disfluency in centenarians. *Journal of Communication Disorders, 35*, 383–392.

Shriberg, L. D., and Kent, R. D. (1995). *Clinical Phonetics* (2nd Ed). Needham Heights, MA.: Allyn and Bacon.

Sinclair, D.C. (1972). Musacles and fasciae. In G.J. Romanes (Ed.) *Cunninghan's textbook of anatomy.* London: Oxford University Press.

Stevens, P. (1960). Spectra of fricative noise in human speech. *Language and Speech, 49*, 32–49.

Stevens, K.N. (1998). *Acoustic Phonetics.* Cambridge, MA.: MIT Press.

Takemoto, H. (2001). Morphological analysis of the human tongue musculature for three- dimensional modeling. *Journal of Speech, Language and Hearing Research, 44*, 95–107.

Tanner, D., Culbertson, W., and Secord, W. (1997). *Developmental Articulation and Phonology Profile.* Oceanside, CA.: Academic Communication Associates.

Sinha, U.K., Rhee, J., Alcaraz, N., and Urken, M.L. (2003). Pressure-specifying sensory device: quantitative sensory measurement in the oral cavity and oropharynx of normal adults. *ENT: Ear, Nose & Throat Journal, 82*, 682–690. Retrieved July 6, 2011 from http://web.ebscohost.com/ehost/detail?vid=4&hid=104&sid=60eb4172-97ba-4ea8-b9e2-6514971d0e89%40sessionmgr108

Varparian, H.K., and Kent, R.D. (2007). Vowel acoustic space development in children: a synthesis of acoustic and anatomic data. *Journal of Speech, Language and Hearing Research, 50*, 1510–1545.

Walsh, B. and Smith, A. (2002). Articulatory movements in adolescents: evidence for protracted development of speech motor control processes. *Journal of Speech, Language & Hearing Research, 45*, 1119–1133.

Wohlert, A.B. and Smith, A. (2002). Developmental change in variability of lip muscle activity during speech. *Journal of Speech, Language & Hearing Research 45*, 1077–1087.

Weitzul, S. and Taylor, R.S. (2006). Lip reconstruction. *eMedicine Specialties.* Retrieved June 8, 2011, from http://www.emedicine.com/derm/topic844.htm

Wisner, A., Dufour, E., Messaudi, M., Nedji, A., Marcel, A., Ungeheuer, M., and Rougeot, C. (2006). Human opiorphin, a natural antinociceptive modulator of opiod-dependent pathways. *Procedings of the National Academy of Sciences of the United States of America: Pharmacology, 103*, 17979–17984.

Zur, K.B., Genden, E.M., Urken, M.L. (2004). Sensory topography of the oral cavity and the impact of free flap reconstruction: A preliminary study. *Head and Neck 26*, 884–889. Retrieved June 8, 20118, 2011, from: http://www3.interscience.wiley.com/cgi-bin/fulltext/109607812/HTMLSTART

IMPORTANT TERMS

Acoustic Filter: A device that modifies the amplitude of specific sound frequencies.

Affricate: A plosive phoneme released with strong turbulence at the place of articulation.

Allophone: The instantaneous manifestation of a phoneme.

Alveolar Ridge: The eminence created by the alveoli in either the maxillary or mandibular dental arches. In phonetics, the term *alveolar* refers to the place of articulation posterior to the superior incisor teeth.

Angle of the Mandible: The geometric relationship between the corpus and ramus of the mandible.

Approximant: A member of the class of phonemes characterized by close juxtaposition of articulatory structures and distinctive vocal tract resonances, unique in their capability to serve as syllable releasers or as syllable nuclei; formerly called *semivowels*. Includes /r/, /l/ /j/ and /w/.

Bregma: Intersection of the coronal and sagittal skull sutures.

Calvaria: Bones of the superior skull, covering the cranial cavity.

Canal: A narrow passage through a bone or through the intersection of bones, through which pass arteries, veins, or nerves.

Choanae: Posterior borders of the nasal cavity at which the nasal cavity communicates with the pharynx.

Commissure: The place at which opposing tissue folds connect with one another.

Consonant: A speech sound formed by a relatively closed vocal tract; consonants usually form syllable boundaries.

Contrastive: An acoustic difference between speech sounds sufficient to signal a change in semantic meaning (*synonym: distinctive*).

Cornu: Anatomical name for a bony or cartilaginous projection.

Corpus: The main body of a bone from which processes project.

Cranial Nerves: Twelve pairs of peripheral nerves having their cell bodies in or near the brain.

Cribiform Plate: Perforated superior process of the ethmoid bone through which pass the sensory end organs of the olfactory system.

Diploe: Marrow of the calvaria.

Distinctive: An acoustic difference between speech sounds sufficient to signal a change in semantic meaning (*synonym: contrastive*).

Distinctive Feature: One of a set of binary articulatory or acoustic features derived by various phoneticians in an attempt to contrast phonemes in terms of binary matrices.

Dorsum: The back of an anatomical structure. The dorsum of the tongue is on its superior surface and the dorsum of the nose is its anterior surface.

Facial Primordia: Embryonic processes destined to develop into the distinctive faces.

Facial Skeleton: Lower, anterior part of the skull, forming the foundation of the face.

Foramen: An opening in a bone, through which may pass an artery, vein, or nerve.

Foregut: Embryonic precursor to the distal alimentary and respiratory tracts.

Frenulum: A thin section of tissue connecting a moveable structure with a relatively fixed one and thus restricting the range of motion in the moveable structure.

Fricative: One of a group of consonant speech sounds distinguished by a continuous, aperiodic sound of turbulence created when pulmonic air is forced through a tight vocal tract constriction.

Fricative Source: The aperiodic, continuous sound source that distinguishes fricative consonants.

Glabella: The facial site located between the eye brows.

Glottal: Referring generally to the space between the vocal folds. In phonetics, *glottal* refers to the place of articulation created by articulation of the vocal folds.

Hard Palate: The anterior part of the oral roof having a bony infrastructure created by the maxillae and palatine bones.

Hypernasal Speech: Speech distinguished by too much nasal resonance.

Hyponasal Speech: Speech distinguished by too little nasal resonance.

Incisive Foramen: Anatomical landmark allowing passage of the sphenopalatine arteries and nasopalatine nerves, and separating the primary palate, or prepalate, from the secondary palate.

International Phonetic Association (IPA): A global organization dedicated to the study of speech sounds in all languages.

Jugae: Bony ridges in the dental arches where the premolar teeth are located.

Labial: Referring to the lips. In phonetics, labial phonemes are formed by juxtaposition of both lips or by the inferior lip and superior incisors.

Lambda: The intersection of the lambdoid and sagittal sutures in the calvaria.

Lamina Propria: A layer of loose connective tissue underlying and supporting the epithelium of hollow organs.

Laryngopharynx: The most inferior part of the pharynx containing the larynx and protecting the respiratory system from intrusion by swallowed material.

Mandible: The lower jaw.

Maxilla: The upper jaw.

Meatus: An anatomical tube or passage.

Muscles of Facial Expression: Superficial muscles of the anterior skull.

Muscles of Mastication: Muscles that create mandibular movement in relationship to the maxilla to mechanically prepare food for digestion.

Nares: The anterior openings of the nasal passages.

Nasal: Referring to the nose. In phonetics, one of a group of phonemes articulated with an open velopharyngeal sphincter.

Nasal Septum: The anatomical wall separating the halves of the nasal cavity.

Nasopharynx: The most superior part of the pharynx immediately posterior to the nasal cavity.

Neurocranium: That part of the skull that houses the brain, auditory, and olfactory end organs.

Open Vowels: Phonemes articulated with the most open vocal tract postures.

Oropharynx: That part of the pharynx located immediately posterior to the oral cavity, bordered superiorly by the nasopharynx and inferiorly by the laryngopharynx.

Palatal: Referring to the roof of the mouth. In phonetics, a term used to describe a group of phonemes articulated wit the tongue juxtaposed against the oral roof.

Papillae: Soft tissue projections form the dorsum of the tongue.

Philtrum: The area of the upper lip sculpted by fusion of the median nasal process of the frontonasal prominence and the maxillary processes.

Phonatory Source: The quasiperiodic driving speech sound source familiarly known as "The Voice."

Phoneme: A group of speech sounds, similar in their articulatory and phonetic characteristics that contrast with other such groups to distinguish spoken meaning.

Phonemic: Acoustic and perceptual characteristics of a phone sufficient to distinguish meaning when compared to other phones used in a language (*synonym: contrastive; distinctive*).

Phonetics: The study of speech sounds.

Phonological System: A psycholinguistic schema consisting of the inventory of acceptable phonemes and combinations of those phonemes.

Place-Manner-Voicing: A simple phonemic organizational system using location of oral or pharyngeal constriction, management of the breath stream and inclusion of the phaontory source as the organizational criteria.

Plosive Source: A transient, aperiodic driving speech sound source.

Plosives: A group of phonemes distinguished by complete vocal tract closure and subsequent release.

Prepalate: The anterior part of the oral roof containing the incisors.

Process: A bony or cartilaginous projection.

Pteryon: The intersection of the frontal, parietal, sphenoid, and temporal bones on the lateral aspect of the calvaria.

Ramus: The posterior part of the mandible by which it articulates with the temporal bone.

Reynold's Number: A coefficient of fricative articulation quantifying the relationship of air viscosity to its inertia and taking into account the contributions of aperture width at the articulatory constriction.

Rounded: A phoneme articulated with any degree of labial constriction.

Rugae: Elevated striations on the oral roof.

Sagittal Suture: The articulation of the parietal bones.

Secondary Palate: The posterior section of the bony palate, just anterior to the velum.

Sella Turcica: A superior projection of the sphenoid body cranial cavity. It is the site of the optic chiasma.

Sinus: Spaces between the outer and inner laminae of the facial skeleton, surrounding the nasal cavity.

Soft Palate: The posterior, soft tissue part of the oral roof that elevates during speech to separate the oropharynx from the nasopharynx.

Speech Articulation: Movements of vocal tract structures in relationship to one another in such a way as to modify pulmonary airflow and acoustic resonating characteristics, thus creating speech sounds.

Squamous: A flat part of a bone.

Stomodeum: The opening of the foregut in the embryo.

Suture: A synarthrodial articulation of skull bones.

Syllable: The minimal motor speech segment consisting of a combination of speech articulatory movements to create variations in airflow and acoustic parameters most often consisting of an initial closed vocal tract posture, followed by a relatively open one, and followed by another closed phase as the speaker moves on to the next syllable.

Uvula: The tip of the velum.

Velar: In phonetics, a group of phonemes articulates by juxtaposition of the back of the tongue with the soft palate.

Velopharyngeal Sphincter: The muscular gateway between the oropharynx and nasopharynx.

Velum: A soft tissue projection posterior to the palate capable of separating the oropharynx and nasopharynx.

Viscerocranium: That part of the skull containing the outer and middle ears and the structures of the distal alimentary and respiratory tracts including the jaws.

Vocal Tract: The distal part of the respiratory tract beginning at the vocal folds and extending to the lips or nares used to create the sounds of speech.

Vowel: A phoneme created by alteration of the shape of a relatively open vocal tract; vowels usually form the nuclei of syllables.

Waldeyer's Ring: A formation of lymphoidal tissue surrounding the mid pharyngeal lumen. The lymphoidal tissues are called tonsils.

X-XI Complex: The combination of nerve fibers from the vagus and accessory cranial nerves supplying motor innervation to the pharynx.

CHAPTER 7
Deglutition

CHAPTER PREVIEW

This chapter addresses mastication and swallowing, beginning with the intake of material to be swallowed into the oral cavity and concluding with its passage into the stomach. We begin with definitions and move on to describe the digestive system in form and function, providing a functional basis for study of the subsequent material.

Included in this chapter is an examination of the relationship of swallowing to speech, providing a rational basis for treatment overlap in the clinical setting. The chapter also contains a section on basic nutrition science to form a basis for the speech-language pathologist to interact with practitioners in that field.

Following an overall examination of swallowing and its functional details, the chapter details digestive tract anatomy, including skeletal and soft tissue components and innervation as far as the entrance to the stomach. Examination of the lower digestive tract is omitted because of its infrequent involvement with communication and its disorders. Developmental anatomy and growth of the digestive system cover those problems faced by clinical practitioners in communication sciences and disorders.

Sidebars include information about frequently encountered digestive system disorders and causes of dysphagia.

CHAPTER OUTLINE

Gross Anatomy of Deglutition

The Digestive System

The digestive or *alimentary* system mediates the transfer of nutrient materials to and from the environment for the purpose of maintaining life. To accomplish this, it must reduce material obtained in the environment to a composition that cells can use and allow passage of metabolic byproducts back to the environment.

The process by which food is consumed and waste is eliminated is called *digestion*. This process begins in the oral cavity, into which water and nutrients are introduced, and ends in the rectum, from which waste is expelled. Material to be digested comes in various consistencies, ranging from thin liquids to tough, fibrous bulk. Structures of the digestive tract are created to ingest, break down and assimilate food, and to pass it from one end of the tract to the other. Water and metabolic byproducts work their ways from the digestive system into the bloodstream. They are removed by the *urinary system,* consisting of the kidneys, ureters, urethra (in males), and bladder. Bulky, solid materials are broken down, assimilated and eliminated by the *gastrointestinal system*.

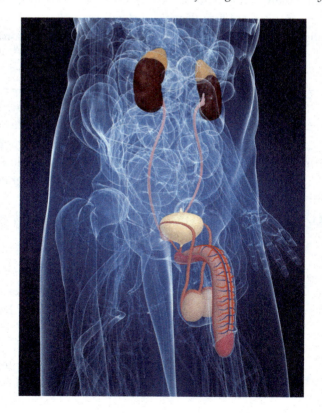

The initial phase of digestion, during which material from the environment is taken into the body, initially prepared for assimilation and transferred to the esophagus is called *deglutition*. The word comes from the Latin *deglutire,* meaning to "*swallow from* or *down*."

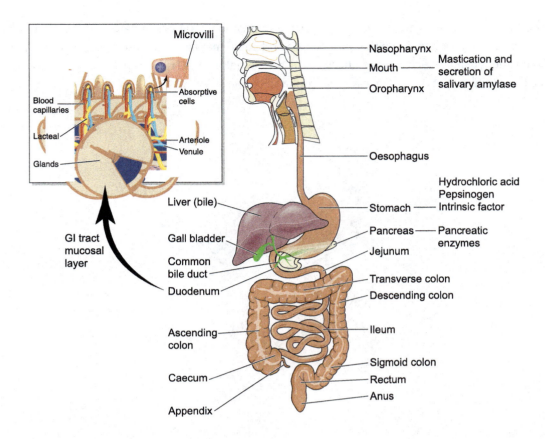

Deglutition is of particular interest to clinical practitioners of speech-language pathology, because it requires sufficiency and competence anatomical structures that also serve vocal tract functions. Speech-language pathologists, by virtue of their knowledge of the form and functions of vocal tract structures, as well as their expertise in behavioral therapeutic techniques, have adopted treatment of deglutition disorders into their scope of practice.

The digestive system begins at the mouth, at the lips, or *labial sphincter*, and ends at the caudal end of the *rectum*, at the *anus*. It consists mainly of a tube, or canal, wide in some places, and narrow in others, surrounded by glands and large organs that add digestive chemicals to the system and help reduce the size of food molecules to a useable size. Clinical speech-language pathologists are concerned mainly with the structures and functions of the mechanism of deglutition. These are the mouth, pharynx, esophagus, and, occasionally, the stomach. The mouth and pharynx will be covered grossly in this chapter, having been discussed in detail in Chapter 6.

Oral Structures for Deglutition

The tube of the digestive tract begins at the oral opening. This primary digestive cavity is also a secondary respiratory tract aperture, used for inspiration and expiration when nasal cavity patency is not sufficient for respiratory demands and, of course, the locus for most of the speech articulators. The oral cavity's triple roles in respiration, speech articulation, and digestion creates situation in which two of the functions must take lesser priorities to the third. Thus, in cases where speech is articulated normally, respiration and deglutition must be compromised, or when respiration is to be accomplished normally, deglutition and speech is sublimated. Normal mastication and swallowing likewise inhibits respiration and speech.

Two muscular lips at the oral cavity's distal end are anatomically designed to open for admission of food and other materials and to close in a *labial sphincter* to keep material within the oral cavity so it can be further prepared for assimilation or to grip around a nipple or other object from which liquid or solid food is to be ingested. The labial sphincter is highly mobile and highly receptive to tactile stimulation. Paralysis or paresis of the labial sphincter can result in *sialorrhea*, or drooling, inhibiting proper food and liquid management and causing social stigma.

Between the labial sphincter, inner cheeks, and the dental arches is the *oral vestibule*. It is U-shaped and surrounds the lateral aspect of the oral cavity in front and on the sides.

Within the oral cavity are structures for breaking down large nutrient masses and for moving them and liquids toward the central parts of the digestive tract. The process of mechanically reducing the size of nutrient or other material and mixing it with saliva is *mastication*. Solid material so treated is grossly formed into a mass called a *bolus*. Oral structures for mechanical mastication are the mandible, teeth, palate, and tongue.

The mandible articulates posteriorly and superiorly with the temporal bones at the temporomandibular joints. A temporomandibular joint is a synovial "ball and socket" joint, consisting of a rounded *condylar* process, the rounded ephysis of the mandibular ramus, inserted in the *mandibular fossa* of the temporal bone. Its free range of motion allows vertical, lateral, and antero-posterior movements of the mandible in relation to the cranium, and many possibilities for grinding a food bolus between the teeth.

The Teeth

In general, the teeth form skeletal structures for cutting, tearing, and grinding solid nutrients. In the adult, there are thirty-two teeth. Anterior teeth cut and tear food, creating portions of the desired size for oral processing, while posterior teeth grind it in to smaller and softer pieces in the actions of *mastication*.

Mastication makes solid material easier to propel into the proximal digestive tract by reducing its size and softening its consistency. In addition, mastication exposes more surface area for chemical interaction with the starch digesting enzymes contained in saliva.

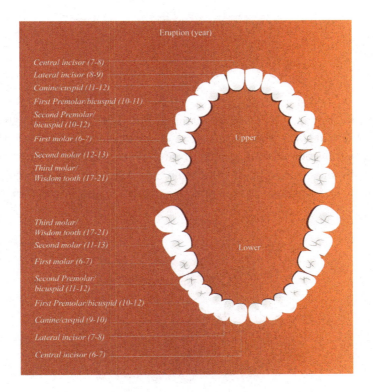

Salivary Glands

Surrounding the oral cavity are glands that secrete *saliva* for the initial chemical treatment of nutrients and for lubrication of the distal canal. These *salivary glands* are exocrine, and transfer their saliva through ducts. The largest of these are the paired *parotid* glands, one of which is located on and around each mandibular ramus. Their ducts, the *parotid ducts*, transfer saliva anteriorly to openings in the buccal walls, at approximately the locations of the second upper molars.

Paired submandibular and sublingual salivary glands flank the midline of the oral floor. Saliva is also secreted from several hundred minor salivary glands in the tongue, *via von Ebner's glands* and from the lips, buccal mucosa, and other epithelial tissues of the oral cavity and oropharynx.

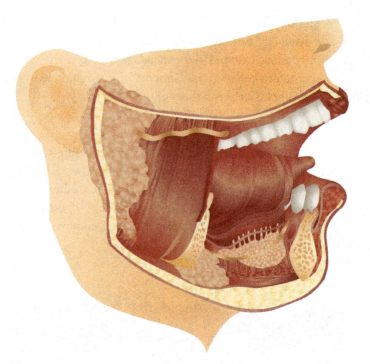

Tongue

As an organ of digestion, the tongue is essential for manipulation of the food bolus. Initially, tongue action helps draw food or liquid into the oral cavity, then, as an organ of mastication, guides material into optimal proximity to the necessary teeth. The transition of food or liquid from the oral cavity to the oropharynx is possible because the tongue moves superiorly and posteriorly. After the main food bolus has passed through the oral cavity, tongue action helps clear the buccal and gingival areas of remaining debris, creating another, smaller bolus for deglutition or for expectoration.

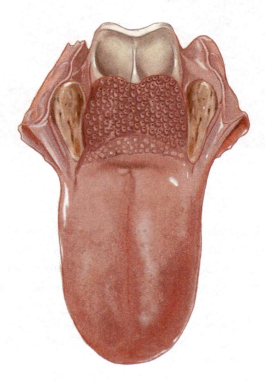

During sucking, the tongue moves anteriorly and posteriorly in the oral cavity. The posterior movement phase of this action, combined with the sealing of the anterior oral cavity by contraction of the labial sphincter and the arching of the tongue's body against the oral roof, creates a piston-like effect, drawing material into the oral cavity between the tongue and the anterior cavity. Success of this function depends upon the creation of negative pressure in the anterior oral cavity during the rearward movement of the tongue. Inferiorly, the tongue is anchored to the oral cavity floor buy a thin, midline *lingual frenulum (frenum)*.

The tissue on the dorsum of the tongue is also the main location for the end organs of taste, and as mentioned above, contains von Ebner's salivary glands. Situated among the wear resistant squamous epithelial cells are specialized and projections called *papillae* (from the Latin, *papulae*: nipple). There are four types of papillae, three of which, the *fungiform, circumvallate*, and *foliate papillae*, are formed to provide foundations for taste organs. The fourth type, the *filliform papillae*, are rough structures, formed to provide mechanical support for mastication and swallowing.

Supporting the *gustatory sense*, or sense of taste, are the *fungiform, circumvallate*, and *foliate* papillae. These contain sensory epithelial cells, or "Taste Buds," that convert physical chemical stimuli into neural action potentials to be interpreted as taste.

The most numerous of these taste-supporting projections are the *fungiform papillae*, distributed evenly on most of the anterior tongue dorsum, with concentrations on the blade and near the apex. Fungiform papilla are named for their vague resemblance to mushrooms, with a narrow base projection upward to a flattened top. Fungiform papilla contain from none to as many as twenty taste buds, with most containing three or four buds. They are sensitive to all five taste types: sweet, sour, salty, bitter, and umami (savory taste).

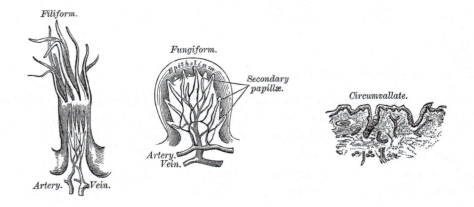

Circumvallate (also called *vallate*) *papillae* are named for the presence of a depressed ring surrounding their bases. Lingual salivary glands, called *Von Ebner's glands*, secrete saliva into this depression, both to assist in nutrient digestion and to flush material away for renewed sensation. Circumvallate papillae are located near the sulcus terminalis and are larger and less numerous than fungiform papillae, numbering from eight to fourteen. Their taste buds are positioned in the sulcus, along the lateral wall, and are especially sensitive to the bitter taste.

Foliate papillae are found on the posterolateral margins of the tongue, near the palatoglossal arch. They extend laterally from the sides of the tongue for a short distance along the posterior edges of the blade and are named for their resemblance to pages of a book.

Filiform papillae provide mechanical support for mastication and swallowing. Named for their rough structure, creating a file-like surface on the tongue, the epithelium at their tips is keratinized to form a tougher, more wear resistant coating. Filiform papillae consist of a core body extending a short distance from the surface of the tongue, then splitting into several slender, pointed projections. They contain no taste buds. Confined to the anterior two thirds of the dorsum, in front of the sulcus terminalis, these most numerous papillae are designed to strip food materials through licking and to guide them into the oropharynx during deglutition.

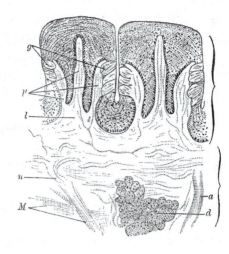

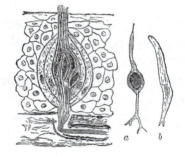

Pharyngeal Structures for Deglutition

The pharynx is the muscular tube connecting the distal openings of the digestive/respiratory tract with the esophagus and trachea. Of the three pharyngeal divisions, only the oropharynx is a normal passageway for both digestion and respiration. Material to be swallowed moves from the oral cavity to the oropharynx and into the esophagus without entering the nasopharynx or laryngopharynx.

The pharynx is lined with epithelial tissue to clean, lubricate, and protect the interior of the tube. Overlying the epithelial tissue lining are two layers of muscles. Internal pharyngeal muscle fibers run in roughly vertical directions, contracting to adjust the positions of the larynx, velum, and tongue within. Fibers of the outer muscles are sphincteric, coursing in a circle around the pharyngeal lumen and contracting in the wavelike action of *peristalsis* to propel material from the distal digestive tract its internal organs.

The hyoid bone and laryngeal cartilages maintain patency of the pharyngeal lumen for respiratory purposes. The esophageal lumen is far less rigid, and generally conforms to the shape of material passing through.

Internally, the pharynx is lined with two types of epithelium, *respiratory epithelium* and *squamous epithelium*. The respiratory epithelium lining the nasopharynx and most of the laryngopharynx is composed of cells that clean, protect, and lubricate the pharyngeal passage. It is a composite of ciliated, pseudostratified, columnar goblet cells and cuboidal basal cells. Stratified, squamous epithelium lines the portions of the pharynx that undergo mechanical wear from transiting materials. These include the oropharynx and medial surfaces of the laryngeal valves.

Vertical muscles act on the posterior tongue, velum, Eustachian tubes, hyoid bone, and larynx, drawing these structures together to propel and guide material from the oral cavity to the esophagus during deglutition. During swallowing, the velopharyngeal sphincter is normally closed, restraining the passage of material into the nasopharynx. An exception to this occurs in the infant, who, by virtue of a horizontal relationship between the laryngopharynx and oropharynx, can breathe nasally during extended nursing sessions. The posterior part of the tongue draws rearward and elevates to contact the stiffened velum, simultaneously propelling material into the oropharynx and sealing the oral cavity against reentry. Circular pharyngeal constrictor muscles contract in sequence to propel material from the oropharyngeal isthmus to the esophagus.

The laryngopharynx, also called the *hypopharynx*, is normally not involved in the passage of swallowed material. On the contrary, as swallowed material passes from the oropharynx to the esophagus, a crucial stage of swallowing involves its proper routing past the laryngeal vestibule and into the esophagus by action of extrinsic and intrinsic laryngeal musculature. Coordinated contraction of laryngeal elevators and aryepiglottic musculature acting on the epiglottis and aryepiglottic folds seal the laryngeal vestibule against penetration while pharyngeal constrictors act in concert to propel material toward the esophagus.

There are, however, several depressions into which swallowed material can be trapped along the way from the oral cavity to the esophagus. Between the anterior edge of the epiglottis and the posterior portion, or root, of the tongue, three folds are created by the eminences of the *median hyoepiglottic ligament*, a connective tissue ligature between the epiglottis and hyoid bone, and the lateral pharyngeal walls. Between the medial and lateral of these *glossoepiglottic folds* are two sulci: the *(glossoepiglottic) valleculae* (from Latin: *valles*: "valley"). These depressions may become traps for swallowed material under certain conditions. In addition to the glossoepiglottic valleculae, two *pyriform sinuses* flank the laryngeal vestibule. The pyriform sinuses are created by the folding of epithelium between the upper edges of the thyroid laminae and the aryepiglottic folds and may also become filled with material passing from the oropharynx to the esophagus.

The Esophagus

The esophagus connects the pharynx and the stomach. It is a muscular tube, with no skeleton of its own. The distal end of the esophagus is at the level of the sixth cervical vertebra and its proximal end opens into the stomach at the level of the twelfth thoracic vertebra. The esophagus is located posterior to the trachea and the heart, and penetrates the diaphragm through the *esophageal hiatus* at the level of the tenth thoracic vertebra.

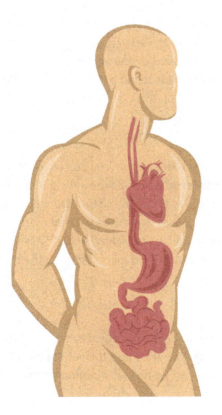

ESOPHAGEAL SPHINCTER DYSFUUNCTIONS AND DYSPHAGIA

"GERD" or *GastroEsophageal Reflux Disease* is a serious, chronic exacerbation of GER, or gastroesophageal reflux. GER is a condition wherein partially digested material from the stomach re-enters the esophagus. It is generally caused by an incompetent lower esophageal sphincter. Gastric material can be carries superiorly to the distal opening of the esophagus and enter the laryngopharyx under some conditions.

In addition to the risk of aspiration by long-term intrusion of gastric contents, GERD poses the threat of tissue erosion by action of digestive acids on esophageal and pharyngeal epithelium.

Achalasia is a swallowing disorder caused by failure of the lower esophageal sphincter to relax and allow food or liquid to enter the stomach. The result is the impounding of swallowed material the esophagus, where it can cause irritation and possibly re-enter the upper airway. Aspiration and tissue erosion are common complications of achalasia.

The esophagus passes from the cervical region to the abdomen, lying posterior to the trachea, heart, and great blood vessels, and anterior to the spinal column. The lumen of the esophageal tube is slightly narrowed at its beginning, in the hypopharynx, again, where it passes near the aorta, next where the left main stem bronchus crosses, and finally where it passes through the diaphragm. However, the esophageal tube is generally flat until a bolus of food passes through it. As the esophagus passes down toward the stomach, it curves along the sagittal kyphotic curvature of the thoracic spinal column. A small left deviation occurs as it enters the thorax and another as it joins the stomach.

Two muscular rings form sphincters at the superior and inferior ends of the esophagus. The superior sphincter is sometimes referred to as the *upper esophageal sphincter (UES)*, located at the level of the sixth cervical vertebra, and the inferior one as the *lower esophageal sphincter (LES)*, located at the level of the tenth thoracic vertebra. Both sphincters' primary functions are to contract to help keep swallowed material from being regurgitated into the lower airway, or through relaxation, to allow material from the oropharynx to pass through the esophagus and into the stomach.

The upper esophageal sphincter is formed mostly by the *cricopharyngeus* muscle. Its more distal fibers are continuous with those of the inferior pharyngeal constrictor muscle. This striated muscle can be controlled voluntarily, and is sometimes contracted to produce an alternate sound source for speech. The *esophageal voice* is created by forcing impounded air in the cervical esophagus through the sphincter formed by the cricopharyngeus muscle (upper esophageal sphincter). The source is quasiperiodic and is primarily employed by individuals having laryngectomies as an alternate phonatory source in *alaryngeal speech*.

Proximally, a circular formation of smooth muscle forms the lower esophageal sphincter at the entrance to the stomach, just inferior to the esophageal hiatus of the thoracic diaphragm. Its action, controlled by the autonomic nervous system and sometimes assisted by the surrounding diaphragm, is to keep swallowed material from re-entering the esophagus once it has passed into the gastric pouch.

Physiology of Deglutition

Deglutition involves movement of food and liquid from the oral cavity to the esophagus. It is the initial part of the overall process of digestion, and although some include the anticipation and procurement of nutrients and liquids in this process (Leopold and Kagel, 1997) as a means of studying the whole process of food and liquid ingestion for clinical therapeutic purposes, the process as it involves the oral cavity and pharynx may be thought of in three phases: an *oral phase*, a *pharyngeal phase*, and an *esophageal phase*. These phases, or stages, as they are sometimes called, can take place sequentially or concurrently, depending upon the bulk and consistency of the material to be swallowed.

Oral Phase

Deglutition begins with the reduction of the bulk and consistency of intake material to a state in which it may be more easily transferred to the lower digestive tract and further chemically assimilated. This phase of deglutition is often referred to as the *oral stage*, and can be further examined in two phases: an *oral preparatory* phase and a *transfer* phase.

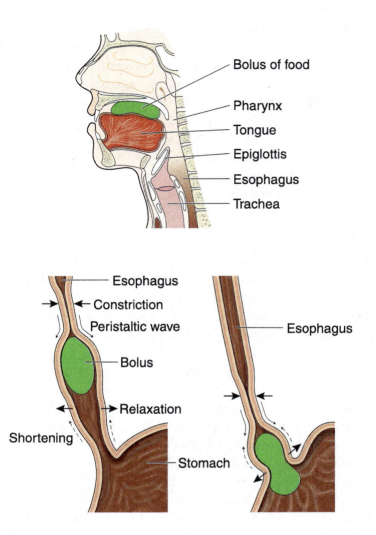

The oral preparatory phase of deglutition begins with the introduction of material to be swallowed into the oral cavity. In order for this to happen, the oral opening, or *labial sphincter*, must be spread through contractions of the muscles that oppose the orbicularis oris and through depression of the mandible, if the bulk of the material so necessitates.

After material is introduced into the oral cavity, the labial sphincter is tightened to keep it within anteriorly, whereas the velopharyngeal sphincter is tightened to keep material from infiltrating the nasopharynx and nasal cavity. The labial sphincter is also used for gripping and sealing delivery apparatuses, such as maternal or artificial nipples, or drinking straws, when material enters the oral cavity by sucking.

Bulky material, such as meat, vegetables, or nuts requires *mastication*, or chewing, to begin the reduction of their bulk and to expose more surface area to action by the enzymes in saliva. Mastication also provides additional time for the oral gustatory (taste) sensory organs to receive and propagate sensations to the central nervous system.

Mastication involves vertical and rotary movements of the mandible, combined with movements of the tongue. Vertical movements occur only in the sagittal plane, and simply bring the mandibular and maxillary teeth into juxtaposition, with material between them being crushed. The vertical movements of mastication may begin with a shearing action, whereby food external to the oral cavity is sliced from a larger body by the incisors. Incisive shearing may be accompanied by a forward mandibular gliding to better bring the lower incisors into play with the upper incisors.

Rotary movements are used to accomplish the grinding, bulk reduction and surface area exposure of mastication. Muscles of mastication move the mandible laterally, forward and backward, and up and down, creating a circular or rotary pattern of movement in the coronal and transverse plane. The tongue, lips, and cheeks position material between the posterior teeth, where it is ground and reduced in bulk for the next stage of swallowing. The tongue produces saliva and contains the majority of the sensory end organs for the sense of taste on its dorsum. Other taste receptors are located in the palate, epiglottis, and upper esophagus.

The surface of the tongue plays another important role in mastication. *Filiform papillae* on the anterior two-thirds of the tongue's dorsum, in front of the sulcus terminalis, create a file-like surface on the tongue, and help strip food materials through licking. Named for their rough structure, the epithelium at their tips is keratinized to form a tougher, more wear resistant coating. Filiform papillae provide mechanical support for mastication and swallowing. Their structures consist of a core body extending a short distance from the surface of the tongue, which then splits into several slender, pointed projections.

Salivation and Digestion

The chemical breakdown or digestion of nutrients begins in the oral cavity during the oral preparatory phase of deglutition. Part of this process is the result of the chemical enzymes in saliva. *Salivary amylase* (formerly called *Ptyalin*) is a protein constituent of saliva that reacts with complex starches (polysaccharides) to chemically break them down into less complex carbohydrates, maltose and dextrin. *Salivary lipase*, also a constituent of saliva, acts on lipids (fats) to begin their digestion in the oral cavity. Other components of saliva are mucin, which lubricates the aerodigestive tract, and water, which softens the material.

Saliva is secreted by three pairs of large salivary glands and by numerous small lingual salivary glands during the oral phase of deglutition. Its chemical composition is a mixture of water, enzymes, electrolytes, mucous, several antibacterial compounds, and opiorphin, a recently discovered pain inhibiter (Wisner, Dufour, Messaoudi, Nejdi, Marcel, Ungeheuer, and Rougeot, 2006). This complex liquid begins the breakdown of food and lubricates the oral cavity and the rest of the digestive tract to ease passage of materials within.

The sublingual, submandibular, and parotid salivary digestive glands secrete saliva, which is then transferred into the oral cavity through ducts located beneath the tongue in the sublingual fold and in the buccal walls. Saliva from the paired *sublingual salivary glands* passes into the oral cavity through about a dozen openings called *Rivinus' ducts*. Several Rivinus' ducts unite at the anterior portion of the sublingual fold to form a larger salivary delivery tube, *Bartholin's duct*, which drains the anterior portion of the sublingual gland. Two *submandibular salivary glands* are located posterior to the sublingual glands, near the hyoid corpus, and also provide saliva for the oral cavity. Their saliva is secreted through *Wharton's ducts* opening anterior to Bartholin's ducts within the sublingual fold. The paired *parotid salivary glands* are the largest salivary glands. These glands are wrapped around the mandibular rami with their free ends extending posteriorly as far as the mastoid processes and external auditory meatuses. They secrete saliva through the *parotid ducts*, also called *Stenson's ducts*, which extend from the glands along the surface of the masseter muscle, then turn medially to penetrate the buccinator muscles and oral mucosa. The ducts open into the oral cavity through the buccal walls, at the level of the second upper molars.

Lingual salivary glands, also called *Von Ebner's glands*, secrete saliva into an annular depression at the base of each large *circumvallate (also called vallate) papilla*, located in the posterior area of the tongue's dorsum and near the *foliate papillae* on the posterolateral surfaces (Spielman, D'Abundo, Field, and Schmale, 1993). In this context, saliva assists in nutrient digestion and flushes material away from the taste bud for renewed sensation.

The final action of the oral phase is the transfer of material into the oropharynx. It begins when the decision is made to swallow all or part of the material in the oral cavity. During this phase, lingual muscles draw the tongue posteriorly in the oral cavity, and the faucial arches dilate. All or part of the material to be swallowed is pushed by the back of the tongue into the oropharynx and deglutition begins its second phase. Material can remain in the oral cavity for further oral preparation.

Bolus Consistency

The size and consistency of the bolus is to be considered as a factor when studying the swallowing process. A *bolus* is a mass of material which passes through a body passageway, particularly the digestive tract. In the study of deglutition, the term refers to a mass of material to be swallowed.

Naturally, a bolus which is bulky and hard requires more preparation than one that is thin. Thin liquids, on the other hand, may move through the digestive tract without muscle intervention, and indeed, may require greater restraint to prevent them from entering the airway than do thicker boluses.

Pharyngeal Phase

The pharyngeal phase of swallowing is the most crucial for airway protection. During this phase, material enters the oropharynx, having been propelled by the tongue, and bypasses the laryngopharynx on its way to the esophagus.

Initially, the individual usually has at least some degree of voluntary control of deglutition during the pharyngeal phase. However, the further into the oropharynx the material moves, the less voluntary control may be held by the one swallowing. As a complete act, swallowing is normally mediated by a *swallowing reflex*, generated in the brainstem. The swallowing reflex is expressed as a complex concert of muscular inhibitions and excitations, beginning with peristaltic movements of the pharyngeal constrictors and concluding with relaxation of the lower esophageal sphincter.

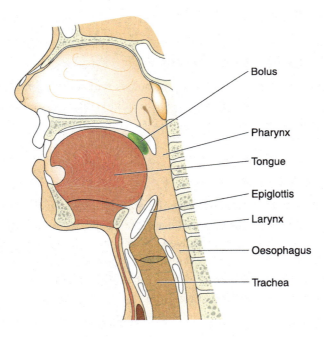

To move swallowed material from the oral cavity in to the oropharynx, the whole body of the tongue draws rearward, pushing the bolus or volume of liquid with its back surface. The sequence of pharyngeal muscular contractions occurs very quickly, usually in less than 1 second (Raj, Goyal, and Mashimo, 2006).

During this quick deglutition sequence, the posterior faucial arch tightens around the retracted tongue and the posterior oropharyngeal wall develops a *Passavant's ridge* (named after German surgeon Philip Gustav Passavant, 1815–1893), by contraction of the superior pharyngeal constrictor to approximate the posterior tongue and help propel the material into the pharynx.

At the same time, the velopharyngeal sphincter tightens to resist nasopharyngeal penetration, and the laryngopharynx moves superiorly and anteriorly, drawn by contractions of the suprahyoid cervical "strap" muscles. With the upward movement of the larynx, the laryngeal vestibule is sealed by action of the aryepiglottic muscles, which cinch the aryepiglottic folds and draw the epiglottis, arytenoid, corniculate, and cuneiform cartilages together. The downward movement of the epiglottis flattens the glossoepiglottic folds, emptying their contents. Deeper in the laryngopharynx, the ventricular and vocal folds adduct, creating a triple valve or "Pharyngeal Squeeze" to protect the airway from intrusion of unwanted material (Flaherty, Seltzer, Campbell, Weisskoff, and Gilbert, 1995; Aviv, Spitzer, Cohen, Ma, Belafsky, and Close, 2002; Olthoff, Schiel, and Kruse, 2004). It must be remembered that the laryngeal entrance opens first in an anterior direction, and its cavity then continues in an inferior direction. With the elevation and sealing of the laryngeal inlet, the esophagus, which has its superior end at the lower level of the laryngeal inlet, is positioned to receive swallowed material as it bypasses the closed laryngeal inlet.

Esophageal Phase

The esophageal phase of swallowing is almost entirely involuntary, being part beginning with the autonomic stimulation of the cricopharyngeal opening and continuing with the peristaltic movement and involuntary stimulation of the lower esophageal sphincter. As mentioned earlier, contractions of the striated cricopharyngeus muscle can be initiated voluntarily, but contractions during the swallowing reflex are usually involuntary. Esophageal musculature is striated at the superior end and gradually gives way to smooth muscle at the inferior end.

Critical to the esophageal phase is normal functioning of the upper esophageal sphincter (UES) and lower esophageal sphincter (LES). The upper sphincter is composed of the striated cricopharyngeus muscle and can be under voluntary control. The muscle that forms the LES is smooth.

The Relationship of Deglutition and Speech

Deglutition and speech involve coordinated action of the bulbar and respiratory musculature to perform their distinct systemic functions. Although movement patterns and origins differ, deglutition and speech share certain anatomic structures. It follows that certain disorders, particularly those of neurological origins, involve concurrent speech and swallowing dysfunction.

Structural Overlap

Deglutition, by definition, involves moving material from the oral and pharyngeal cavities into the esophagus. The peripheral speech mechanism is most often described as being of one substance with the distal respiratory tract, but it is also accepted that the distal digestive tract is involved as well. Indeed, the distal part of both systems is often described as the *aerodigestive* tract, a term which includes both respiratory and digestive applications.

Structures used in both deglutition and speech include the oral cavity and its contents, including those most external structures of the oral cavity, the lips. It is through movements and juxtapositions of oral structures most phonemes are articulated and food is consumed and initially prepared for swallowing. A final oral deglutition function is propelling the material into the oropharynx.

The velopharyngeal sphincter and the oropharynx also have common speech and deglutition functions (Edwards, 1992; Logemann, 1995). During deglutition, the velopharyngeal sphincter must seal the nasal cavity to prevent the infiltration of material into the nasopharynx, whereas the faucial arches of the oral isthmus form a sphincter with the posterior tongue to prevent material from re-entering the oral cavity. The oropharynx serves as a final conduit for swallowed material into the esophagus for deglutition and as a resonating chamber for speech. The speech function of the velopharyngeal sphincter is also to separate the nasopharynx and nasal cavity from the rest of the vocal tract in order to regulate the amount of nasal resonance in the speech acoustic. Alternately, the velopharyngeal sphincter is opened, allowing a normal amount of nasal resonance when it is desired. Pharyngeal fricatives (/ʕ/ and /ħ/), not standard in English phonology, are articulated in the oropharynx by juxtaposing the tongue root against the pharyngeal wall.

The esophagus begins at the lower extent of the oropharynx, at which point, the laryngopharynx diverts the airway in an anterior direction for a short extent before continuing inferiorly to the lower airway. Here, the structural duality of aerodigestive tract for speech and deglutition purposes no longer applies.

Physiological Similarities between Swallowing and Speech Articulation

In function, swallowing and speech both involve coordinated contraction of similar muscle groups supporting the same anatomical structures in the upper airway. For example, anterior-to-posterior lingual movements push the bolus toward the oropharynx during deglutition. Similar movements produce dental-alveolar-postalveolar and palatal approximants ("glides") to back vowel target syllabic nuclei during speech. Intrinsic laryngeal muscle groups that adduct the vocal folds during vowel articulation, sonant consonant articulation and glottal consonant articulation also function sequentially to secure airway closure as a bolus passes through the pharynx. As mentioned above, the velopharyngeal valving action that couples or uncouples the nasal cavity to the vocal tract prevents penetration of food and liquid into the nasopharynx and allows reduction of relative intraoral air pressure for sucking also changes vocal tract resonance for nasal or non-nasal consonant articulation. Further, contractions of extrinsic laryngeal muscle groups to elevate the larynx and seal the airway during deglutition are activated to varying extents during pitch increases (Zemlin, 1998).

Neurologically, there are additional similarities between speech and swallowing. In the central nervous system, diverse centers of brain activity appear to exert central control over both swallowing and speech movements (Mosier, Wen-Chu, Maldjian, Shah, and Modi 1999; Suzuki, et al 2003; and Marin, et al (2004)). Overlapping brain centers for both functions include primary motor and sensory cortices, with a significant laterality effect tending to favor the left (or dominant) cerebral hemisphere, motor association cortices, temporal, cingulate, and insular cortical areas and the internal capsule. Activity in the thalamus, insula, basal nuclei, and cerebellum has also been reported (Mosier and Bereznaya, 2001), and little evidence has supported cerebral cortical areas dedicated to swallowing as distinct from other tongue movements.

Peripheral nervous system support for speech and swallowing includes the lower motor neuron and first order sensory neuron contributions of the cranial and spinal nerves. While there may be a much larger direct role for parasympathetic and sympathetic innervation of common structures in swallowing, the role of salivary secretion in the maintenance of vocal tract lubricity is commonly recognized.

Physiological Differences between Swallowing and Speech Articulation

The most obvious physiological difference between speech and swallowing is systemic. Speech employs the oral, nasal, and pharyngeal cavities as respiratory structures while swallowing employs them as alimentary structures.

NEUROPATHOLOGY OF SPEECH AND DEGLUTITION

Speech and swallowing disorders may occur simultaneously in several neurological diseases, since muscle groups associated with speech also function in the oral and pharyngeal stages of swallowing (Kennedy, Pring, and Fawcus, 1993). It is logical to expect dysarthria and neurogenic dysphagia to occur concurrently.

Duffy (1995) found dysarthria to be the most frequently presented acquired neurogenic communication disorder in over 3,000 patients. Miller and Langmore (1994) identified neurological disorders as the cause of most dysphagia. Almost all of the subjects of Darley, Aronson, and Brown's (1975) subjects presented dysphagia as well as dysarthria, no matter what their neuropathological classifications. Logemann (1983) observed diminished lateral and vertical lingual range of motion, reduced buccal tension, and limited rotary mandibular movement in oral stage neurogenic dysphagia. Dobie (1978) noted reduced, uncoordinated lingual movements and decreased oral sensation that affected bolus formation in the oral preparatory phase, both of which are essential to normal speech. Among a group of individuals with neurogenic dysphagia, Miller (1982) observed reduced lingual elevation, lingual range of motion, disorganized anterior to posterior lingual patterns, and limited mandibular movement in mastication during the oral phase of the swallow.

Swallowing problems in the oral phase may include difficulty with mastication, with managing and initiating bolus transfer and with poor retention of intraoral material. These swallowing functions depend upon lingual and mandibular mobility, and so does speech articulation.

The vertical lingual movement required to maneuver a bolus along the palatal (maxillary) vault and to maneuver food from the buccal sulci is also essential to dental, alveolar, postalveolar, or velar consonant production as well as for close vowel articulation.

Mandibular elevation is required to bring the tongue high enough for it to contact the upper oral cavity and to bring the upper and lower dental arches into proximity. Such proximity is necessary for mastication and also for lingua-palatal and labial juxtaposition in the articulation of consonants and some vowels.

Pharyngeal phase swallowing disorders may be concomitant with dysarthria, resulting from the same injury or disease process that disrupts efficient muscular control. The involuntary swallowing reflex includes coordinated velopharyngeal and laryngeal closure combined with rhythmic and sphincteric, pharyngeal constrictor action. Neuromuscular impairment of the swallowing reflex can allow food or liquid to penetrate the laryngopharynx, compromising the airway (Buchholz, 1994). Kilman and Goyal (1976) described a delayed or absent swallowing reflex, inadequate velopharyngeal closure, and reduced pharyngeal peristalsis as aspects of the pathophysiology of dysphagia. There is good evidence that the presence of a "gag" reflex, triggered by touch at the oropharyngeal wall, is not a prerequisite for swallowing without airway infiltration, but that pharyngeal sensation, particularly in the laryngopharynx, is essential (Davies, Kidd, Stone, and MacMahon, 1995; Aviv, Spitzer, Cohen, Ma, Belafsky, and Close, 2002).

Velopharyngeal closure and laryngeal elevation and closure are well-known functions of both swallowing and speech. Thus, the same neurological disease that reduces strength and coordination of the pharyngeal musculature active in swallowing may also alter nasopharyngeal resonance and affect laryngeal phonatory function. The swallowing reflex may be defective without significant dysarthria (Kennedy, Pring, and Fawcus, 1993), such as might occur in impairment of the afferent aspect of the reflex.

Neuromuscular differences between speech and swallowing involve their behavioral origins and variations in movement patterns. Behavioral origins of speech and deglutition arise from their diverse physiological foundations. These origins have clinical import because they influence patients' therapeutic compliance and, hence, the success of treatment. Initiation and continuation of speech movement patterns involve the interplay of diverse central affective motivations and neurolinguistic substrates with sensorimotor functions. Once speech is underway, the speaker consciously maintains speech neuromuscular subsystem coordination and modifies ongoing patterns and intensities of contraction through afferent system feedback (Mysak, 1976, Darley, *et al.*, 1975). It is apparent, then, that most speech movements originate voluntarily and continue at least partly through conscious self-monitoring. Swallowing origins are less well understood (Logemann, 1995), but appear to involve an interplay of conscious and unconscious motivations and movements. Voluntary intake and preparation of the food bolus begin a chain of swallowing events. Once underway, deglutition becomes successively more reflexive and ultimately autonomic (Zemlin, 1988, Tanner and Culbertson, 1999).

Speaking and swallowing movement patterns are distinct in their directionality and rhythm. Speech movements are usually egressive, with air flowing out of the body, whereas swallowing is always ingressive. Speech movement patterns vary with the language of the speaker, but include movements that vary widely in range, direction, and velocity. Swallowing patterns are much more stereotyped and sequential, varying relatively little in action (Kennedy and Kent, 1985). Speaking and swallowing are occasionally concurrent, and momentary interruptions of speech are common while a speaker clears saliva from the mouth. Autonomic esophageal phases of deglutition can continue while the mouth is occupied with speaking.

Respiratory Considerations

Structural and functional similarities aside, the primary biological and life-sustaining role of the upper aerodigestive tract is to make intake of food, liquid, and air safe. The valves in the oral cavity and pharynx exist to admit as well as to prevent admission of selected substances. Of these, food and liquid are probably the most important materials to keep out of the lower airway. Choking was listed by the U.S. National Safety Council as being among the top five causes of accidental death in 2007 (National Safety Council, 2007).

Even when material in the lower airway does not cause asphyxiation, the introduction of bacteria thriving on food and liquid material either from the oropharynx or from regurgitation of gastric contents poses the risk of *aspiration pneumonia*. Normal mechanisms of defense against aspiration are normally functioning swallowing and cough reflexes (Lee, 2004).

Nutrition Science

Clinicians who treat patients with dysfunctions of deglutition must be aware of basic human nutritional needs. These needs are studied individually and statistically by the *nutrition scientist, nutritionist* or *dietician,* and questions regarding the nutritional or hydration status of patients can most often be answered through consultation with these members of the rehabilitation team.

With training at the baccalaureate level or beyond, nutrition science professionals plan meals and supervise their preparation and serving to students, including those with special needs, and patients in hospitals or nursing homes. Most schools and similar public institutions have a dietician, whereas the *clinical nutritionist* is a professional who addresses the nutritional needs of hospitalized patients. The scope of nutrition science extends beyond that of merely recommending what foods a patient needs and is well beyond the training of the typical speech-language pathologist.

The Body, Nutrients, and Health

Human nutrition can be categorized into two broad groups: essential nutrients and non-essential nutrients. *Essential nutrients* are those that must be supplied in the diet, whereas *non-essential nutrients* are those that the body can produce from raw materials it has consumed. Children and some patients have special metabolic conditions that can alter the list of essential requirements. In those cases, additions to the essential list are considered *conditionally essential* (Truswell and Mann, 2007).

Essential nutrients include essential fatty acids (lipids), essential amino acids (proteins), certain vitamins, and dietary minerals. Water, sunlight, and oxygen, not generally considered to be nutrients, are also required and cannot be created by the body.

Nutrient Sources

Essential fatty acids are consumed by eating fatty meats, dairy products, vegetable seeds, including nuts, fish oils, and plant leaves. Essential amino acids are broken down from proteins formed in longer polypeptide chains and consumed from animal tissues, dairy products, eggs, beans, cereals, vegetable leaves, and soy products. Essential

vitamins and dietary minerals occur naturally in foods and can be supplemented pharmaceutically. Exposure to short periods of sunlight is often recommended for natural vitamin D synthesis.

Hydration

Although water is not considered a nutrient, it comprises most of the substance of the body and is essential to bodily function. Sufficient water must be taken from the environment through *hydration*, lest the individual become *dehydrated*. Signs of dehydration include dry skin, decreased and dark urine output, confusion, headache, and muscle weakness (Mayo Clinic, 2007).

Nutritional Needs Across the Lifespan

The need for energetic, protein supplying foods is at its greatest during the growth and development periods of life, from birth through adolescence. During this period, when bones are growing calcium is also a much needed dietary component. Growth rate during the first five years of life is rapid, tapering off after the first year but continuing at a diminished rate through adolescence and into adulthood. Nutritional needs are at their peaks during adolescence (Whitney, Cataldo, and Rolfes, 2002), particularly among active people.

Nutritional needs taper off during adulthood and particularly in the long-lived individual Whitney, Cataldo, and Rolfes (2002) estimated that adult nutritional needs decrease on the average at about 5% per decade. However, this does not diminish the importance of adequate nourishment in long-lived individuals. The need for energy from carbohydrates and for protein from low-calorie sources remains, and malnutrition may be a real danger in circumstances where the individual is either unwilling or unable to secure appropriate nutrition.

Another important nutritional consideration for long-lived individuals is the changes in the anatomy and physiology of the distal digestive tract. In many older individuals, dentition becomes a problem. Missing teeth or entirely edentulous dental arches make mastication difficulties not uncommon. With the lack of supporting lower dentition, the angle of the mandible increases to almost the same degree as that found in the infant (Joseph, 1976).

Neuromuscular control of swallowing appears to be compromised in advancing age in some individuals. Humbert, *et al.* (2009) compared the swallowing reflexes of twelve young subjects with those of eleven older subjects and concluded that neurophysiological changes were evident, resulting in cerebral as well as neuromuscular distinctions between the groups.

Skeletal Framework of Deglutition

The skeletal framework for deglutition is the same as the one for most of the peripheral speech mechanism, namely the bones of the skull and thorax. The hyaline and elastic cartilages that keep the laryngopharynx and lower respiratory tract patent are not parts of the digestive system. The epiglottis, formed of elastic cartilage, provides a substantial barrier against substances entering the laryngeal vestibule as they pass inferiorly on the way to the esophagus.

Skeletal Support for Deglutition

Skeletal support for deglutition can be considered externally and internally. External skeletal structures support and protect the soft tissues of the digestive tract, and internal support provides patency as well as rigid surfaces against which to process solid material.

External Support

The distal digestive tract is protected by the bones of the facial skeleton, the hyoid bone, and the cervical spine. Facial skeleton bones include the mandible and maxillae. Damage or destruction of these bones severely compromises the integrity of the soft internal oral and pharyngeal tissues and is a common cause of dysphagia.

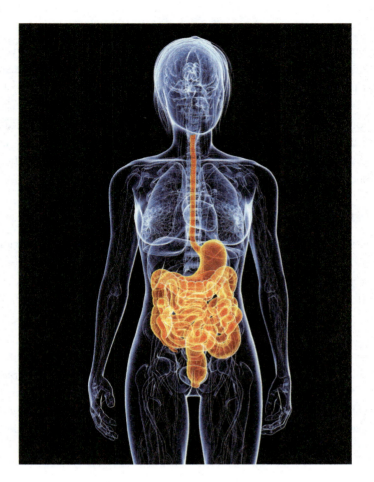

In the neck, skeletal structures play a much more crucial role. Mechanical injury of the hyoid bone can compromise patency of the oropharynx, resulting in obstruction of the areodigestive tract. Cervical spine fracture, in addition to creating the possibility of accompanying spinal cord and cranial nerve injury (Wolf and Meiners, 2003), can affect the patency of the oropharynx as well. Typically, these patients will have a tracheostomy to open and maintain the lower airway during recovery, and modifications such as *tube feeding* may also be required for nutrition and hydration. *Enteral* tube feeding deliver nutrients and liquids directly into the stomach or small intestine, bypassing the oral cavity and neck structures, and *parenteral* tube feeding delivers nutrients and water directly into the bloodstream.

In the thorax, the ribs, sternum, and spinal column may be considered protective structures for all the soft tissues, including the esophagus.

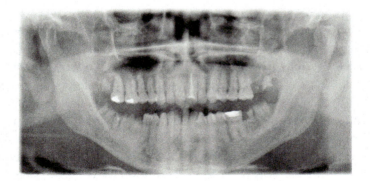

Internal Support

Internal support for deglutition is provided by the bones of the oral and oropharyngeal cavities. The bones of the oral cavity provide structure to support mastication and provide patency to the distal digestive tract. These include the alveolar and palatine processes of the maxillae, the horizontal processes of the palatine bones, the mandible, and the hyoid bone.

Damage or destruction of the oral cavity bones compromises the ability of the oral cavity to processes the oral phase of deglutition, and damage to the hyoid bone can affect the patency of the distal digestive tract. Damage to the mandible or maxillae is fairly common and decreases or eliminates the ability of the paired upper and lower dentition to masticate solid material.

Dentition and Deglutition

Both upper and lower dentition is required to perform proper mastication, and the various teeth are designed to perform mastication in sequential stages. The anterior teeth are called incisors because their thin, sharp shape is best for slicing smaller portions of food from a larger body. Posterior teeth are flattened and stout to provide a surface and withstand the pressures of grinding and increasing the surface area of food.

Food and Dental Health

As is the case with any art of the body, proper nutrition is important to keep teeth functional. Insufficient dietary intake can compromise the structure of teeth and make them brittle or soft. The nutrition scientist should be consulted for advice regarding nutritional requirements for dental health. Further, excess accumulation of substances found in food or drink can promote the deterioration of dental crowns and ultimately destroy the entire tooth. Dental specialists, including doctors of dental science and dental hygienists should be consulted regarding the physical and health status of dentition.

Edentulous Mastication

The individual without teeth in the alveolar processes is *edentulous*. Complete or partial dentures can be fitted, depending upon individual needs, and a surprising amount of mastication can occur without teeth, through use of the gingival surfaces.

Muscles of Deglutition

Muscles involved deglutition include the lower muscles of facial expression, the intrinsic and extrinsic tongue muscles, muscles of mandibular elevation, oropharyngeal muscles, and upper esophageal sphincter muscles. These muscles control the labial, oropharyngeal, velopharyngeal, and esophageal sphincters, move the mandible and tongue during mastication, and propel material from the oral cavity into the esophagus.

Facial Muscles

Facial muscles used in deglutition are those that open and close the labial sphincter. These are the *lower muscles of facial expression,* and are those with attachments at or below the zygomatic bones.

The orbicularis oris and the paired incisivus superior and incisivus inferior tighten and close the labial sphincter, containing material within the oral cavity. *Orbicularis oris* is a complex sphincteric muscle, with fibers that circle the oral opening. Its fibers are continuous with those of many other facial muscles, including those that act in opposition. Contraction of orbicularis oris fibers tightens the opening of the labial sphincter and seals the oral opening during sucking, mastication, or transferring material form the oral cavity to the oropharynx. *Incisivus superior* muscles arise on the maxillary alveolar processes, at the eminences created by the cuspid teeth (*canine jugae*) and insert from behind into the fibers of the orbicularis oris muscle, at both sides of the upper lip, near the

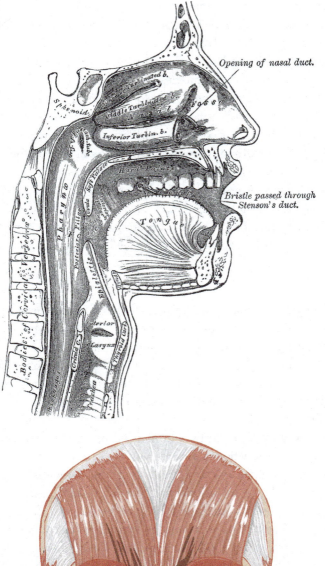

Opening of nasal duct.

Bristle passed through Stenson's duct.

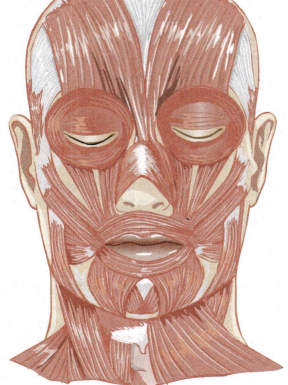

angles of the mouth. *Incisivus inferior* muscles originate at the eminences created by the lower lateral incisors and insert into obicularis oris fibers at the oral angles (Zemlin, 1998; Bardeen, 1907). Contraction of either incisivus muscle fibers draws the lips tight against the alveolar processes of either the maxillae or mandible.

Muscles that oppose tightening of the labial sphincter open the lips to allow admission of material to be swallowed or to bare the teeth for biting. These muscles originate at various facial sites radiating around the oral opening and have fibers that interdigitate with those of the orbicularis oris muscle at angular or horizontal angles. For details regarding their origins and insertions, as well as their specific functions, see Chapter 6.

The labial *levators* (from Latin *levare*: to lift) oppose orbicularis oris by drawing the upper lip superiorly, toward the nose and cheeks. This opens the labial sphincter and allows material to enter the oral cavity. Muscles that draw the upper lip superiorly and help open the labial sphincter include *levator labii superioris, levator labii superioris alaeque nasi, zygomaticus minor, zygomaticus major,* and *levator anguli oris.*

Movement of the lower lip away from the upper lip is also part of a normal labial sphincter opening. In anatomical position, gravity can assist or even effect labial opening if unopposed by orbicularis oris. The lower lip is depressed by contraction of the paired *labial depressors, depressor labii inferioris, depressor anguli oris,* and the *platysma.*

Oral Muscles

The role of oral muscles in deglutition is to position solid material for mastication, to elevate and depress the mandible for mastication, and to transfer the material to be swallowed from the oral cavity into the oropharynx. Maintenance of the bolus for mastication is accomplished largely by intrinsic and extrinsic tongue muscles, assisted by the cheek muscles. Intrinsic tongue muscles are named for the direction of their fibers: *superior* and *inferior longitudinal,* which elevate, retroflex (superior longitudinal) or depress (inferior longitudinal) the tongue apex, *vertical,* which compresses the tongue in the vertical dimension, and *transverse,* which compresses the tongue in the horizontal dimension. These muscles act in concert to produce great flexibility of movement within the oral cavity, and thus, the tongue can move from left to right and curl upward and down to clear the buccal sulci and palatal vault. Intrinsic tongue muscles function in concert with the extrinsic tongue muscles for licking.

Contractions of extrinsic tongue muscles move the body of the tongue to either position its tip and blade advantageously or to propel material into the oropharynx. The tongue body is moved anteriorly by contraction of the paired *genioglossus* muscles to position the blade and tip of the tongue forward or even out of the oral cavity for licking, etc. These are opposed by two pairs of muscles that move the tongue body posteriorly for retraction of its tip back into the oral or for transfer of food and liquid into the oropharynx, *styloglossus,* and *palatoglossus. Mylohyoid* and *hyoglossus* draw the body of the tongue inferiorly toward the mandibular corpus and the hyoid bone.

Tongue movements are augmented by the paired *buccinator* muscles, deep within the cheeks. Buccinator contractions provide resistance to tongue movements and assist with positioning material to be swallowed within the oral cavity.

Muscles of mandibular elevation provide the biting and grinding movements needed for mastication. The *temporalis, masseter,* and *pterygoid* (lateral and medial) muscles originate at various skull sites and insert on the mandible to elevate it and move the mandible from left to right and forward and backward. Refer to Chapter Six for details regarding origins and insertions of the oral musculature.

Pharyngeal Muscles

Pharyngeal muscles act in concert and in sequence to begin the peristaltic wave by closing around the tongue root as material to be swallowed passes from the oral cavity into the oropharynx on its way to the esophagus. They also seal the passage between the oropharynx and nasopharynx to prevent infiltration of material into the superior pharyngeal cavity. Muscles of the laryngopharynx contract to seal the airway and prevent material from entering the respiratory tract, whereas several of the vertical cervical strap muscles elevate the larynx and hyoid bone during the process. Finally, the inferior *palatoglossus* and *palatopharyngeus* form the muscular foundations of the *faucial arches*, also called the *fauces*. These muscular vaults mark the borders of the oral isthmus, separating the oral cavity from the oropharynx, and, with the body of the posterior tongue, contract to prevent material from reentering the oral cavity. Following quickly upon contractions of the faucial arches, sequential, wave-like contractions of the *superior*, *middle*, and *inferior pharyngeal constrictor* muscles push material inferiorly and into the upper esophagus. Here, the upper esophageal sphincter, formed predominantly by the lower fibers of the inferior constrictor muscle, identified separately as the *cricopharyngeu*s muscle, relaxes and dilates the upper esophageal lumen.

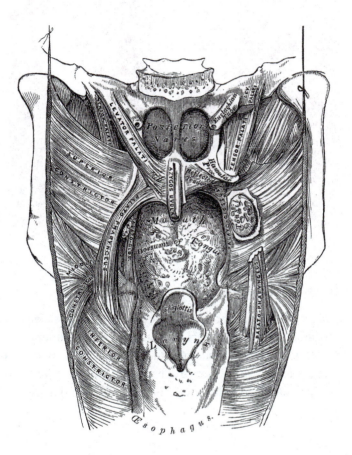

Coincident with contractions of the sphincteric pharyngeal muscles during deglutition and bolus transfer to the esophagus are contractions of the vertically oriented *suprahyoid* and some of the *extrinsic laryngeal* muscles. These elevate the larynx so that the laryngeal additus approximates the lowered epiglottis. The epiglottis, in turn, is drawn posteriorly and inferiorly by the aryepiglottic muscles, forming the uppermost seal of the laryngeal cavity.

Although they are not anatomically parts of the digestive system, several of *intrinsic laryngeal muscles* contract in unison with contractions of oropharyngeal muscles. Their contractions ensure against infiltration of swallowed materials into the lower airway. Contraction of the aryepiglottic muscle, coincident with upward displacement of the laryngeal body, seals the entrance to the laryngopharynx by cinching the aryepiglottic folds. These folds are strengthened by a cartilaginous infrastructure formed by the epiglottis, the cuneiform and corniculate cartilages, and by the apices of the arytenoid cartilages.

Two seals are formed within the laryngeal cavity and inferior to the laryngeal entrance by adductions of the *ventricular folds* and *vocal folds*. The ventricular folds are adducted by contractions of the *ventricular muscles* and the *oblique arytenoid* muscles, and the vocal folds are adducted by contractions of the *oblique arytenoid, transverse arytenoid*, and *lateral cricoarytenoid* muscles.

Esophageal Muscles

The upper esophageal sphincter, also called the cricopharyngeus muscle may also be considered a continuation of the striated fibers of the lower pharyngeal sphincter. Its relaxation allows food and liquid to enter the esophagus, and its contraction prevents gastric material form regurgitating into the pharynx.

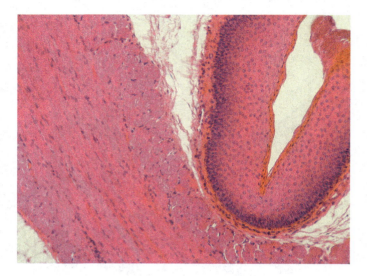

Cytology and Histology of the Digestive Tract

In general, the tissues of the digestive tract are arranged in four laminae. These layers are, from internal to external, epithelium, a submucosal lamina propria, external musculature, and adventitia. Tissues of the distal digestive tract, that is, those of the oral cavity and the oropharynx, are distinguished from those of the lower digestive tract by their more complex musculature and by the presence of lymphoidal tissues. Non-oral digestive tract tissues will be treated from internal to external. Oral and pharyngeal histology is covered in detail in Chapter 6.

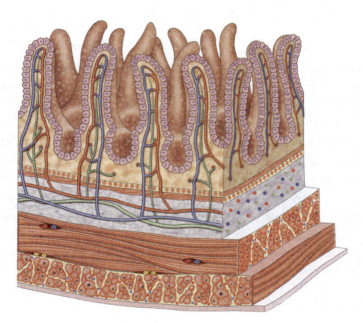

Lining the lumen of the general digestive tube is a layer of *epithelium*. Note that this layer is the surface tissue of the inside, or lumen, of the esophageal tube. In the upper or distal levels of the canal, this layer is stratified, squamous epithelium, suited anatomically to the abrasion created when swallowed material passes through. Squamous epithelium lines the pharynx and esophagus, since swallowed material passes through both tubes. At lower levels of the digestive tract, that is, from the stomach on, the internal epithelium becomes columnar, providing the appropriate anatomy for secretion and absorption.

The next of the four tissue layers, superficial to the inner epithelium, is a layer of *lamina propria*, which supports the esophageal epithelium. This loose connective tissue provides physical structure for the tract while allowing it to be flexible as material passes through. Its flexibility also allows peristaltic muscular contractions to move foods and fluids deeper into the digestive tract.

Superficial to the lamina propria is a layer of external musculature. This musculature layer is striated and capable of voluntary contractions above the upper esophagus, and gives way to smooth, involuntary musculature at levels approaching the stomach. The lower esophageal sphincter is one of smooth muscle. The smooth musculature consists of an inner sphincteric or circular layer and a superficial longitudinal layer.

Providing protection and more structure, and forming the outer covering of the digestive tract, is a connective tissue layer called either the *adventitia* or *serosa*. At upper levels, that is, in the cervical and thoracic regions, the outer layer is referred to as the adventitia. Here it consists of a layer of loose connective tissue. Within the abdominal cavity, the outer layer is protected by a layer of simple squamous epithelium and is called serosa.

The Esophagus

Four layers of tissue form the esophageal tube in an arrangement similar to that of the digestive tract in general. A *fibrous layer* (adventitia) forms the outside, covering the intermediate *muscular layer*. Deep to the muscular layer, and forming the basement membrane of the internal layer is the *submucous layer (lamina proporia)*. The inner lining is a *mucous layer* formed of non-keratinized, stratified squamous epithelium, designed for wear by material passing within. The submucosal lamina propria between the squamous tissue lining and the muscular layer contains glands formed by groups of goblet mucous secreting cell. These groups project ducts through the squamous tissue and secrete mucous into the esophageal lumen to lubricate and facilitate the passage of swallowed material.

Esophageal musculature is a complex mixture of striated and smooth muscle fibers. An external layer of muscle tissue forms the outer wall of the organ, covering an inner layer of sphincteric muscle tissue. This musculature is striated at the upper one-third and smooth at the lower one-third, reflecting the transition from voluntary to involuntary control. The musculature in between is a mixture of striated and smooth muscle. An inner layer of smooth muscle tissue, the *muscularis mucosae*, is located internally, just deep to the epithelial lining of the tube. Its constant agitation action keeps the internal tissues of the esophagus free from debris (Huang and Chang, 2004).

Neurology of Deglutition

The act of seeking food or liquid may be triggered by autonomic, visceral impulses, with behavioral as well as nutritional origins, while the act itself, including acquisition of food and placing it in the mouth, is normally voluntary and conscious. Thus, neurological functions in swallowing are largely voluntary and conscious at first, becoming more involuntary and unconscious as the swallowed material proceeds through the esophagus and beyond.

This is not to downplay the role of psychological, sometimes unconscious, central nervous system mechanisms involved in the process of procuring and assimilating food or water. Psychological urges evoked by external or internal stimuli create the appetite, and must be considered when studying the individual's actions centering on food and water intake. Salivation, for example, increases as the memory of food is triggered by smell, sight, sound, or touch of certain foods (Tortora and Grabowski, 1996).

Salivation and other digestive responses can be triggered by both parasympathetic and sympathetic branches of the peripheral nervous system. Normally, parasympathetic impulses keep a steady stream of saliva flowing in the digestive tract, decreasing during sleep, with 1 to 1.5 liters produced daily. Sympathetic activation tends to suppress salivary secretions.

The oral phase of deglutition is also initially voluntary and conscious, with the exception of the glandular responses of salivary glands. Musculature of the oral cavity and oropharynx can be contracted under voluntary control, and sensation emanating from these areas reaches consciousness. It must be noted that involuntary and complex patterns of facial, oral and pharyngeal muscles contractions are common, having diverse central origins delivered through common peripheral pathways, and that swallowing patterns can be either volitional or unconscious. Mosier and Bereznaya (2001) described these seemingly diverse movement pattern origins as arising from parallel neural networks.

Voluntary contractions of pharyngeal muscles, including those of the oropharynx, laryngopharynx, and distal esophagus, have central origins and are delivered to the muscles via the accessory nerves. Involuntary contractions, such as occur during peristalsis, are delivered via the vagus and glossopharyngeal nerves.

Muscles of the face facilitate deglutition by opening the external entrance to the oral cavity to permit material to be inserted therein and by sealing that entrance to retain material in the oral cavity during mastication. These muscles receive their motor innervation via maxillary and mandibular terminal branches of the facial nerve (VII).

Muscles of mastication receive their motor nerve supply via the trigeminal (V), vagus (X) and hypoglossal (XII) cranial nerves. Those that elevate, extend, and rotate the mandible, including the temporalis, masseter, and pterygoid muscles, receive their motor innervation from the mandibular branch of the trigeminal nerve. Intrinsic and extrinsic tongue muscles, which assist mastication by maintaining bolus position, receive their motor innervation via the hypoglossal nerve.

The velopharyngeal sphincter usually closes during deglutition to inhibit infiltration of material into the nasopharynx. The levator velipalatini muscles, which draw the velum upward and toward the posterior pharyngeal wall when contracted, receive their voluntary motor innervation via the accessory nerve (XI), distributed with those of the pharyngeal branch of the vagus nerve (X). The tensor velipalatini muscles, which contract to add stiffness to the soft palate, receive their motor supply via the maxillary branch of the trigeminal nerve (V).

Conscious sensations of touch, pain, and proprioception are conveyed from the face and anterior oral cavity to the central nervous system via the trigeminal nerve (V). Moving proximally, sensation from the posterior one-third of the oral cavity, from the oropharynx and from the superior esophagus are conveyed to the central nervous system via the nerves of the pharyngeal plexus, including the glossopharyngeal (IX), vagus (X), and accessory (XI) nerves. Conscious sensations associated with taste are also conveyed to the central nervous system via cranial sensory ganglia associated with the facial, glossopharyngeal, and vagus cranial nerves (VII, IX, and X).

Sympathetic and parasympathetic nervous system reflex arcs play an essential role in digestive tract function. Primary visceral sensory first-order neurons monitor conditions in the upper digestive tract and deliver feedback to the central nervous system at the brain stem and spinal cord levels. There, blood sugar, oxygen, carbon dioxide, and the presence of toxins are monitored and an appropriate response generated. Input is conveyed via *cranial sensory neurons*, with cell bodies aggregated in *cranial sensory ganglia* associated with the facial, glossopharyngeal, and vagus cranial nerves (VII, IX, and X), and by terminal branches of the cervical sympathetic trunk. It should be noted that similar input is monitored at lower digestive tract levels, as well, via receptors at lower segmental levels.

Afferent information triggers secretions of mucosal epithelium and salivary glands as needed. Efferent nervous system activation is conveyed back to the muscles, mucosa, and salivary glands via parasympathetic efferent neurons associated with the cranial nerves and via sympathetic efferent neurons of the cervical sympathetic trunk. The functional tension between parasympathetic and sympathetic nervous systems is particularly variable in digestive system actions, with parasympathetic regulations for resting functions frequently switching with sympathetic arousal functions during the course of food and liquid intake activity.

In the central nervous system, nuclei and tracts associated with the trigeminal, facial, glossopharyngeal, vagus, accessory, and hypoglossal cranial nerves provide coordination and projection of peripheral nerve activation to higher levels. These cranial nerve nuclei are located mostly in the pons and medulla oblongata (the trigeminal ganglion is located lust outside the mid pons), with projections to the thalamus, hypothalamus, basal nuclei, cerebellum, and cerebral cortex. In the case of afferent somesthetic stimulation via the trigeminothalamic tracts, peripheral, first order neurons synapse with second order neurons at the nucleus of the descending tract of V at cervical spinal cord levels as far caudally as the C-4 segment.

In the brain, several diverse centers of activity appear to exert central control over swallowing. Mosier, Wen-Chu, Maldjian, Shah, and Modi (1999) used functional magnetic resonance imaging (fMRI) techniques and eight subjects to observe activity in the cerebral hemispheres. Later, Suzuki, *et al.* (2003) and Marin, *et al.* (2004) performed similar investigations, using functional magnetic resonance imaging with eleven and fourteen subjects, respectively. These investigators observed activity in the primary motor and sensory cortices (pre- and post-central gyri, respectively) with a significant laterality effect tending to favor the left cerebral hemisphere. They reported additional cortical activity in motor association cortices, and in the temporal, cingulate, and insular cortical areas as well as in the internal capsule. Mosier and Bereznaya (2001) also noted activity in the thalamus, insula, basal nuclei, and cerebellum. Interestingly, neither study located cerebral cortical areas dedicated to swallowing as distinct from other tongue movement.

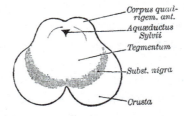

Developmental Anatomy of the Digestive System

Embryological development of the digestive system parallels that of the speech articulatory mechanism. Major external and internal respiratory and digestive tract structures begin to develop during the fourth week of gestation.

At the cephalic end of the embryo, ventral folding creates an enclosure for the primitive distal alimentary tract when part of the yolk sack is separated and incorporated into the inner embryo. The primitive *prochordal plate*, at which endodermal and ectodermal germ tissue layers meet, folds under the head and creates the *buccopharyngeal* (also called *oropharyngeal*) membrane. This membrane separates the frontal end of the cranium from the first branchial arch. Shortly after this event, the respiratory tract and digestive tract separate to follow their own developmental courses.

At their distal ends, the digestive (alimentary) tract and the respiratory tract are composed of the same structures in the same way they remain after birth. These structures include the nose, mouth, and pharynx.

An endodermal *foregut* develops at the distal end of the digestive tract and extends in a caudal direction. In it will develop structures common to the digestive system and respiratory system, but it will continue developing mainly as a foundation of the digestive system, extending and differentiating into specialized digestive tissues as far as the duodenum, a digestive structure beyond the stomach. At its rostral end, however, the foregut remains separated from the future oral cavity by the buccopharyngeal membrane.

The trachea and esophagus begin as the same tube, the foregut, until the programmed death of cells, *apoptosis*, coordinated with ventral invagination of the foregut, creates a *tracheoesophageal septum* during the first month after fertilization. The exact timing and sequence of this process is still under investigation (His, 1885, Zaw-Tun, 1982, Williams, Quan, and Beasley, 2003).

Meanwhile, at the very rostral and of the embryo, a shallow depression appears the outer ectodermal covering. This depression is the *stomodeum,* and it deepens steadily, pressing inward toward the buccopharyngeal membrane on its way to join the foregut. It is destined to become the mouth and nose, with the facial primordial fusing to separate the facial features and their oral and nasal cavities forming the distal respiratory and digestive tracts.

The buccopharyngeal membrane ruptures on about the 24th day, bringing the primitive gut into communication with the fluid filled *amniotic cavity*, enveloping the embryo.

Rostral digestive structures, being also parts of the respiratory and vocal tracts, develop from the cephalic end of the embryo and from the branchial (pharyngeal or gill) arches grooves and pouches. The first of these arches appears at the cranial end of the primitive foregut after about twenty-two days of gestation, and the last five will develop in a cranio-caudal sequence, with the fifth arch disappearing in human beings.

From the cephalic (head) end and from the first and second branchial arches develop jaws and superficial facial features that open and close the labial sphincter, allowing admission and retention of material. These develop as *facial primordia*, distinct tissue regions, and appear grow and differentiate on the ventral surfaces of the head and branchial arches. Specific facial primordial called the *frontonasal prominence*, and paired *maxillary* and *mandibular prominences* appear, enlarge and grow towards one another from above, below, and from either side, fusing where they meet. Proliferation and migration of their mesenchyme surrounds the stomodeal opening, forming the external and internal mouth by the end of the fifth gestational week.

The upper and lower jaws, maxilla and mandible, their associated digestive tract structures and several other skull bones, will grow from divisions of the first branchial arch, also called the *mandibular arch* or *Meckel's arch*. As the first arch develops, two distinct cartilaginous divisions appear. From the more caudal of the two, *Meckel's cartilage*, will come the maxilla and the mandible.

Muscles of facial expression develop from the second branchial arch. The buccinators, deep muscles of the cheeks, provide tension between the oral commissures, anteriorly, and the pterygopharyngeal raphe in the posterior oral region, also derives form the second arch, as do posterior belly of the digastric and the stylohyoid muscles, active during laryngeal elevation associated with swallowing. The cranial nerve associated with the second arch is the facial nerve (VII), and it provides the common final pathway for motor control from the central nervous system.

Muscles of mastication, even though they are ultimately located caudal to those of facial expression, evolve from the mesoderm of the first arch. Muscles of mastication include the temporalis, masseter, and pterygoid muscles. Likewise, the tensors velipalatini, which, as their name suggests, stiffen the soft palate, are derivatives of first arch mesoderm. These pared muscles are active during swallowing as a means of helping separate the nasopharynx from the oropharynx and to provide a more rigid surface against which to hold the tongue during deglutition. Other muscles active during swallowing originate from the first branchial arch. Contractions of the anterior belly of the digastric and the mylohyoid muscles draw the hyoid bone superiorly during deglutition. Muscles of the first branchial arch receive their motor and sensory innervation from the trigeminal nerve (V).

Most muscles of velopharyngeal sphincter and posterior oral cavity originate from the fourth branchial arch and receive their motor innervation form the vagus and accessory nerves (X-XI). The levator velipalatini and superior pharyngeal constrictors that close the velopharyngeal sphincter, and the palatoglossus and palatopharyngeus muscles, that pull it open when gravity isn't sufficient, form a muscular sling around the opening between the oropharynx and nasopharynx. These are assisted during deglutition by the paired tensor velipalatini muscles, mentioned above.

Only one pair of muscles associated with swallowing originates from the third branchial arch. These are the stylopharyngeus muscles, originating at the bases of each styloid process, below and medial to the external auditory meatuses, and inserting between the middle and inferior pharyngeal constrictors. Contractions of the stylopharyngeus muscles draw the pharyngeal walls upward and rearward, carrying with them the larynx, toward the styloid process of the temporal bone. Motor innervation is from the glossopharyngeal nerve (IX).

Digestive tract structures differentiate and grow caudally, the foregut eventually communicating with the *midgut* and *hindgut*. Caudal foregut structures will include the stomach and duodenum. Midgut tissues will develop into the small intestines and first two-thirds of the colon. The hindgut will become the remainder of the colon, the rectum, and the upper part of the anal canal.

CHAPTER SUMMARY

This chapter has focused on structures and functions related to swallowing. Many speech-language pathologists have a large part of their clinical caseloads devoted to assessment and treatment of swallowing disorders, so the information contained in this chapter will find much use among the practitioner. With definitions used throughout listed at the beginning for ease of use, the test started with a description of the gross anatomy of the mechanism sued for swallowing. The more distal of these structures, particularly the oral cavity and the oropharynx are also used for respiration and for speech. A discussion of the muscular esophagus followed as the chapter reached the limit of its coverage, since speech-language pathologists rarely deal directly with disorders of the stomach and other more caudal structures of the digestive system. The chapter covered the physiology of deglutition, with discussions of both voluntary and involuntary functions, next. Comparison and contrast of the forms and functions of the structures and mechanisms for speech and for swallowing followed. These concepts were followed by a brief coverage of nutrition science and its importance for those clinicians treating patients with swallowing disorders. Descriptions of the hard and soft tissues of the digestive tract form the labial sphincter to the lower esophageal sphincter were followed by descriptions of the neurology of deglutition and, finally the developmental anatomy of the digestive tract.

REFERENCES AND SUGGESTED READING

Aviv J.E., Spitzer J., Cohen M., Ma, G., Belafsky, P., Close, L.G. (2002). Laryngeal adductor reflex and pharyngeal squeeze as predictors of laryngeal penetration and aspiration. *Laryngoscope, 112,* 338–341.

Buchholz, D. (1994). Dysphagia associated with neurological disorders. *Acta Oto-Rhino Laryngologica Belgica, 48,* 143–155.

Crystal, D. (1969). *Prosodic systems and intonations in English.* London: Cambridge University Press, 4–7.

Darley, F., Aronson, A., & Brown, J. (1975). *Motor Speech Disorders.* Philadelphia: W.B. Sanders, 35–65.

Davies AE, Kidd D, Stone SP, MacMahon J. (1995). Pharyngeal sensation and gag reflex in healthy subjects. *Lancet, 345,* 48748–8.

DeVault, K.R. & Castell, D.O. (2004). Updated guidelines for the diagnosis and treatment of gastroesophageal reflux disease. *American Journal of Gastroenterology, 94,* 1434–1442.

Dobie, R. (1978). Rehabilitation of swallowing disorders. *AFP, 17,* 84–95.

Duffy, J. (1995). *Motor Speech Disorders.* St. Louis, MO: Mosby-Year Book, 8.

Edwards, H. (1992). *Applied phonetics: The sounds of American English.* San Diego: Singular, 19–23.

Fisher, H., & Logemann, J. (1971). *The Fisher-Logemann Test of Articulatory Competence.* Boston: Houghton-Mifflin.

Goyal, R. K. and Mashimo, H. (May, 2006). Physiology of oral, pharyngeal, and esophageal motility. *GI Motility online.* Retrieved June 9, 2011, from http://www.nature.com/gimo/contents/pt1/full/gimo1.html

Hamilton W, McMinn R: Digestive System. In: Hamilton W (Ed). *Textbook of Human Anatomy* (2nd ed.). St. Louis: C.V. Mosby, 1977, pp: 357–358.

Hamdy, S. (2006). Role of cerebral cortex in the control of swallowing. GI Motility online doi:10.1038/gimo8 Retrieved June 9, 2011, from http://www.nature.com/gimo/contents/pt1/full/gimo8.html

His, W. (1885). Anatomie menschlicher embryonen. III. *Zur Gischichte der Organe* (pp. 12–19). Leipzig: Vogel. Cited in Zaw-Tun, H. A. (1982). The tracheo-esophageal septum- fact or fantasy? Origin and development of the respiratory primordium and esophagus. *Acta Anatomia. 114*(1), 1–21.

Huang, Shih-Che, Chang, Bee-Song. (2004). Endothelin causes contraction of human esophageal muscularis mucosae through interaction with both ETA and ETB receptors. *Regulatory Peptides, 117* (3) 179–186.

Humbert I.A., Fitzgerald, M.E., McLaren, D.G., Johnson, S., Porcaro, E., Kosmatka, K., Hind, J., Robbins, J. (2009). Neurophysiology of swallowing: effects of age and bolus type. *Neuroimage, 44,* 982–991. Epub Oct 28, 2008.

International Phonetic Association (2005). *Reproduction of the International Phonetic Alphabet.* Retrieved June 9, 2011, from http://www.arts.gla.ac.uk/IPA

Joseph, J, Locomotor System. In: Hamilton W (Ed). *Textbook of Human Anatomy* (2nd ed.). St. Louis: C.V. Mosby, 1977, p. 81.

Kennedy, G., Pring, T., & Fawcus, R. (1993). No place for motor speech acts in assessment of dysphagia? Intelligibility and swallowing difficulties in stroke and Parkinson's disease patients. *European Journal of Disorders of Communication, 28,* 213–226.

Kennedy J., & Kent R. (1985). Anatomy and physiology of deglutition and related functions. *Seminars in Speech and Language, 6,* 257–272.

Kilman, W., & Goyal, R. (1976). Disorders of pharyngeal and upper esophageal sphincter motor function. *Archives of Internal Medicine, 126,* 592–601.

Lee, J. (2004). Aspiration Pneumonia. *eMedicine Specialities.* Retrieved July 6, 2011, from http://www.emedicine .com/Radio/topic57.htm#section~AuthorsandEditors

Leopold, N. A. and Kagel, M.C. (1997). Dysphagia—ingestion or deglutition?: a proposed paradigm. *Dysphagia, 12,* 202–206.

Logemann, J. (1995). Dysphagia: evaluation and treatment. *Folia Phoniatrica Logopedia, 47,* 140–164.

Logemann, J. (1983). *Evaluation and treatment of swallowing disorders.* San Diego, CA: College Hill Press, 70–73.

Marin, R.E., MacIntosh, B.J., Smith, R.C., Barr, A.M., Stevens, T.K., Gati J.S. and Menon, R.S. (2004). Cerebral Areas processing swallowing and tongue movements are overlapping but distinct. *Journal of Neurophysiology, 92,* 2428–2443.

Miller, AJ. (1982). Deglutition. *Physiological Reviews, 62,* 129–184.

Miller, R. & Langmore, S. (1994). Treatment efficacy for adults with oropharyngeal dysphagia. *Archives of Physical Medicine and Rehabilitation, 75,* 1256–1262.

Mosier, K.M., Wen-Chu, L., Maldjian, J., Shah, R. and Modi, B. (1999). Lateralization of cortical function in swallowing: a functional MR imaging study. *American Journal of Radiology, 20,* 1520–1526.

Mosier, K, and Bereznaya, I. (2001). Parallel cortical networks for volitional control of swallowing in humans. *Experimental Brain Research, 140,* 280–289.

Mysak, E. (1976). *Pathologies of speech systems.* Baltimore, MD: Williams and Wilkins, 38–45.

National Safety Council Press Release. (June 7, 2007). Accidental deaths increasing at alarming rate poisonings, overdoses, seeing greatest gains. Retrieved June 9, 2011, from http://www.nsc.org/Pages/Accidental DeathsIncreasingatAlarmingRate.aspx

Olthoff, A., Schiel, R., Kruse, E. (2004). The supraglottic nerve supply: an anatomic study with clinical implications. *Laryngoscope. 117,* 1930–1933.

Spielman, A. I., D'Abundo, S., Field, R. B. and Schmale, H. (1993). Protein analysis of human von Ebner saliva and a method for its collection from the foliate papillae. *Journal of Dental Research, 72,* 1331–1335.

Suzuki, M., Asada, Y., Ito, J., Hayashi, K., Inoue, H., and Kitano, H. (2003). Activation of cerebellum and basal ganglia on volitional swallowing detected by functional magnetic resonance imaging. *Dysphagia, 18,* 71–77.

Swanson, L.W. (1988). The neural basis of motivated behavior. *Acta Morphologica* Neerl Scandinavia 26, 165–176.

Tanner, D., & Culbertson, W. (1999). *Quick Assessment for dysphagia.* Oceanside, CA: Academic Communication Associates.

Tortora, G.J. and Grabowski, R. (1996). *Principles of Anatomy and Physiology* (8th ed.). New York: Harper Collins.

Tiffany, W., & Carrell, J. (1977). *Phonetics: theory and application*. New York: McGraw-Hill Book Company, 105.

Weinstein BE: Presbycusis. In: Katz, J. (Ed) (1994). *Handbook of Clinical Audiology* (4th ed.). Baltimore, MD: Williams and Wilkins, pp. 568–584.

Whitney, E.N., Cataldo, C.B. and Rolfes, S.R. (2002). *Understanding Normal and Clinical Nutrition* (6th ed.). Stamford, Ct.: Wadsworth/Thomson Learning.

Williams, A. K., & Quan, Q. B., & Beasley, S. W. (2003). Three-dimensional imaging clarifies the process of tracheoesophageal separation in the rat. *Journal of Pediatric Surgery. 38*, 173–177.

Wisner, A., Dufour, E., Messaudi, M., Nedji, A., Marcel, A., Ungeheuer, M., and Rougeot, C. (2006). Human opiorphin, a natural antinociceptive modulator of opiod-dependent pathways. *Procedings of the National Academy of Sciences of the United States of America: Pharmacology, 103*, 17979–17984.

Zaw-Tun, H. A. (1982). The tracheo-esophageal septum-fact or fantasy? Origin and development of the respiratory primordium and esophagus. *Acta Anatomia. 114*, 1–21.

Zemlin, W. (1998). *Speech and Hearing Science: Anatomy and Physiology* 4th ed. Englewood Cliffs, NJ: Prentice-Hall, 280–281.

IMPORTANT TERMS

Aerodigestive Tract: Anatomic passage common to both the respiratory and digestive systems.

Alimentary System: Organ system which assimilates nutrients from the environment for body maintenance; also called the digestive tract.

Anus: Muscular ring at the caudal extent of the alimentary system controlling passage of feces from the rectum into the environment.

Bolus: A mass of material to be swallowed.

Deglutition: Transfer of material from the oral cavity to the esophagus. The word comes from the Latin *deglutire,* meaning to "swallow from or down."

Digestion: Breaking down of nutrients into units that can be absorbed.

Edentulous: Lacking teeth.

Enteral Tube Feeding: Method of delivering nutrition and liquids directly into the stomach or small intestine, bypassing the oral cavity and neck structures.

Esophageal Hiatus: Opening in the diaphragm to allow passage of the esophagus; located roughly at the level of the tenth thoracic vertebra.

Esophageal Voice: Substitute for the phonatory speech source created by tightening the crocopharyngeus muscle and forcing esophageal air through the narrow opening so created.

Esophagus: Muscular tube connecting the oropharynx and the stomach.

Essential Nutrients: Those that must be supplied in the diet.

Exocrine Glands: Glands that transfer their secretions, such as saliva, through ducts.

Foregut: Embryonic endodermal tube that develops at the distal end of the digestive tract and extends in a caudal direction; in it will develop structures common to the digestive system and respiratory system.

Hydration: Consumption of water.

Lower Esophageal Sphincter (LES): Muscular ring controlling passage of material between the esophagus and the stomach.

Mastication: Chewing.

Meckel's Cartilage: Embryonic skeletal component of the first branchial arch; from it will come the maxilla and the mandible.

Non-Essential Nutrients: Nutrients that the body can produce from raw materials it has consumed.

Nutrition Scientist: Clinician who treats patients by determining and supplying their basic human nutritional needs; also called nutritionist or dietician.

Labial Sphincter: Muscular ring around the entrance to the digestive tract; the lips.

Parenteral Tube Feeding: Method of delivering nutrients and water directly into the bloodstream.

Peristalsis: Rythmic sequential contractions of the sphincteric digestive tract muscles to propel material from the distal digestive tract its internal organs.

Pyriform Sinuses: Pockets flanking the laryngeal vestibule. The pyriform sinuses are created by created by the folding of epithelium between the upper edges of the thyroid laminae and the aryepiglottic folds and may become filled with material passing from the oropharynx to the esophagus.

Salivary Glands: Glands located in and around the oral cavity that secrete saliva to support digestion and swallowing.

Sialorrhea: Excessive and/or uncontrolled salivation; "drooling."

Swallowing Phases: Distinct repetitive events in the consumption of nutrients and their passage into the stomach.

Swallowing Reflex: A complex concert of muscular inhibitions and excitations, beginning with peristaltic movements of the pharyngeal constrictors and concluding with relaxation of the lower esophageal sphincter; generated in the brainstem.

Upper Esophageal Sphincter (UES): Muscular ring controlling passage of material between the oropharynx and esophagus; located at the level of the sixth cervical vertebra.

Valleculae (singular: valleculae): Depressions inform one another by the eminence created by the median hyoepiglottic ligament. The valleculae may become traps for swallowed material under certain conditions; from Latin: *valles:* "valley".

Von Ebner's Glands: Salivary glands contained in the epithelium of the oral cavity, surrounding the circumvallate and foliate papillae of the tongue.

Index